SHEAR DEFORMABLE BEAMS AND PLATES
Relationships with Classical Solutions

Elsevier Science Internet Homepage
http://www.elsevier.nl (Europe)
http://www.elsevier.com (America)
http://www.elsevier.co.jp (Asia)

Consult the Elsevier homepage for full catalogue information on all books, journals and electronic products and services.

Elsevier Titles of Related Interest

CHAN & TENG
ICASS '99, Advances in Steel Structures.
(2 Volume Set)
ISBN: 008-043015-5

DUBINA
SDSS '99 - Stability and Ductility of Steel Structures.
ISBN: 008-043016-3

FRANGOPOL, COROTIS & RACKWITZ
Reliability and Optimization of Structural Systems.
ISBN: 008-042826-6

FUKUMOTO
Structural Stability Design.
ISBN: 008-042263-2

MAKELAINEN
ICSAS '99 Int Conf on Light-Weight Steel and Aluminium Structures.
ISBN: 008-043014-7

RIE & PORTELLA
Low Cycle Fatigue and Elasto-Plastic Behaviour of Materials.
ISBN: 008-043326-X

SRIVASTAVA
Structural Engineering World Wide 1998 (CD-ROM Proceedings with Printed Abstracts Volume, 702 papers).
ISBN: 008-042845-2

USAMI & ITOH
Stability and Ductility of Steel Structures.
ISBN: 008-043320-0

VOYIADJIS *ET AL.*
Damage Mechanics in Engineering Materials.
ISBN: 008-043322-7

VOYIADJIS & KATTAN
Advances in Damage Mechanics: Metals and Metal Matrix Composites.
ISBN: 008-043601-3

Related Journals
Free specimen copy gladly sent on request. Elsevier Science Ltd, The Boulevard, Langford Lane, Kidlington, Oxford, OX5 1GB, UK

Advances in Engineering Software
CAD
Composite Structures
Computer Methods in Applied Mechanics & Engineering
Computers and Structures
Construction and Building Materials
Engineering Failure Analysis
Engineering Fracture Mechanics
Engineering Structures
Finite Elements in Analysis and Design

International Journal of Mechanical Sciences
International Journal of Plasticity
International Journal of Solids and Structures
International Journal of Fatigue
Journal of Applied Mathematics and Mechanics
Journal of Constructional Steel Research
Journal of Mechanics and Physics of Solids
Mechanics of Materials
Mechanics Research Communications
Structural Safety
Thin-Walled Structures

To Contact the Publisher
Elsevier Science welcomes enquiries concerning publishing proposals: books, journal special issues, conference proceedings, etc. All formats and media can be considered. Should you have a publishing proposal you wish to discuss, please contact, without obligation, the publisher responsible for Elsevier's civil and structural engineering publishing programme:

Mr Ian Salusbury
Senior Publishing Editor
Elsevier Science Ltd
The Boulevard, Langford Lane
Kidlington, Oxford
OX5 1GB, UK

Phone: +44 1865 843425
Fax: +44 1865 843920
E.mail: i.salusbury@elsevier.co.uk

General enquiries, including placing orders, should be directed to Elsevier's Regional Sales Offices – please access the Elsevier homepage for full contact details (homepage details at the top of this page).

SHEAR DEFORMABLE BEAMS AND PLATES
Relationships with Classical Solutions

C.M. Wang
*Department of Civil Engineering,
The National University of Singapore,
10 Kent Ridge Crescent, Singapore 119260*

J.N. Reddy
*Department of Mechanical Engineering,
Texas A&M University, College Station,
Texas 77843-3123, USA*

K.H. Lee
*Department of Mechanical and Production Engineering,
The National University of Singapore,
10 Kent Ridge Crescent, Singapore 119260*

2000

ELSEVIER

Amsterdam - Lausanne - New York - Oxford - Shannon - Singapore - Tokyo

ELSEVIER SCIENCE Ltd
The Boulevard, Langford Lane
Kidlington, Oxford OX5 1GB, UK

© 2000 Elsevier Science Ltd. All rights reserved.

This work is protected under copyright by Elsevier Science, and the following terms and conditions apply to its use:

Photocopying
Single photocopies of single chapters may be made for personal use as allowed by national copyright laws. Permission of the Publisher and payment of a fee is required for all other photocopying, including multiple or systematic copying, copying for advertising or promotional purposes, resale, and all forms of document delivery. Special rates are available for educational institutions that wish to make photocopies for non-profit educational classroom use.

Permissions may be sought directly from Elsevier Science Global Rights Department, PO Box 800, Oxford OX5 1DX, UK; phone: (+44) 1865 843830, fax: (+44) 1865 853333, e-mail: permissions@elsevier.co.uk. You may also contact Global Rights directly through Elsevier's home page (http://www.elsevier.nl), by selecting 'Obtaining Permissions'.

In the USA, users may clear permissions and make payments through the Copyright Clearance Center, Inc., 222 Rosewood Drive, Danvers, MA 01923, USA; phone: (978) 7508400, fax: (978) 7504744, and in the UK through the Copyright Licensing Agency Rapid Clearance Service (CLARCS), 90 Tottenham Court Road, London W1P 0LP, UK; phone: (+44) 171 631 5555; fax: (+44) 171 631 5500. Other countries may have a local reprographic rights agency for payments.

Derivative Works
Tables of contents may be reproduced for internal circulation, but permission of Elsevier Science is required for external resale or distribution of such material.
Permission of the Publisher is required for all other derivative works, including compilations and translations.

Electronic Storage or Usage
Permission of the Publisher is required to store or use electronically any material contained in this work, including any chapter or part of a chapter.

Except as outlined above, no part of this work may be reproduced, stored in a retrieval system or transmitted in any form or by any means, electronic, mechanical, photocopying, recording or otherwise, without prior written permission of the Publisher.
Address permissions requests to: Elsevier Global Rights Department, at the mail, fax and e-mail addresses noted above.

Notice
No responsibility is assumed by the Publisher for any injury and/or damage to persons or property as a matter of products liability, negligence or otherwise, or from any use or operation of any methods, products, instructions or ideas contained in the material herein. Because of rapid advances in the medical sciences, in particular, independent verification of diagnoses and drug dosages should be made.

First edition 2000

Library of Congress Cataloging in Publication Data

British Library Cataloguing in Publication Data
A catalogue record from the British Library has been applied for.

Library of Congress Cataloging-in-Publication Data

Wang, C. M.
  Shear deformable beams and plates : relationships with classical solutions / C.M. Wang, J.N. Reddy, K.H. Lee.
    p. cm.
  Includes bibliographical references and index.
  ISBN 0-08-043784-2 (hardcover)
   1. Plates (Engineering)--Mathematical models. 2. Girders--Mathematical models. 3. Shear (Mechanics) 4. Deformations (Mechanics) 5. Mathematical analysis. I. Reddy, J. N. (John N.) II. Lee, K. H. III. Title.

TA660.P6 W27 2000
624.1'7765--dc21
                                                                          00-035437
ISBN: 0 08 043784 2

⊚ The paper used in this publication meets the requirements of ANSI/NISO Z39.48-1992 (Permanence of Paper).
Printed in The Netherlands.

Contents

Preface xiii

1 Introduction 1
    1.1 Preliminary Comments 1
    1.2 An Overview of Plate Theories 3
    1.3 Present Study 5
    Problems 6

Part 1: Beams

2 Bending of Beams 11
    2.1 Beam Theories 11
        2.1.1 Introduction 11
        2.1.2 Euler–Bernoulli Beam Theory (EBT) 14
        2.1.3 Timoshenko Beam Theory (TBT) 17
        2.1.4 Reddy–Bickford Beam Theory (RBT) 20
    2.2 Relationships Between EBT and TBT 24
        2.2.1 General Comments 24
        2.2.2 Simply Supported (SS) Beams 25
        2.2.3 Clamped-Free (CF) Beams 25
        2.2.4 Free-Clamped (FC) Beams 25
        2.2.5 Clamped-Simply Supported (CS) Beams 26
        2.2.6 Simply Supported-Clamped (SC) Beams 26
        2.2.7 Clamped (CC) Beams 26
        2.2.8 Summary of Relationships 27
    2.3 Relationships Between EBT and RBT 28

    2.4 Examples 31
        2.4.1 Simply Supported Beam 32
        2.4.2 Cantilever Beam 34

    2.5 Summary 36

    Problems 37

3 Shear-Flexural Stiffness Matrix 39

    3.1 Introduction 39

    3.2 Summary of Relationships 41
        3.2.1 Relationships Between TBT and EBT 41
        3.2.2 Relationships Between RBT and EBT 41
        3.2.3 Relationships Between Simplified RBT and EBT 42

    3.3 Stiffness Matrix 44

    3.4 Frame Structure - An Example 48

    3.5 Concluding Remarks 50

    Problems 52

4 Buckling of Columns 55

    4.1 Introduction 55

    4.2 Relationship Between Euler–Bernoulli
        and Timoshenko Columns 56
        4.2.1 General Relationship 56
        4.2.2 Pinned-Pinned Columns 58
        4.2.3 Fixed-Fixed Columns 60
        4.2.4 Fixed-Free Columns 61

    4.3 Relationship Between Euler–Bernoulli and
        Reddy–Bickford Columns 64
        4.3.1 General Relationship 64
        4.3.2 Pinned-Pinned Columns 69
        4.3.3 Fixed-Fixed Columns 70
        4.3.4 Fixed-Free Columns 71
        4.3.5 Pinned-Pinned Columns with End Rotational
            Springs of Equal Stiffness 72

    4.4 Concluding Remarks 74

    Problems 75

5 Tapered Beams 77

5.1 Introduction 77

5.2 Stress Resultant-Displacement Relations 78

5.3 Equilibrium Equations 78

5.4 Deflection and Force Relationships 79
    5.4.1 General Relationships 79
    5.4.2 Simply Supported (SS) Beams 80
    5.4.3 Clamped-Free (CF) Beams 81
    5.4.4 Free-Clamped (FC) Beams 81
    5.4.5 Clamped (CC) Beams 82
    5.4.6 Clamped-Simply Supported (CS) Beams 82
    5.4.7 Simply Supported-Clamped (SC) Beams 83
    5.4.8 An Example 84

5.5 Symmetrically Laminated Beams 85

5.6 Concluding Remarks 86

Problems 86

Part 2: Plates

6 Theories of Plate Bending 89

6.1 Overview of Plate Theories 89

6.2 Classical (Kirchhoff) Plate Theory (CPT) 92
    6.2.1 Equations of Equilibrium 92
    6.2.2 Boundary Conditions 94
    6.2.3 Governing Equations in Terms of the Deflection 98

6.3 First-Order Shear Deformation Plate Theory (FSDT) 100
    6.3.1 Equations of Equilibrium 100
    6.3.2 Plate Constitutive Equations 102
    6.3.3 Governing Equations in Terms of Displacements 103

6.4 Third-Order Shear Deformation Plate Theory (TSDT) 105
    6.4.1 Equations of Equilibrium 105
    6.4.2 Plate Constitutive Equations 107

Problems 108

7 Bending Relationships for Simply Supported Plates 111

7.1 Introduction 111

7.2 Relationships Between CPT and FSDT 112

7.3 Examples 116
    7.3.1 Simply Supported, Uniformly Loaded, Equilateral Triangular Plate 116
    7.3.2 Simply Supported, Uniformly Loaded, Rectangular Plate 117

7.4 Relationships Between CPT and TSDT 118
    7.4.1 Introduction 118
    7.4.2 Governing Equations 119
    7.4.3 The Kirchhoff Plate Theory (CPT) 123
    7.4.4 Relationships Between the Theories 124
    7.4.5 An Accurate Simplified Relationship 125
    7.4.6 An Example 127

7.5 Closure 128

Problems 129

8 Bending Relationships for Lévy Solutions 133

8.1 Introduction 133

8.2 Governing Equations 134
    8.2.1 Introduction 134
    8.2.2 Stress Resultant-Displacement Relations 135
    8.2.3 Equilibrium Equations 137

8.3 Bending Relationships 137
    8.3.1 General Relationships 137
    8.3.2 SSSS Plates 143
    8.3.3 SCSC Plates 144
    8.3.4 SFSF Plates 145
    8.3.5 SCSS Plates 147
    8.3.6 SFSS Plates 148

8.4 Numerical Results 148
    8.4.1 SCSC Plates 149
    8.4.2 SFSF Plates 151

Problems 152

9 Bending Relationships for Circular and Annular Plates 153

9.1 Governing Equations 153

9.2 Relationships Between CPT and FSDT 156
    9.2.1 General Relationships 156
    9.2.2 Examples 162

9.3 Relationships Between CPT and TSDT 165
    9.3.1 General Relationships 165
    9.3.2 An Example 169

9.4 Closure 171

Problems 171

10 Bending Relationships for Sectorial Plates 177

10.1 Introduction 177

10.2 Formulation 178
    10.2.1 The Kirchhoff Plate Theory (CPT) 178
    10.2.2 The Mindlin Plate Theory (FSDT) 180
    10.2.3 Governing Equations 181

10.3 Exact Bending Relationships 182
    10.3.1 General Relationships 182
    10.3.2 SSS Sectorial Plates 185
    10.3.3 SSC Sectorial Plates 186
    10.3.4 SSF Sectorial Plates 186

10.4 Examples 187
    10.4.1 SSS Plates 187
    10.4.2 SSC Plates 188
    10.4.3 SSF Plates 188
    10.4.4 Numerical Results 189

10.5 Conclusions 192

Problems 192

11 Buckling Relationships 195

11.1 Polygonal Plates 195
    11.1.1 Governing Equations 195
    11.1.2 Relationships Between CPT and FSDT 199
    11.1.3 Relationships Between CPT and TSDT 200

11.2 Circular Plates 205
    11.2.1 Governing Equations 205
    11.2.2 Relationship Between CPT and FSDT 207
    11.2.3 Relationship Between CPT and TSDT 208
    11.2.4 Numerical Results 210

11.3 Sectorial Mindlin Plates 211
    11.3.1 Governing Equations 211
    11.3.2 Buckling Load Relationship 215

Problems 218

12 Free Vibration Relationships 223

12.1 Introduction 223

12.2 Relationships Between CPT and FSDT 226
    12.2.1 General Relationship 226
    12.2.2 Numerical Results 229

12.3 Relationships Between CPT and TSDT 235

12.4 Concluding Remarks 241

Problems 241

13 Relationships for Inhomogeneous Plates 243

13.1 Deflection Relationships for Sandwich Plates 243
    13.1.1 Introduction 243
    13.1.2 Governing Equations for Kirchhoff Plates 244
    13.1.3 Governing Equations for Sandwich Mindlin Plates . 245
    13.1.4 Relationship Between Sandwich and
           Kirchhoff Plates 248
    13.1.5 Examples 250
    13.1.6 Relationship Between Sandwich ....
           and Solid Mindlin Plates 252

13.2 Deflection Relationships for Functionally Graded
    Circular Plates 253
    13.2.1 Introduction 253
    13.2.2 Formulation 254
    13.2.3 Relationships Between CPT and FSDT 256
    13.2.4 Relationships for Various Boundary Conditions 260
    13.2.5 Illustrative Examples 264

13.3 Buckling Load Relationships for Sandwich
    Mindlin Plates..269
    13.3.1 Governing Equations.............................269
    13.3.2 Buckling Load Relationship......................271
13.4 Free Vibration Relationships for Sandwich Plates........272
    13.4.1 Governing Equations.............................272
    13.4.2 Free Vibration Relationship.....................275
13.5 Summary..276

References...279

Subject Index..293

Preface

There exist many books on the theory and analysis of beams and plates. Most of the books deal with the classical (Euler-Bernoulli/Kirchhoff) theories but few include shear deformation theories in detail. The classical beam/plate theory is not adequate in providing accurate bending, buckling, and vibration results when the thickness-to-length ratio of the beam/plate is relatively large. This is because the effect of transverse shear strains, neglected in the classical theory, becomes significant in deep beams and thick plates. In such cases, shear deformation theories provide accurate solutions compared to the classical theory.

Equations governing shear deformation theories are typically more complicated than those of the classical theory. Hence it is desirable to have exact relationships between solutions of the classical theory and shear deformation theories so that whenever classical theory solutions are available, the corresponding solutions of shear deformation theories can be readily obtained. Such relationships not only furnish benchmark solutions of shear deformation theories but also provide insight into the significance of shear deformation on the response. The relationships for beams and plates have been developed by the authors and their colleagues over the last several years. However, this valuable information is dispersed in the literature. Therefore, the goal of this monograph is to bring together these relationships for beams and plates in a single volume.

The book is divided into two parts. Following the introduction, Part 1 consists of Chapters 2 to 5 dealing with beams, and Part 2 consists of Chapters 6 to 13 covering plates. Problems are included at the end of each chapter to use, extend, and develop new relationships. The book is suitable as a reference by engineers and scientists working in industry and academia. An introductory course on mechanics of materials and elasticity should prove to be helpful but not necessary because a review of the basics is included in the relevant chapters.

The authors gratefully acknowledge the support and encouragement of their respective universities in carrying out the collaborative research and the writing of this book. It is a pleasure to acknowledge the help of the following colleagues in proof reading of the preliminary manuscript: Kok-Keng Ang, Goy-Teck Lim, and Yang Xiang. Special thanks go to Poh Hong, Aruna and See Fong for their love and patience while their husbands were occupied with writing this book.

C. M. Wang
Singapore

J. N. Reddy
College Station, Texas

K. H. Lee
Singapore

Chapter 1

Introduction

1.1 Preliminary Comments

The primary objective of this book is to study the relationships between the solutions of classical theories of beams and plates with those of the shear deformation theories. Shear deformation theories are those in which the effect of transverse shear strains is included. Relationships are developed for bending, buckling, and free vibration solutions.

A *plate* is a structural element with plane form dimensions that are large compared to its thickness and is subjected to loads that cause bending deformation in addition to stretching. In most cases, the thickness is no greater than one-tenth of the smallest in-plane dimension. Because of the smallness of the thickness dimension, it is often not necessary to model the plate using 3D elasticity equations. Beams are one-dimensional counterparts of plates.

The governing equations of beams and plates can be derived using either vector mechanics or energy and variational principles. In vector mechanics, the forces and moments on a typical element of the plate are summed to obtain the equations of equilibrium or motion. In energy methods, the principles of virtual work or their derivatives, such as the principles of minimum potential energy or complementary energy, are used to obtain the equations. While both methods can give the same equations, the energy methods have the advantage of providing information on the form of the boundary conditions.

Beam and plate theories are developed by assuming the form of the displacement or stress field as a linear combination of unknown functions and the thickness coordinate. For example, in plate theories we assume

$$\varphi_i(x,y,z,t) = \sum_{j=0}^{N}(z)^j \varphi_i^j(x,y,t) \qquad (1.1.1)$$

where φ_i is the ith component of displacement or stress, (x,y) are the in-plane coordinates, z is the thickness coordinate, t denotes the time, and φ_i^j are functions to be determined.

When φ_i are displacements, the equations governing φ_i^j are determined by the principle of virtual displacements

$$\delta W \equiv \delta U + \delta V = 0 \qquad (1.1.2a)$$

or its dynamic version, i.e., Hamilton's principle

$$\int_{t_1}^{t_2} (\delta K - \delta U - \delta V)\, dt = 0 \qquad (1.1.2b)$$

where $(\delta U, \delta V, \delta W, \delta K)$ denote the virtual internal (strain) energy, virtual potential energy due to applied loads, the total virtual work done, and virtual kinetic energy, respectively. These quantities are determined in terms of the actual stresses and virtual strains, which depend on the assumed displacement functions φ_i and their variations. For plate structures, the integration over the domain of the plate is represented as the product of integration over the plane of the plate and integration over the thickness of the plate (volume integral=integral over the plane × integral over the thickness). This is possible due to the explicit nature of the assumed displacement field in the thickness coordinate. Thus, we can write

$$\int_{Vol.} (\cdot)\, dV = \int_{-\frac{h}{2}}^{\frac{h}{2}} \int_{\Omega_0} (\cdot)\, d\Omega\, dz \qquad (1.1.3)$$

where h denotes the thickness of the plate and Ω_0 denotes the undeformed mid-plane of the plate, which is assumed to coincide with the xy−plane. Since all undetermined variables are explicit functions of the thickness coordinate, the integration over plate thickness is carried out explicitly, reducing the problem to a two-dimensional one. Hence, the Euler–Lagrange equations associated with Eq. (1.1.2a,b) consist of differential equations involving the dependent variables $\varphi_i^j(x,y,t)$ and the thickness-averaged stress resultants $R_{ij}^{(m)}$ per unit length:

$$R_{ij}^{(m)} = \int_{-\frac{h}{2}}^{\frac{h}{2}} (z)^m \sigma_{ij}\, dz \qquad (1.1.4)$$

The stress resultants can be written in terms of φ_i with the help of the assumed constitutive equations and strain-displacement relations. More complete development of this procedure is presented in the forthcoming chapters.

The same approach is used when φ_i denote stress components, except that the basis of the derivation of the governing equations is the principle of virtual forces. In the present book, stress-based theories will receive very little attention. Readers interested in stress-based theories may consult the book by Panc (1975).

1.2 An Overview of Plate Theories

The simplest *plate theory* of bending is the classical plate theory (CPT). In the case of pure bending, the displacement of the CPT is given by (see Reddy 1984b, 1997a, 1999a)

$$u(x,y,z,t) = -z\frac{\partial w_0}{\partial x}$$
$$v(x,y,z,t) = -z\frac{\partial w_0}{\partial y}$$
$$w(x,y,z,t) = w_0(x,y,t) \qquad (1.2.1)$$

where (u,v,w) are the displacement components along the (x,y,z) coordinate directions, respectively, and w_0 is the transverse deflection of a point on the mid-plane (i.e., $z = 0$). The displacement field (1.2.1) implies that straight lines normal to the xy-plane before deformation remain straight and normal to the mid-surface after deformation. The Kirchhoff assumption amounts to neglecting both transverse shear and transverse normal strain effects, i.e., deformation is due entirely to bending.

The next theory in the hierarchy of refined theories is the *first-order shear deformation theory* (or FSDT) (Mindlin 1951 and Reddy 1984b, 1999a), which is based on the displacement field

$$u(x,y,z,t) = z\phi_x(x,y,t)$$
$$v(x,y,z,t) = z\phi_y(x,y,t)$$
$$w(x,y,z,t) = w_0(x,y,t) \qquad (1.2.2)$$

where ϕ_x and $-\phi_y$ denote rotations about the y and x axes, respectively. The FSDT extends the kinematics of the classical plate theory by including a gross transverse shear deformation in its kinematic assumptions, i.e., the transverse shear strain is assumed to be constant with respect to the thickness coordinate. In the first-order shear deformation theory, shear correction factors are introduced to correct for the discrepancy between the actual transverse shear force distributions and those computed using the kinematics relations of the FSDT. The shear correction factors depend not only on the geometric parameters, but also on the loading and boundary conditions of the plate. In both the CPT and FSDT, the plane-stress state assumption is used and the plane-stress reduced form of the constitutive law is used.

Second- and higher-order plate bending theories employ higher-order polynomials in the expansion of the displacement components through the thickness of the plate. The higher-order theories introduce additional unknowns that are often difficult to interpret in physical terms. The second-order theory with transverse inextensibility is based on the displacement field

$$u(x,y,z,t) = z\phi_x(x,y,t) + z^2\psi_x(x,y,t)$$
$$v(x,y,z,t) = z\phi_y(x,y,t) + z^2\psi_y(x,y,t)$$
$$w(x,y,z,t) = w_0(x,y,t) \tag{1.2.3}$$

There are a number of third-order theories in the literature, and a review of these theories is given by Reddy (1997a). The third-order shear deformation plate theory (TSDT) of Reddy (1984a, 1984b, 1997a, 1999a) is based on the displacement field

$$u(x,y,z,t) = z\phi_x(x,y,t) + z^3\left(-\frac{4}{3h^2}\right)\left(\phi_x + \frac{\partial w_0}{\partial x}\right)$$
$$v(x,y,z,t) = z\phi_y(x,y,t) + z^3\left(-\frac{4}{3h^2}\right)\left(\phi_y + \frac{\partial w_0}{\partial y}\right)$$
$$w(x,y,z,t) = w_0(x,y,t) \tag{1.2.4}$$

The displacement field accommodates a quadratic variation of transverse shear strains (and hence stresses) and the vanishing of transverse shear stresses at the top and bottom surfaces of a plate. Thus there is no need to use shear correction factors in a third-order theory. Third-order theories provide a slight increase in accuracy relative to the FSDT solution, at the expense of an increase in computational effort.

In addition to its inherent simplicity and low computational cost, the FSDT often provides sufficiently accurate description of the global response for thin to moderately thick plates, e.g., maximum deflections, critical buckling loads, and free vibration frequencies and associated mode shapes. Therefore, it is of interest to determine the deflections, buckling loads, and natural frequencies of plates using the FSDT.

1.3 Present Study

Often, the higher-order beam/plate theories require solutions of more complicated governing equations. In view of the fact that solutions of classical beam and plate theories are available for a vast number of problems and the familiarity of engineers with these solutions, it is desirable to have relationships between solutions of higher-order theories and those of the classical theories. This book presents relationships between the solutions of the classical and shear deformation theories of beams and plates. The relationships for deflections, buckling loads and natural frequencies enable one to obtain the solutions of the shear deformation plate theories for specific problems and thereby reduce the effort of solving the complicated equations of shear deformation theories.

The book is divided into two major parts. Part 1 deals with beams and Part 2 is devoted to plates. Part 1 contains four chapters namely Chapters 2 to 5, and Part 2 covers Chapters 6 to 13.

Following this introduction, a review of beam theories and the relationships between the Euler–Bernoulli beam theory (EBT), Timoshenko beam theory (TBT) and Reddy–Bickford beam theory (RBT) are presented in Chapter 2. The relationships are used to develop the shear-flexural stiffness matrix in Chapter 3, which allows the analysis of shear deformable continuous beams and frames. Chapter 4 is devoted to the development of buckling load and vibration frequency relationships. Bending relationships for tapered beams are presented in Chapter 5.

A derivation of the governing equations of the classical, first-order, and third-order plate theories for static bending is presented in Chapter 6. Bending relationships are presented in Chapter 7 for simply supported polygonal plates, Chapter 8 for rectangular Lévy plates, Chapter 9 for circular and annular plates, and Chapter 10 for sectorial and annular sectorial plates. Chapter 11 is devoted to buckling load relationships, while Chapter 12 covers frequency relationships for free vibration. Finally, Chapter 13 contains bending, buckling, and vibration relationships of sandwich and functionally graded plates. Exercise problems are included at the end of each chapter, and references cited in these chapters are placed in alphabetical order at the end of the book.

Problems

1.1 Starting with a linear distribution of the displacements through the beam thickness in terms of unknown functions (w_0, F_1, F_2)

$$u(x, z) = zF_1(x), \quad w(x, z) = w_0(x) + zF_2(x) \tag{i}$$

determine the functions (F_1, F_2) in terms of w_0 such that the following Euler–Bernoulli hypothesis holds:

$$\frac{\partial w}{\partial z} = 0, \quad \frac{\partial u}{\partial z} = -\frac{\partial w}{\partial x} \tag{ii}$$

1.2 Starting with the displacement field

$$u(x, z) = z\phi(x) + z^2\psi(x) + z^3\theta(x), \quad w(x, z) = w_0(x) \tag{i}$$

determine the functions (ψ, θ) in terms of w_0 and ϕ such that the transverse shear stress vanishes at $z = \pm\frac{h}{2}$:

$$\sigma_{xz}(x, -\frac{h}{2}) = 0, \quad \sigma_{xz}(x, \frac{h}{2}) = 0 \tag{ii}$$

where h is the thickness of the beam.

1.3 Consider the following equations of equilibrium of 2-D (xz-plane) elasticity in the absence of body forces [$\sigma_{xx} = \sigma_{xx}(x, z)$, $\sigma_{xz} = \sigma_{xz}(x, z)$]:

$$\frac{\partial \sigma_{xx}}{\partial x} + \frac{\partial \sigma_{xz}}{\partial z} = 0 \tag{i}$$

$$\frac{\partial \sigma_{xz}}{\partial x} + \frac{\partial \sigma_{zz}}{\partial z} = 0 \tag{ii}$$

Integrate the above equations with respect to z over the interval $(-h/2, h/2)$ and express the result in terms of the forces N_{xx} and Q_x

$$N_{xx} = b \int_{-\frac{h}{2}}^{\frac{h}{2}} \sigma_{xx} \, dz, \quad Q_x = b \int_{-\frac{h}{2}}^{\frac{h}{2}} \sigma_{xz} \, dz \tag{iii}$$

where b is the width and h is the thickness of the beam. Use the following boundary conditions:

$$\sigma_{xz}(x, y, -\frac{h}{2}) = 0, \quad \sigma_{xz}(x, y, \frac{h}{2}) = 0$$
$$\sigma_{zz}(x, y, -\frac{h}{2}) = q_b, \quad \sigma_{zz}(x, y, \frac{h}{2}) = q_t \qquad (iv)$$

Next, multiply equations (i) and (ii) with z, integrate with respect to z over the interval $(-h/2, h/2)$, and express the result in terms of the moment M_{xx} and shear force Q_x

$$M_{xx} = b \int_{-\frac{h}{2}}^{\frac{h}{2}} z\sigma_{xx} \, dz \qquad (v)$$

Eliminate Q_x from the final equations.

PART 1
BEAMS

Chapter 2

Bending of Beams

Presented in this chapter are the various beam theories, progressing from the simple Euler–Bernoulli beam theory to the first-order shear deformation beam theory of Timoshenko and finally to the third-order beam theory of Reddy and Bickford. The latter two beam theories allow for the effect of transverse shear deformation which has been neglected in the Euler–Bernoulli beam theory. Using the principle of minimum potential energy, the governing equilibrium equations and boundary conditions have been derived for transversely loaded, uniform beams on the basis of the kinematic assumptions of the aforementioned beam theories. In view of the mathematical similarity of the governing equations and on the basis of load equivalence, exact relationships between the bending solutions of these three beam theories are derived. These relationships enable the conversion of the well-known Euler–Bernoulli beam solutions to their shear deformable beam counterparts. Examples are given to illustrate the use of these relationships.

2.1 Beam Theories

2.1.1 Introduction

There are a number of beam theories that are used to represent the kinematics of deformation. To describe the various beam theories, we introduce the following coordinate system. The x-coordinate is taken along the length of the beam, z-coordinate along the thickness (the height) of the beam, and the y-coordinate is taken along the width of the beam. In a general beam theory, all applied loads and geometry are such that the displacements (u, v, w) along the coordinates (x, y, z) are only functions of the x and z coordinates. Here it is further assumed that the displacement v is identically zero.

The simplest beam theory is the *Euler–Bernoulli beam theory* (EBT), which is based on the displacement field

$$u^E(x, z) = -z \frac{dw_0^E}{dx} \tag{2.1.1a}$$

$$w^E(x, z) = w_0^E(x) \tag{2.1.1b}$$

where w_0 is the transverse deflection of the point $(x, 0)$ of a point on the mid-plane (i.e., $z = 0$) of the beam and the superscript 'E' denotes the quantities in the Euler–Bernoulli beam theory. The displacement field in Eq. (2.1.1) implies that straight lines normal to the mid-plane before deformation remain straight and normal to the mid-plane after deformation, as shown in Figure 2.1.1a. These assumptions amount to neglecting both transverse shear and transverse normal strains.

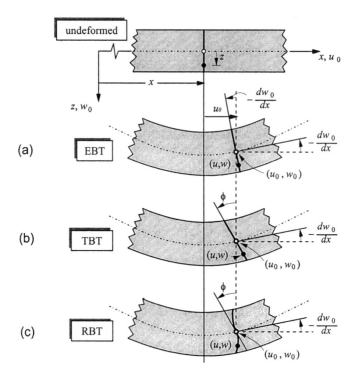

Figure 2.1.1. Deformation of a typical transverse normal line in various beam theories (u_0 denotes displacement due to in-plane stretching, which is not considered here).

The next theory in the hierarchy of beam theories is the Timoshenko beam theory (TBT) [e.g., Timoshenko (1921)], which is based on the displacement field

$$u^T(x, z) = z\phi^T(x) \tag{2.1.2a}$$
$$w^T(x, z) = w_0^T(x) \tag{2.1.2b}$$

where ϕ denotes the rotation of the cross section (see Figure 2.1.1b) and the superscript 'T' denotes the quantities in the Timoshenko beam theory. In the Timoshenko beam theory the normality assumption of the Euler–Bernoulli beam theory is relaxed and a constant state of transverse shear strain (and thus constant shear stress computed from the constitutive equation) with respect to the thickness coordinate is included. The Timoshenko beam theory requires shear correction factors to compensate for the error due to this constant shear stress assumption. As stated earlier, the shear correction factors depend not only on the material and geometric parameters but also on the loading and boundary conditions.

In higher-order theories, the Euler–Bernoulli hypothesis is further relaxed by removing the straightness assumption. Theories higher than third order are seldom used because the accuracy gained is so little that the effort required to solve the equations is not justified.

A second-order theory with transverse inextensibility is based on the displacement field

$$u(x, z) = z\phi(x) + z^2\psi(x) \tag{2.1.3a}$$
$$w(x, z) = w_0(x) \tag{2.1.3b}$$

where ϕ now represents the slope $\partial u/\partial x$ at $z = 0$ (see Figure 2.1.1c) of the deformed line that was straight in the undeformed beam, and ϕ and ψ together define the quadratic nature of the deformed line. Similarly, a third-order beam theory [see Jemielita (1975), Levinson (1981), Bickford (1982), Reddy (1984a,b), Heyliger and Reddy (1988)] is based on the displacement field

$$u^R(x, z) = z\phi^R(x) + z^2\psi^R(x) + z^3\theta^R(x) \tag{2.1.4a}$$
$$w^R(x, z) = w_0^R(x) \tag{2.1.4b}$$

where the superscript R denotes the quantities in the Reddy beam theory.

The following displacement field can be found in the works of Jemielita (1975), and a similar displacement field was used by Levinson (1980, 1981), Bickford (1982), Reddy (1984a,b), and Heyliger and Reddy (1988):

$$u^R(x,z) = z\phi^R(x) - \alpha z^3 \left(\phi^R + \frac{dw_0^R}{dx}\right) \qquad (2.1.5a)$$

$$w^R(x,z) = w_0^R(x) \qquad (2.1.5b)$$

where $\alpha = 4/(3h^2)$. The displacement field accommodates a quadratic variation of the transverse shear strain (and hence shear stress) and the vanishing of transverse shear strain (and hence shear stress) on the top and bottom planes of a beam. Thus, there is no need to use shear correction factors in the third-order beam theory. Levinson (1981) used a vector approach to derive the equations of equilibrium, which are essentially the same as those of the Timoshenko beam theory. Bickford (1982) and Reddy (1984a,b) independently derived variationally consistent equations of motion associated with the displacement field (2.1.5a,b). Bickford's work was limited to isotropic beams, while Reddy's study considered laminated composite plates. The third-order laminated plate theory of Reddy (1984a,b) was specialized by Heyliger and Reddy (1988) to study linear and nonlinear bending and vibrations of isotropic beams. For other pertinent works on third-order theory of beams, the reader may consult the textbooks of Reddy (1984b, 1997a, 1999a) and references therein.

2.1.2 Euler–Bernoulli Beam Theory (EBT)

The virtual strain energy δU of a beam is given by

$$\delta U = \int_0^L \int_A \sigma_{xx} \delta \varepsilon_{xx} \, dA dx \qquad (2.1.6)$$

where δ is the variational symbol, A the cross-sectional area of the uniform beam, L the length of the beam, σ_{xx} the axial stress, and ε_{xx} the normal strain. Note that the strain energy associated with the shearing strain is zero in the Euler–Bernoulli beam theory.

Using the linear strain-displacement relation [see Eq. (2.1.1a)]

$$\varepsilon_{xx} = \frac{\partial u^E}{\partial x} = -z \frac{d^2 w_0^E}{dx^2} \qquad (2.1.7)$$

in Eq. (2.1.6), we obtain

$$\delta U = -\int_0^L M_{xx}^E \frac{d^2 \delta w_0^E}{dx^2}\, dx \qquad (2.1.8)$$

where M_{xx}^E is the bending moment

$$M_{xx}^E = \int_A z\sigma_{xx}\, dA \qquad (2.1.9)$$

Assuming that the transverse load $q(x)$ acts at the centroidal axis of the beam and that there are no other applied loads, the virtual potential energy of the load q is given by

$$\delta V = -\int_0^L q \delta w_0^E\, dx \qquad (2.1.10)$$

The principle of virtual displacements states that *if a body is in equilibrium, then the total virtual work done, $\delta W = \delta U + \delta V$, is zero.* Thus, we have

$$\delta W = -\int_0^L \left(M_{xx}^E \frac{d^2 \delta w_0^E}{dx^2} + q \delta w_0^E \right) dx = 0 \qquad (2.1.11)$$

Integration by parts of the first term in Eq. (2.1.11) twice leads to

$$\int_0^L \left(-\frac{d^2 M_{xx}^E}{dx^2} - q \right) \delta w_0^E\, dx + \left[M_{xx}^E \frac{d \delta w_0^E}{dx} - \frac{d M_{xx}^E}{dx} \delta w_0^E \right]_0^L = 0 \qquad (2.1.12)$$

Since δw_0 is arbitrary in $(0 < x < L)$, we obtain the equilibrium equation

$$-\frac{d^2 M_{xx}^E}{dx^2} = q \quad \text{for } 0 < x < L \qquad (2.1.13)$$

It is useful to introduce the shear force Q_x^E and rewrite the equilibrium equation (2.1.13) in the following form

$$-\frac{d M_{xx}^E}{dx} + Q_x^E = 0, \quad -\frac{d Q_x^E}{dx} = q \qquad (2.1.14)$$

The form of the boundary conditions of the Euler–Bernoulli theory is provided by the boundary expression in Eq. (2.1.12). It is clear that either the displacement w_0^E is known or the shear force $Q_x^E \equiv dM_{xx}^E/dx$ is specified at a point on the boundary. In addition, either the slope dw_0^E/dx is specified or the bending moment M_{xx}^E is known at a boundary point. Thus, we have

$$\text{Specify}: \quad \begin{Bmatrix} w_0^E \\ \dfrac{dw_0^E}{dx} \end{Bmatrix} \quad \text{or} \quad \begin{Bmatrix} Q_x^E \equiv \dfrac{dM_{xx}^E}{dx} \\ M_{xx}^E \end{Bmatrix} \quad (2.1.15)$$

Note that specifying w_0^E or dw_0^E/dx is known as an essential boundary condition while specifying Q_x^E or M_{xx} is known as the natural boundary condition. In mechanics, an essential boundary condition is known as the kinematic or geometric boundary condition while the natural boundary condition is known as the statical or force boundary condition.

Using Hooke's law, we can write

$$\sigma_{xx} = E_x \varepsilon_{xx} = -E_x z \frac{d^2 w_0^E}{dx^2} \quad (2.1.16)$$

where E_x is the modulus of elasticity. Thus, we have

$$M_{xx}^E = \int_A z\sigma_{xx}\, dA = -D_{xx}\frac{d^2 w_0^E}{dx^2} \quad (2.1.17)$$

where $D_{xx} = E_x I_{yy}$ is the flexural rigidity of the beam and $I_{yy} = \int_A z^2 dA$ the second moment of area about the y-axis. The substitution of Eq. (2.1.17) into Eqs. (2.1.13) and (2.1.15) yields

$$\frac{d^2}{dx^2}\left(D_{xx}\frac{d^2 w_0^E}{dx^2}\right) = q \quad \text{for } 0 < x < L \quad (2.1.18)$$

$$\text{Specify}: \quad \begin{Bmatrix} w_0^E \\ \dfrac{dw_0^E}{dx} \end{Bmatrix} \quad \text{or} \quad \begin{Bmatrix} Q_x^E = -\dfrac{d}{dx}\left(D_{xx}\dfrac{d^2 w_0^E}{dx^2}\right) \\ M_{xx}^E = -D_{xx}\dfrac{d^2 w_0^E}{dx^2} \end{Bmatrix} \quad (2.1.19)$$

Some standard boundary conditions associated with the Euler–Bernoulli beam theory are given below:

Simple support: The transverse displacement w_0^E is prescribed as zero and the transverse shear force $Q_x^E \equiv dM_{xx}^E/dx$ is unknown. In addition, the bending moment M_{xx}^E should be specified while the slope dw_0^E/dx is not specified.

Clamped: The transverse deflection w_0^E as well as the slope dw_0^E/dx are specified to be zero. The shear force Q_x^E and bending moment M_{xx}^E are unknown.

Free: The transverse deflection w_0^E as well as the slope dw_0^E/dx are not specified. The shear force Q_x^E and bending moment M_{xx}^E should be specified.

Elastically supported: The shear force is given by $Q_x^E = -k_1 w_0^E$ at the support, where k_1 is the spring constant of the elastic support (assumed to be linear). If, in addition, a rotational spring is there, the bending moment is then $M_{xx}^E = k_2(dw_0^E/dx)$, where k_2 is the torsional spring constant.

The bending solutions for the Euler–Bernoulli beam under the transverse load q may be readily obtained by integrating the fourth-order differential equation (2.1.18) and using two boundary conditions from (2.1.19) at each end of the beam to evaluate the integration constants.

2.1.3 Timoshenko Beam Theory (TBT)

In view of the displacement field given in Eq. (2.1.2), the strain-displacement relations are given by

$$\varepsilon_{xx} = \frac{\partial u^T}{\partial x} = z\frac{d\phi^T}{dx} \tag{2.1.20a}$$

$$\gamma_{xz} = \frac{\partial u^T}{\partial z} + \frac{\partial w^T}{\partial x} = \phi^T + \frac{dw_0^T}{dx} \tag{2.1.20b}$$

Note that the transverse shear strain is nonzero. Hence, the virtual strain energy δU includes the virtual energy associated with the shearing strain, i.e.

$$\begin{aligned}\delta U &= \int_0^L \int_A (\sigma_{xx}\delta\varepsilon_{xx} + \sigma_{xz}\delta\gamma_{xz})\,dA\,dx \\ &= \int_0^L \int_A \left[\sigma_{xx} z\frac{d\delta\phi^T}{dx} + \sigma_{xz}\left(\delta\phi^T + \frac{d\delta w_0^T}{dx}\right)\right] dA\,dx \\ &= \int_0^L \left[M_{xx}^T \frac{d\delta\phi^T}{dx} + Q_x^T\left(\delta\phi^T + \frac{d\delta w_0^T}{dx}\right)\right] dx \end{aligned} \tag{2.1.21}$$

Here, σ_{xx} is the normal stress, σ_{xz} the transverse shear stress, and M_{xx}^T and Q_x^T are the bending moment and shear force, respectively

$$M_{xx}^T = \int_A z\sigma_{xx}\, dA, \quad Q_x^T = \int_A \sigma_{xz}\, dA \qquad (2.1.22)$$

As before, assuming that the transverse load $q(x)$ acts at the centroidal axis of the Timoshenko beam, the virtual potential energy of the transverse load q is given by

$$\delta V = -\int_0^L q(x)\delta w_0^T\, dx \qquad (2.1.23)$$

Substituting the expressions for δU and δV into $\delta W = \delta U + \delta V$, and carrying out integration by parts to relieve δw_0^T and $\delta \phi^T$ of any differentiation, we obtain

$$\begin{aligned}
0 &= \int_0^L \left[M_{xx}^T \frac{d\delta\phi^T}{dx} + Q_x^T \left(\delta\phi^T + \frac{d\delta w_0^T}{dx}\right) - \delta w_0^T q \right] dx \\
&= \int_0^L \left[\left(-\frac{dM_{xx}^T}{dx} + Q_x^T\right)\delta\phi^T + \left(-\frac{dQ_x^T}{dx} - q\right)\delta w_0^T \right] dx \\
&\quad + \left[M_{xx}^T \delta\phi^T + Q_x^T \delta w_0^T \right]_0^L \qquad (2.1.24)
\end{aligned}$$

Setting the coefficients of δw_0^T and $\delta \phi^T$ in $0 < x < L$ to zero, the following equilibrium equations are obtained [c.f. Eq. (2.1.14)] :

$$-\frac{dM_{xx}^T}{dx} + Q_x^T = 0, \quad -\frac{dQ_x^T}{dx} = q \qquad (2.1.25)$$

The boundary conditions of the Timoshenko beam theory are of the form

$$\text{Specify}: \quad \begin{Bmatrix} w_0^T \\ \phi^T \end{Bmatrix} \quad \text{or} \quad \begin{Bmatrix} Q_x^T \\ M_{xx}^T \end{Bmatrix} \qquad (2.1.26)$$

Using the constitutive relations

$$\sigma_{xx} = E_x \varepsilon_{xx}, \quad \sigma_{xz} = G_{xz}\gamma_{xz} \qquad (2.1.27)$$

we can express the bending moment and shear force in terms of the generalized displacements (w_0^T, ϕ^T)

$$M_{xx}^T = \int_A z\sigma_{xx}\, dA = D_{xx}\frac{d\phi^T}{dx} \qquad (2.1.28)$$

$$Q_x^T = K_s \int_A \sigma_{xz}\, dA = K_s A_{xz}\left(\phi^T + \frac{dw_0^T}{dx}\right) \qquad (2.1.29)$$

where

$$D_{xx} = \int_A E_x z^2\, dA = E_x I_{yy}, \quad A_{xz} = \int_A G_{xz}\, dA = G_{xz} A \qquad (2.1.30)$$

and K_s is the shear correction factor that has been introduced to compensate for the error caused by assuming a constant transverse shear stress distribution through the beam depth. The usual approaches for estimating the shear correction factors are either by matching the high frequency spectrum of vibrating beams (e.g., Mindlin and Deresiewicz 1954, Stephen 1982) or by using approximation procedures and simplifying assumptions within the linear theory of elasticity (e.g., Cowper 1966 and Bert 1973).

Substituting for M_{xx}^T and Q_x^T from Eqs. (2.1.28) and (2.1.29) into (2.1.25) and (2.1.26), we obtain the governing equations and boundary conditions in terms of the generalized displacements:

$$-\frac{d}{dx}\left(D_{xx}\frac{d\phi^T}{dx}\right) + K_s A_{xz}\left(\phi^T + \frac{dw_0^T}{dx}\right) = 0 \qquad (2.1.31)$$

$$-\frac{d}{dx}\left[K_s A_{xz}\left(\phi^T + \frac{dw_0^T}{dx}\right)\right] = q \qquad (2.1.32)$$

for $0 < x < L$ and

$$\text{Specify}: \quad \begin{Bmatrix} w_0^T \\ \phi^T \end{Bmatrix} \quad \text{or} \quad \begin{Bmatrix} K_s A_{xz}\left(\phi^T + \frac{dw_0^T}{dx}\right) \\ D_{xx}\frac{d\phi^T}{dx} \end{Bmatrix} \qquad (2.1.33)$$

at the boundary.

The common boundary conditions associated with the Timoshenko beam theory are given below:

Simple support: The transverse displacement w_0^T is prescribed as zero and the transverse shear force Q_x^T is unknown. In addition, the bending moment M_{xx}^T should be specified while the rotation ϕ^T is not specified.

Clamped: The transverse deflection w_0^T as well as the rotation ϕ^T are specified to be zero. The shear force Q_x^T and bending moment M_{xx}^T are unknown.

Free: The transverse deflection w_0^T as well as the rotation ϕ^T are not specified. The shear force Q_x^T and bending moment M_{xx}^T should be specified.

Elastically supported: The shear force is given by $Q_x^T = -k_1 w_0^T$ at the support, where k_1 is the spring constant of the elastic support (assumed to be linear). If, in addition, a rotational spring is there, the bending moment is equal to $M_{xx}^T = -k_2 \phi^T$, where k_2 is the torsional spring constant.

The equilibrium equations (2.1.25) of the Timoshenko beam theory may be combined to obtain

$$\frac{d^2 M_{xx}^T}{dx^2} = \frac{dQ_x^T}{dx} = -q \quad \text{or} \quad \frac{d^2}{dx^2}\left(D_{xx}\frac{d\phi^T}{dx}\right) = -q \qquad (2.1.34)$$

$$\frac{d}{dx}\left(K_s A_{xz}\frac{dw_0^T}{dx}\right) = -q - \frac{d}{dx}\left(K_s A_{xz}\phi^T\right) \qquad (2.1.35)$$

These equations can be readily integrated to determine ϕ^T first and w_0^T next.

2.1.4 Reddy–Bickford Beam Theory (RBT)

The strain-displacement relations of the Reddy–Bickford beam theory are given by [see Eqs. (2.1.5a,b)]

$$\varepsilon_{xx} = \frac{\partial u^R}{\partial x} = z\frac{d\phi^R}{dx} - \alpha z^3\left(\frac{d\phi^R}{dx} + \frac{d^2 w_0^R}{dx^2}\right) \qquad (2.1.36a)$$

$$\gamma_{xz} = \frac{\partial u^R}{\partial z} + \frac{\partial w^R}{\partial x} = \phi^R + \frac{dw_0^R}{dx} - \beta z^2\left(\phi^R + \frac{dw_0^R}{dx}\right) \qquad (2.1.36b)$$

where

$$\alpha = \frac{4}{3h^2}, \quad \beta = 3\alpha = \frac{4}{h^2} \qquad (2.1.37)$$

Hence, the virtual strain energy δU becomes

$$\delta U = \int_0^L \int_A (\sigma_{xx}\delta\varepsilon_{xx} + \sigma_{xz}\delta\gamma_{xz})\, dAdx$$

$$= \int_0^L \int_A \left\{ \sigma_{xx} \left[z\frac{d\delta\phi^R}{dx} - \alpha z^3 \left(\frac{d\delta\phi^R}{dx} + \frac{d^2\delta w_0^R}{dx^2} \right) \right] \right.$$

$$\left. + \sigma_{xz} \left[\left(1 - \beta z^2\right) \left(\delta\phi^R + \frac{d\delta w_0^R}{dx} \right) \right] \right\} dAdx$$

$$= \int_0^L \left[\left(M_{xx}^R - \alpha P_{xx}\right) \frac{d\delta\phi^R}{dx} - \alpha P_{xx}\frac{d^2\delta w_0^R}{dx^2} \right.$$

$$\left. + \left(Q_x^R - \beta R_x\right) \left(\delta\phi^R + \frac{d\delta w_0^R}{dx} \right) \right] dx \qquad (2.1.38)$$

where M_{xx}^R and Q_x^R are the usual bending moment and shear force

$$M_{xx}^R = \int_A z\sigma_{xx}\, dA, \quad Q_x^R = \int_A \sigma_{xz}\, dA \qquad (2.1.39)$$

and P_{xx} and R_x are the higher-order stress resultants

$$P_{xx} = \int_A z^3\sigma_{xx}\, dA, \quad R_x = \int_A z^2\sigma_{xz}\, dA \qquad (2.1.40)$$

It is important to note that unlike the Timoshenko beam theory, there is no need to use a shear correction factor in the Reddy–Bickford beam theory. This is due to the fact that the transverse shear strain is quadratic through the thickness of the beam. The virtual potential energy of the transverse load q is given by

$$\delta V = -\int_0^L q(x)\delta w_0^R\, dx \qquad (2.1.41)$$

Applying the principle of virtual displacements, $\delta W = \delta U + \delta V = 0$, we obtain

$$0 = \int_0^L \left[\left(M_{xx}^R - \alpha P_{xx}\right) \frac{d\delta\phi^R}{dx} - \alpha P_{xx}\frac{d^2\delta w_0^R}{dx^2} \right.$$

$$\left. + \left(Q_x^R - \beta R_x\right) \left(\delta\phi^R + \frac{d\delta w_0^R}{dx} \right) - q\delta w_0^R \right] dx$$

$$= \int_0^L \left[\left(-\frac{d\hat{M}_{xx}^R}{dx} + \hat{Q}_x^R \right) \delta\phi^R + \left(-\alpha\frac{d^2 P_{xx}}{dx^2} - \frac{d\hat{Q}_x^R}{dx} - q \right) \delta w_0^R \right] dx$$

$$+ \left[\hat{M}_{xx}^R \delta\phi^R + \left(\alpha\frac{dP_{xx}}{dx} + \hat{Q}_x^R \right) \delta w_0^R - \alpha P_{xx}\frac{d\delta w_0^R}{dx} \right]_0^L \qquad (2.1.42)$$

where
$$\hat{M}_{xx}^R = M_{xx}^R - \alpha P_{xx}, \quad \hat{Q}_x^R = Q_x^R - \beta R_x \qquad (2.1.43)$$

Setting the expressions associated with the arbitrary δw_0^R and $\delta \phi^R$ in $0 < x < L$ to zero, we obtain the following equilibrium equations for the Reddy–Bickford beam theory:

$$-\frac{d\hat{M}_{xx}^R}{dx} + \hat{Q}_x^R = 0 \qquad (2.1.44)$$

$$-\alpha \frac{d^2 P_{xx}}{dx^2} - \frac{d\hat{Q}_x^R}{dx} = q \qquad (2.1.45)$$

The form of the boundary conditions for the Reddy–Bickford theory is

$$\text{Specify}: \left\{ \begin{array}{c} w_0^R \\ \frac{dw_0^R}{dx} \\ \phi^R \end{array} \right\} \quad \text{or} \quad \left\{ \begin{array}{c} \hat{V}_x^R \equiv \alpha \frac{dP_{xx}}{dx} + \hat{Q}_x^R \\ \alpha P_{xx} \\ \hat{M}_{xx}^R \end{array} \right\} \qquad (2.1.46)$$

where the $\hat{V}_x^R$ is the effective shear force. For the Reddy–Bickford beam theory, it can be seen that the boundary conditions require the specification of the primary (or kinematic) variables w_0^R, dw_0^R/dx, ϕ^R or else the secondary (force) variables $\hat{V}_x^R$, P_{xx}, $\hat{M}_{xx}^R$ are equal to zero. Note that the Reddy–Bickford beam theory has three boundary conditions at each end of the beam, unlike both Euler–Bernoulli and Timoshenko beam theories which have only two boundary conditions at each end of the beam. The total of six boundary conditions are required because the Reddy–Bickford beam theory is a sixth-order theory while the other two beam theories are fourth-order theories.

The stress resultant-displacement relations for the RBT are given by

$$M_{xx}^R = \int_A z\sigma_{xx}\, dA = \hat{D}_{xx}\frac{d\phi^R}{dx} - \alpha F_{xx}\frac{d^2 w_0^R}{dx^2} \qquad (2.1.47)$$

$$P_{xx} = \int_A z^3 \sigma_{xx}\, dA = \hat{F}_{xx}\frac{d\phi^R}{dx} - \alpha H_{xx}\frac{d^2 w_0^R}{dx^2} \qquad (2.1.48)$$

$$Q_x^R = \int_A \sigma_{xz}\, dz = \hat{A}_{xz}\left(\phi^R + \frac{dw_0^R}{dx}\right) \qquad (2.1.49)$$

$$R_x = \int_A z^2 \sigma_{xz}\, dz = \hat{D}_{xz}\left(\phi^R + \frac{dw_0^R}{dx}\right) \qquad (2.1.50)$$

where

$$\hat{D}_{xx} = D_{xx} - \alpha F_{xx}, \quad \hat{F}_{xx} = F_{xx} - \alpha H_{xx} \quad (2.1.51a)$$
$$\hat{A}_{xz} = A_{xz} - \beta D_{xz}, \quad \hat{D}_{xz} = D_{xz} - \beta F_{xz} \quad (2.1.51b)$$

$$(A_{xx}, D_{xx}, F_{xx}, H_{xx}) = \int_A (1, z^2, z^4, z^6) E_x \, dA \quad (2.1.52a)$$
$$(A_{xz}, D_{xz}, F_{xz}) = \int_A (1, z^2, z^4) G_{xz} \, dA \quad (2.1.52b)$$

Expressing the equations of equilibrium in terms of the displacements w_0 and ϕ, we have

$$-\frac{d}{dx}\left(\bar{D}_{xx}\frac{d\phi^R}{dx} - \alpha\hat{F}_{xx}\frac{d^2 w_0^R}{dx^2}\right) + \bar{A}_{xz}\left(\phi^R + \frac{dw_0^R}{dx}\right) = 0$$
$$(2.1.53)$$

$$-\alpha\frac{d^2}{dx^2}\left(\hat{F}_{xx}\frac{d\phi^R}{dx} - \alpha H_{xx}\frac{d^2 w_0^R}{dx^2}\right) - \frac{d}{dx}\left[\bar{A}_{xz}\left(\phi^R + \frac{dw_0^R}{dx}\right)\right] = q$$
$$(2.1.54)$$

where

$$\bar{A}_{xz} = \hat{A}_{xz} - \beta\hat{D}_{xz}, \quad \bar{D}_{xx} = \hat{D}_{xx} - \alpha\hat{F}_{xx}, \quad \bar{F}_{xx} = \hat{F}_{xx} - \alpha\hat{H}_{xx} \quad (2.1.55)$$

The common boundary conditions associated with the Reddy–Bickford beam theory are given below:

Simple support: The transverse displacement w_0^R is prescribed as zero and the transverse shear force $\hat{Q}_x^R$ is unknown. In addition, the bending moment $\hat{M}_{xx}^R$ and the higher-order stress resultant αP_{xx} should be specified while the rotation ϕ^R and the slope dw_0^R/dx are not specified.

Clamped: The transverse deflection w_0^R as well as the rotation ϕ^R and the slope dw_0^R/dx are specified to be zero. The shear force $\hat{Q}_x^R$, bending moment $\hat{M}_{xx}^R$, and the higher-order stress resultant αP_{xx} are unknown.

Free: The transverse deflection w_0^R, the rotation ϕ^R, and the slope dw_0^R/dx are not specified. The shear force $\hat{Q}_x^R$, bending moment $\hat{M}_{xx}^R$, and the higher-order stress resultant αP_{xx} should be specified.

Elastically supported: The shear force is given by $\hat{Q}_x^R = -k_1 w_0^R$ at the support, where k_1 is the spring constant of the elastic support (assumed to be linear). If a rotational spring is also there, the bending moment is then $\hat{M}_{xx}^R = -k_2 \phi^R$, where k_2 is the torsional spring constant. In addition, the higher-order stress resultant αP_{xx} is zero.

2.2 Relationships Between EBT and TBT

2.2.1 General Comments

The objective of this section is to establish relationships between the bending solutions (i.e., deflection, rotation, bending moment, and shear force) of the Timoshenko beam theory (TBT) in terms of the corresponding quantities of the Euler–Bernoulli beam theory (EBT). The relationships are established using the load equivalence, as shown below (see Wang 1995a).

It is clear from Eqs. (2.1.14), (2.1.17), (2.1.25), and (2.1.28) that the shear forces, bending moments and the slopes of the two beams are related by

$$Q_x^T = Q_x^E + C_1 \tag{2.2.1}$$

$$M_{xx}^T = M_{xx}^E + C_1 x + C_2 \tag{2.2.2}$$

$$\phi^T = -\frac{dw_0^E}{dx} + C_1 \frac{x^2}{2D_{xx}} + C_2 \frac{x}{D_{xx}} + C_3 \frac{1}{D_{xx}} \tag{2.2.3}$$

Substitution of Eq. (2.2.3) into Eq. (2.1.29), using Eqs. (2.1.25) and (2.2.2), and then integrating with respect to x yields the following deflection relationship:

$$w_0^T = w_0^E + \frac{M_{xx}^E}{K_s A_{xz}} + C_1 \left(\frac{x}{K_s A_{xz}} - \frac{x^3}{6 D_{xx}} \right) - C_2 \frac{x^2}{2 D_{xx}} - C_3 \frac{x}{D_{xx}} - C_4 \frac{1}{D_{xx}} \tag{2.2.4}$$

where C_1, C_2, C_3, C_4 are constants of integration. These constants are to be determined using the boundary conditions of the particular beam. For free (F), simply supported (S) and clamped (C) ends, the boundary conditions are given by

$$\mathbf{F}: M_{xx}^E = M_{xx}^T = Q_x^E = Q_x^T = 0 \tag{2.2.5}$$

$$\mathbf{S}: w_0^E = w_0^T = M_{xx}^E = M_{xx}^T = 0 \tag{2.2.6}$$

$$\mathbf{C}: w_0^E = w_0^T = \frac{dw_0^E}{dx} = \phi^T = 0 \tag{2.2.7}$$

In the following subsections, the evaluation of the constants C_i is illustrated for single-span beams of various combinations of end conditions. For brevity, the beams will be designated by two letters. The first letter indicates the type of end condition at the left end ($x = 0$) and the second letter refers to the end condition at the right end ($x = L$). Also, for convenience, the non-dimensional shear parameter Ω is introduced, as follows:

$$\Omega = \frac{D_{xx}}{K_s A_{xz} L^2} = \frac{E_x}{K_s G_{xz}} \left(\frac{r}{L}\right)^2 \tag{2.2.8}$$

where $r = \sqrt{I_{yy}/A}$ is the radius of gyration and L/r denotes the slenderness ratio of the beam.

2.2.2 Simply Supported (SS) Beams

The boundary conditions for simply supported beams are given by Eq. (2.2.6) for $x = 0$ and $x = L$. The substitution of these boundary conditions into Eqs. (2.2.2) and (2.2.4) gives the following values of the constants:

$$C_1 = C_2 = C_3 = C_4 = 0 \tag{2.2.9}$$

2.2.3 Clamped-Free (CF) Beams

The boundary conditions for clamped-free beams are given by Eq. (2.2.7) for $x = 0$ and by Eq. (2.2.5) for $x = L$. The substitution of these boundary conditions into Eqs. (2.2.1) to (2.2.4) gives the following values of the constants:

$$C_1 = C_2 = C_3 = 0 \quad \text{and} \quad C_4 = M^E_{xx}(0)\Omega L^2 \tag{2.2.10}$$

2.2.4 Free-Clamped (FC) Beams

The boundary conditions for free-clamped beams are given by Eq. (2.2.5) for $x = 0$ and by Eq. (2.2.7) for $x = L$. The substitution of these boundary conditions into Eqs. (2.2.1) to (2.2.4) gives the following:

$$C_1 = C_2 = C_3 = 0 \quad \text{and} \quad C_4 = M^E_{xx}(L)\Omega L^2 \tag{2.2.11}$$

2.2.5 Clamped-Simply Supported (CS) Beams

The boundary conditions for clamped-simply supported beams are given by Eq. (2.2.7) for $x = 0$ and by Eq. (2.2.6) for $x = L$. The substitution of these boundary conditions into Eqs. (2.2.2) to (2.2.4) gives the following:

$$C_1 = \frac{3\Omega}{(1+3\Omega)L}M_{xx}^E(0), \quad C_2 = -C_1, \quad C_3 = 0, \quad C_4 = M_{xx}^E(0)\Omega L^2 \quad (2.2.12)$$

2.2.6 Simply Supported-Clamped (SC) Beams

The boundary conditions for simply supported-clamped beams are given by Eq. (2.2.6) for $x = 0$ and by Eq. (2.2.7) for $x = L$. The substitution of these boundary conditions into Eqs. (2.2.2) to (2.2.4) gives the following:

$$C_1 = -\frac{3\Omega}{(1+3\Omega)L}M_{xx}^E(L), \quad C_2 = 0, \quad C_3 = -C_1, \quad C_4 = 0 \quad (2.2.13)$$

2.2.7 Clamped (CC) Beams

The boundary conditions for clamped beams are given by Eq. (2.2.7) for $x = 0$ and $x = L$. The substitution of these boundary conditions into Eqs. (2.2.3) and (2.2.4) gives the following:

$$C_1 = -\frac{12\Omega}{(1+12\Omega)L}\left[M_{xx}^E(L) - M_{xx}^E(0)\right]$$
$$C_2 = \frac{6\Omega}{(1+12\Omega)}\left[M_{xx}^E(L) - M_{xx}^E(0)\right]$$
$$C_3 = 0, \quad C_4 = M_{xx}^E(0)\Omega L^2 \quad (2.2.14)$$

The results show that for statically determinate beams, the shear force, bending moment, and slope in the two theories remain the same, while the deflection differs. For statically indeterminate beams, the solutions for shear force, bending moment, slope, and deflection predicted by the two theries are not the same.

2.2.8 Summary of Relationships

In view of the foregoing expressions for the constants $C_i, i = 1, ..., 4$, the relationships of the slope, the bending moment, the shear force and the deflection may be obtained respectively from Eqs. (2.2.1) to (2.2.4). The relationships are summarized in Table 2.2.1.

Table 2.2.1 Generalized deflection and force relationships between Timoshenko and Euler–Bernoulli beams.

B. C.	Relationships
SS	$w_0^T(x) = w_0^E(x) + \frac{\Omega L^2}{D_{xx}} M_{xx}^E(x)$ $\phi^T(x) = -\frac{dw_0^E(x)}{dx}$ $M_{xx}^T(x) = M_{xx}^E(x)$ $Q_x^T(x) = Q_x^E(x)$
CF	$w_0^T(x) = w_0^E(x) + \frac{\Omega L^2}{D_{xx}} \left[M_{xx}^E(x) - M_{xx}^E(0) \right]$ $\phi^T(x) = -\frac{dw_0^E(x)}{dx}$ $M_{xx}^T(x) = M_{xx}^E(x)$ $Q_x^T(x) = Q_x^E(x)$
FC	$w_0^T(x) = w_0^E(x) + \frac{\Omega L^2}{D_{xx}} \left[M_{xx}^E(x) - M_{xx}^E(L) \right]$ $\phi^T(x) = -\frac{dw_0^E(x)}{dx}$ $M_{xx}^T(x) = M_{xx}^E(x)$ $Q_x^T(x) = Q_x^E(x)$
CS	$w_0^T(x) = w_0^E(x) + \frac{\Omega L^2}{D_{xx}} \left[M_{xx}^E(x) - M_{xx}^E(0) \right]$ $\qquad + \frac{3\Omega L^2}{D_{xx}(1+3\Omega)} \frac{x}{L} \left(\Omega + \frac{x}{2L} - \frac{x^2}{6L^2} \right) M_{xx}^E(0)$ $\phi^T(x) = -\frac{dw_0^E(x)}{dx} + \frac{3\Omega L}{D_{xx}(1+3\Omega)} \frac{x}{L} \left(1 - \frac{x}{2L}\right) M_{xx}^E(0)$ $M_{xx}^T(x) = M_{xx}^E(x) - \frac{3\Omega}{(1+3\Omega)} \left(1 - \frac{x}{L}\right) M_{xx}^E(0)$ $Q_x^T(x) = Q_x^E(x) + \frac{3\Omega}{(1+3\Omega)L} M_x^E(0)$

(Table 2.2.1 is continued on the next page)

(Table 2.2.1 is continued from the previous page)

SC	$w_0^T(x) = w_0^E(x) + \frac{\Omega L^2}{D_{xx}} M_{xx}^E(x)$ $\quad - \frac{3\Omega L^2}{D_{xx}(1+3\Omega)} \frac{x}{L} \left(\Omega + \frac{1}{2} - \frac{x^2}{6L^2}\right) M_{xx}^E(L)$ $\phi^T(x) = -\frac{dw_0^E(x)}{dx} + \frac{3\Omega L}{2D_{xx}(1+3\Omega)} \left(\frac{x^2}{L^2} - 1\right) M_{xx}^E(L)$ $M_{xx}^T(x) = M_{xx}^E(x) - \frac{3\Omega}{(1+3\Omega)} \frac{x}{L} M_{xx}^E(L)$ $Q_x^T(x) = Q_x^E(x) - \frac{3\Omega}{(1+3\Omega)L} M_{xx}^E(L)$
CC†	$w_0^T(x) = w_0^E(x) + \frac{\Omega L^2}{D_{xx}} \left[M_{xx}^E(x) - M_{xx}^E(0)\right]$ $\quad + \frac{3\mu L^2}{D_{xx}} \frac{x}{L} \left(\frac{2}{3} \frac{x^2}{L^2} - 4\Omega - \frac{x}{L}\right) \left[M_{xx}^E(L) - M_{xx}^E(0)\right]$ $\phi^T(x) = -\frac{dw_0^E}{dx} - \frac{6\mu L}{D_{xx}} \frac{x}{L} \left(\frac{x}{L} - 1\right) \left[M_{xx}^E(L) - M_{xx}^E(0)\right]$ $M_{xx}^T(x) = M_{xx}^E(x) - 6\mu \left(2\frac{x}{L} - 1\right) \left[M_{xx}^E(L) - M_{xx}^E(0)\right]$ $Q_x^T(x) = Q_x^E(x) - \frac{12\mu}{L} \left[M_{xx}^E(L) - M_{xx}^E(0)\right]$

†· $\mu = \frac{\Omega}{(1+12\Omega)}$.

It can be seen from Table 2.2.1 that the bending moments and shear forces are the same for statically determinate Timoshenko and Euler–Bernoulli beams, i.e. SS, CF and FC beams. Also for these beams, the rotation of the Timoshenko beam is equal to the slope of the Euler–Bernoulli beam. For statically indeterminate CS, SC and CC beams, the stress-resultants are not the same, because the compatibility equation involving the effect of transverse shear deformation is required for the solution. The deflection relationships show clearly the effect of shear deformation. The shear-deflection component increases with increasing magnitude of the Euler–Bernoulli moment (or transverse load) and shear parameter.

2.3 Relationships Between EBT and RBT

Here, we develop the relationships between the bending solutions of the Euler–Bernoulli beam theory (EBT) and the Reddy–Bickford beam theory (RBT). At the outset, we note that both EBT and TBT are fourth-order theories whereas the RBT is a sixth-order theory. The order referred to here is the total order of all equations of equilibrium expressed in terms of the generalized displacements. The refined beam theory is governed by a fourth-order equation in w_0 and a second-order equation in ϕ. Therefore, the relationships between the solutions of two different order theories can only be established by solving an additional second-order equation. The relationships are developed between deflections, rotations and stress resultants of the EBT and RBT for an easy comparison between theories.

First we note that Eqs. (2.1.44) and (2.1.45) together yield

$$-\frac{d^2 M_{xx}^R}{dx^2} = q \qquad (2.3.1)$$

Equating the loads in Eqs. (2.1.13) and (2.3.1), and after integration twice, we obtain

$$M_{xx}^R = M_{xx}^E + C_1 x + C_2 \qquad (2.3.2)$$

The stress resultant-displacement relationships in Eqs. (2.1.47) and (2.1.50) can be expressed as

$$M_{xx}^R = \frac{\hat{D}_{xx}}{\hat{A}_{xz}} \frac{dQ_x^R}{dx} - D_{xx} \frac{d^2 w_0^R}{dx^2} \qquad (2.3.3)$$

$$R_x = \frac{\hat{D}_{xz}}{\hat{A}_{xz}} Q_x^R \qquad (2.3.4)$$

$$P_{xx} = \frac{\hat{F}_{xx}}{\hat{A}_{xz}} \frac{dQ_x^R}{dx} - F_{xx} \frac{d^2 w_0^R}{dx^2}$$

$$= \left(\frac{\hat{F}_{xx}}{\hat{A}_{xz}} - \frac{\hat{F}_{xx} \hat{D}_{xx}}{D_{xx} \hat{A}_{xx}} \right) \frac{dQ_x^R}{dx} - \left(\frac{F_{xx}}{D_{xx}} \right) M_{xx}^R \qquad (2.3.5)$$

where the stiffness coefficients with hats and bars were defined in Eqs. (2.1.51a,b) and (2.1.55), and α and β are defined in Eq. (2.1.37).

Replacing P_{xx} and R_x in Eq. (2.1.44) with the expressions in Eqs. (2.3.4) and (2.3.5), we obtain

$$\left(\frac{\hat{D}_{xx}}{D_{xx}} \right) \frac{dM_{xx}^R}{dx} = \frac{\bar{A}_{xz}}{\hat{A}_{xz}} Q_x^R - \alpha \left(\frac{F_{xx} \hat{D}_{xx}}{D_{xx} \hat{A}_{xz}} - \frac{\hat{F}_{xx}}{\hat{A}_{xz}} \right) \frac{d^2 Q_x^R}{dx^2} \qquad (2.3.6)$$

Using Eq. (2.3.2) and simplifying the coefficients, we arrive at

$$\alpha \left(\frac{F_{xx} \hat{D}_{xx}}{D_{xx} \hat{A}_{xz}} - \frac{\hat{F}_{xx}}{\hat{A}_{xz}} \right) \frac{d^2 Q_x^R}{dx^2} - \frac{\bar{A}_{xz}}{\hat{A}_{xz}} Q_x^R + \left(\frac{\hat{D}_{xx}}{D_{xx}} \right) \left(Q_x^E + C_1 \right) = 0 \qquad (2.3.7)$$

Thus, a second-order differential equation must be solved to determine Q_x^R in terms of Q_x^E. Once Q_x^R is known, M_x^R, ϕ^R, and w_0^R can be determined as will be shown shortly.

The effective shear force V_x^R in the Reddy–Bickford beam theory can be computed from

$$\hat{V}_x^R(x) = Q_x^R - \beta R_x + \alpha \frac{dP_{xx}}{dx} = \frac{dM_{xx}^R}{dx}$$
$$= Q_x^E(x) + C_1 \qquad (2.3.8)$$

where Eqs. (2.1.44) and (2.3.2) are used to derive the last equality.

To determine ϕ^R, we use Eq. (2.1.47):

$$D_{xx}\frac{d\phi^R}{dx} = M_{xx}^R + \alpha F_{xx}\left(\frac{d\phi^R}{dx} + \frac{d^2 w_0^R}{dx^2}\right)$$
$$= M_{xx}^E + C_1 x + C_2 + \frac{\alpha F_{xx}}{\hat{A}_{xz}}\frac{dQ_x^R}{dx}$$
$$= -D_{xx}\frac{d^2 w_0^E}{dx^2} + C_1 x + C_2 + \frac{\alpha F_{xx}}{\hat{A}_{xz}}\frac{dQ_x^R}{dx} \qquad (2.3.9)$$

or

$$D_{xx}\phi^R(x) = -D_{xx}\frac{dw_0^E}{dx} + \frac{\alpha F_{xx}}{\hat{A}_{xz}}Q_x^R + C_1\frac{x^2}{2} + C_2 x + C_3 \qquad (2.3.10)$$

where Eqs. (2.1.17) and (2.3.2) are used in arriving at the last equation.

Lastly, we derive the relation between w_0^R and w_0^E. Using Eqs. (2.1.49) and (2.3.10), we can write

$$D_{xx}\frac{dw_0^R}{dx} = -D_{xx}\phi^R(x) + \frac{D_{xx}}{\hat{A}_{xz}}Q_x^R$$
$$= D_{xx}\frac{dw_0^E}{dx} + \frac{\hat{D}_{xx}}{\hat{A}_{xz}}Q_x^R - C_1\frac{x^2}{2} - C_2 x - C_3 \qquad (2.3.11)$$

and integrating with respect to x, we obtain

$$D_{xx}w_0^R(x) = D_{xx}w_0^E(x) + \frac{\hat{D}_{xx}}{\hat{A}_{xz}}\left(\int Q_x^R(\eta)d\eta\right) - C_1\frac{x^3}{6} - C_2\frac{x^2}{2} - C_3 x - C_4 \qquad (2.3.12)$$

This completes the derivation of the relationships between the solutions of the Euler–Bernoulli beam theory and the Reddy–Bickford beam theory. The constants of integration, C_1, C_2, C_3, C_4 appearing in Eqs. (2.3.2), (2.3.8), (2.3.11) and (2.3.12) are determined using the boundary conditions. Since there are six boundary conditions in the Reddy–Bickford theory [see Eq. (2.1.46)], the remaining two boundary conditions are used in the solving of the second-order differential equation (2.3.7). Boundary conditions for various types of supports are defined below, consistent with the kinematic and natural variables [see Eq. (2.1.46)] of the theory:

$$\mathbf{F}:\ Q_x^R - \beta R_x + \alpha \frac{dP_{xx}}{dx} = 0,\ M_{xx}^R - \alpha P_{xx} = 0,\ P_{xx} = 0 \quad (2.3.13)$$

$$\mathbf{S}:\ w_0^R = 0,\ M_{xx}^R - \alpha P_{xx} = 0,\ P_{xx} = 0 \quad (2.3.14)$$

$$\mathbf{C}:\ w_0^R = 0,\ \phi^R = 0,\ \frac{dw_0^R}{dx} = 0 \quad (2.3.15)$$

Since the second-order equation (2.3.7) requires boundary conditions on Q_x^R, we reduce the force boundary conditions in Eqs. (2.3.13) to (2.3.15) to one in terms of Q_x^R:

Free (F): Equations (2.3.13) and (2.3.15) imply

$$\frac{dQ_x^R}{dx} = 0 \quad (2.3.16)$$

Simply supported (S): Equation (2.3.14) implies

$$\frac{dQ_x^R}{dx} = 0 \quad (2.3.17)$$

Clamped (C): Equations (2.3.15) and (2.1.49) imply

$$Q_x^R = 0 \quad (2.3.18)$$

2.4 Examples

Here, we present two examples to derive the solutions of TBT and of RBT using the relationships derived in Sections 2.2 and 2.3 and the solutions of EBT.

2.4.1 Simply Supported Beam

Consider a simply supported beam under uniformly distributed load of intensity q_0. Using the equilibrium equation (2.1.13) and boundary conditions in Eq. (2.1.19), the stress-resultants and the deflection of the Euler–Bernoulli beam are found to be

$$Q_x^E(x) = \frac{q_0}{2}(L - 2x) \tag{2.4.1}$$

$$M_{xx}^E(x) = \frac{q_0}{2}x(L - x) \tag{2.4.2}$$

$$w_0^E(x) = \frac{q_0 L^4}{24 D_{xx}}\left(\frac{x}{L} - \frac{2x^3}{L^3} + \frac{x^4}{L^4}\right) \tag{2.4.3}$$

Using the relationship for simply supported (SS) beams in Table 2.2.1, the corresponding bending solutions for the Timoshenko beam are

$$Q_x^T(x) = Q_x^E(x) = \frac{q_0 L}{2}\left(1 - 2\frac{x}{L}\right) \tag{2.4.4}$$

$$M_{xx}^T(x) = M_{xx}^E(x) = \frac{q_0 L^2}{2}\frac{x}{L}\left(1 - \frac{x}{L}\right) \tag{2.4.5}$$

$$\begin{aligned} w_0^T(x) &= w_0^E(x) + \frac{1}{K_s A_{xz}} M_{xx}^E(x) \\ &= \frac{q_0 L^4}{24 D_{xx}}\left(\frac{x}{L} - 2\frac{x^3}{L^3} + \frac{x^4}{L^4}\right) + \frac{q_0 L^2}{2 K_s A_{xz}}\left(\frac{x}{L} - \frac{x^3}{L^3}\right) \end{aligned} \tag{2.4.6}$$

In the case of the Reddy–Bickford beam, we need to first solve the second-order differential for the transverse shear force. From Eq. (2.3.8) and Eq. (2.4.6), we have

$$\frac{d^2 Q_x^R}{dx^2} - \lambda^2 Q_x^R = -\mu\left[\frac{q_0}{2}(L - 2x) + C_1\right] \tag{2.4.7}$$

where

$$\lambda^2 = \frac{\bar{A}_{xz} D_{xx}}{\alpha(F_{xx}\hat{D}_{xx} - \hat{F}_{xx}D_{xx})}, \quad \mu = \frac{\hat{A}_{xz}\hat{D}_{xz}}{\alpha(F_{xx}\hat{D}_{xx} - \hat{F}_{xx}D_{xx})} \tag{2.4.8}$$

The solution to this differential equation is

$$Q_x^R(x) = C_5 \sinh \lambda x + C_6 \cosh \lambda x + \frac{\mu}{\lambda^2}\left[\frac{q_0}{2}(L - 2x) + C_1\right] \tag{2.4.9}$$

where C_5 and C_6 are constants to be determined, along with C_1, C_2, C_3, C_4, using the boundary conditions.

The boundary conditions for the problem at hand are

$$w_0^E(0) = w_0^E(L) = M_{xx}^E(0) = M_{xx}^E(L) = 0 \qquad (2.4.10)$$
$$w_0^R(0) = w_0^R(L) = M_{xx}^R(0) = M_{xx}^R(L) = P_{xx}(0) = P_{xx}(L) = 0 \qquad (2.4.11)$$

We note from Eq. (2.3.5) that

$$M_{xx}^R(0) = P_{xx}(0) \text{ imply } \frac{dQ_x^R}{dx}(0) = 0 \qquad (2.4.12a)$$

$$M_{xx}^R(L) = P_{xx}(L) = 0 \text{ imply } \frac{dQ_x^R}{dx}(L) = 0 \qquad (2.4.12b)$$

Using the boundary conditions (2.4.10)–(2.4.12), we find that

$$C_1 = C_2 = C_3 = 0, \quad C_4 = \left(\frac{q_0\mu}{\lambda^4}\right)\left(\frac{\hat{D}_{xx}}{\bar{A}_{xz}}\right)$$

$$C_5 = \frac{q_0\mu}{\lambda^3}, \quad C_6 = -\frac{q_0\mu}{\lambda^3}\tanh\left(\frac{\lambda L}{2}\right) \qquad (2.4.13)$$

and the solution becomes

$$Q_x^R(x) = \left(\frac{q_0\mu}{\lambda^3}\right)\left[\sinh\lambda x - \tanh\left(\frac{\lambda L}{2}\right)\cosh\lambda x \right.$$
$$\left. + \frac{\lambda}{2}(L - 2x)\right] \qquad (2.4.14)$$

$$M_{xx}^R(x) = M_{xx}^E(x) = \frac{q_0}{2}x(L - x) \qquad (2.4.15)$$

$$w_0^R(x) = w_0^E(x) + \left(\frac{q_0\mu}{\lambda^4}\right)\left(\frac{\hat{D}_{xx}}{\bar{A}_{xz}D_{xx}}\right)\left[-\tanh\left(\frac{\lambda L}{2}\right)\sinh\lambda x \right.$$
$$\left. + \cosh\lambda x + \frac{\lambda^2}{2}x(L - x) - 1\right] \qquad (2.4.16)$$

For a rectangular cross-section beam, it can be shown that

$$\frac{\hat{D}_{xx}\hat{D}_{xx}}{\bar{A}_{xz}D_{xx}D_{xx}} = \frac{6}{5\bar{A}_{xz}}, \quad \frac{\hat{D}_{xx}}{\bar{A}_{xz}D_{xx}} = \frac{6}{5\bar{A}_{xz}} \qquad (2.4.17)$$

A close examination of Eq. (2.4.16) shows that the Reddy–Bickford beam solution has an effective shear coefficient, based on the coefficient in the expression for $w_0^R(x)$, of $K_s = 5/6$. Of course, the refined third-order beam theory does not require a shear correction factor. Also the shear correction factor for the Timoshenko beam theory can be obtained, for example, by comparing the maximum deflections of the Timoshenko beams with those of the Reddy–Bickford beams.

2.4.2 Cantilever Beam

For a cantilever beam under uniformly distributed load of intensity q_0, the stress-resultants and the deflection of the Euler–Bernoulli beam are found to be

$$Q_x^E(x) = q_0 L \left(1 - \frac{x}{L}\right) \tag{2.4.18}$$

$$M_{xx}^E(x) = -\frac{q_0 L^2}{2}\left(1 - \frac{x}{L}\right)^2 \tag{2.4.19}$$

$$w_0^E(x) = \frac{q_0 L^4}{24 D_{xx}}\left(6\frac{x^2}{L^2} - 4\frac{x^3}{L^3} + \frac{x^4}{L^4}\right) \tag{2.4.20}$$

Using the relationship for clamped-free (CF) beams in Table 2.2.1, the corresponding bending solutions for the Timoshenko beam are

$$Q_x^T(x) = Q_x^E(x) = q_0 L\left(1 - \frac{x}{L}\right) \tag{2.4.21}$$

$$M_{xx}^T(x) = M_{xx}^E(x) = -\frac{q_0 L^2}{2}\left(1 - \frac{x}{L}\right)^2 \tag{2.4.22}$$

$$\begin{aligned}w_0^T(x) &= w_0^E + \frac{1}{K_s A_{xz}}\left[M_{xx}^E(x) - M_{xx}^E(0)\right] \\ &= \frac{q_0 L^4}{24 D_{xx}}\left(6\frac{x^2}{L^2} - 4\frac{x^3}{L^3} + \frac{x^4}{L^4}\right) + \frac{q_0 L^2}{2 K_s A_{xz}}\frac{x}{L}\left(2 - \frac{x}{L}\right)\end{aligned} \tag{2.4.23}$$

In the case of the Reddy–Bickford beam, we need to first solve the second-order differential equation for the transverse shear force. The general solution of Eq. (2.3.7) with Q_x^E as defined in Eq. (2.4.18) is

$$Q_x^R(x) = C_5 \sinh \lambda x + C_6 \cosh \lambda x + \frac{\mu}{\lambda^2}[q_0(L-x) + C_1] \tag{2.4.24}$$

where λ and μ are defined by Eq. (2.4.8).

The boundary conditions for the cantilever beam are

$$w_0^E(0) = \frac{dw_0^E}{dx}(0) = Q_x^E(L) = M_{xx}^E(L) = 0 \quad (2.4.25)$$

$$w_0^R(0) = \frac{dw_0^R}{dx}(0) = \phi^R(0) = \frac{dM_{xx}^R}{dx}(L) = M_{xx}^R(L) = P_{xx}(L) = 0 \quad (2.4.26)$$

We note from Eq. (2.3.5) that

$$M_{xx}^R(L) - P_{xx}(L) = 0 \quad \text{imply} \quad \frac{dQ_x^R}{dx}(L) = 0 \quad (2.4.27)$$

and from Eqs. (2.3.10) and (2.3.12)

$$\frac{dw_0^R}{dx}(0) = \phi^R(0) = \frac{dw_0^E}{dx}(0) = 0 \quad \text{imply} \quad Q_x^R(0) = 0 \quad (2.4.28)$$

Although $Q_x^R(0)$ obtained from the constitutive relations is zero at the clamped edge, the effective shear force of the theory $\hat{V}_x^R$ at $x = 0$ is indeed not zero. It is given by Eq. (2.3.8).

Using the boundary conditions (2.4.10)–(2.4.12), we find that

$$C_1 = C_2 = C_3 = 0, \quad C_4 = \left(\frac{q_0\mu}{\lambda^4}\right)\left(\frac{1 + \lambda L \sinh \lambda L}{\cosh \lambda L}\right)\left(\frac{\hat{D}_{xx}}{\hat{A}_{xz}}\right)$$

$$C_5 = \frac{q_0\mu}{\lambda^3}\left(\frac{1 + \lambda L \sinh \lambda L}{\cosh \lambda L}\right), \quad C_6 = -\frac{q_0 L\mu}{\lambda^3} \quad (2.4.29)$$

and the solution becomes

$$Q_x^R(x) = \left(\frac{q_0\mu}{\lambda^3 \cosh \lambda L}\right)[\sinh \lambda x - \lambda L \cosh \lambda(L-x)] + \frac{q_0\mu}{\lambda^2}(L-x) \quad (2.4.30)$$

$$M_{xx}^R(x) = M_{xx}^E(x) = q_0\left(Lx - \frac{x^2}{2} - \frac{L^2}{2}\right) \quad (2.4.31)$$

$$w_0^R(x) = w_0^E(x) + \left(\frac{q_0\mu}{2\lambda^2}\right)\left(\frac{\hat{D}_{xx}}{\hat{A}_{xz}D_{xx}}\right)(2Lx - x^2)$$

$$+ \left(\frac{q_0\mu}{\lambda^4 \cosh \lambda L}\right)\left(\frac{\hat{D}_{xx}}{\hat{A}_{xz}D_{xx}}\right)[\cosh \lambda x + \lambda L \sinh \lambda(L-x)]$$

$$- \left(\frac{q_0\mu}{\lambda^4}\right)\left(\frac{\hat{D}_{xx}}{\hat{A}_{xz}D_{xx}}\right)\left(\frac{1 + \lambda L \sinh \lambda L}{\cosh \lambda L}\right) \quad (2.4.32)$$

From Eqs. (2.4.30)–(2.4.32), it can be shown that the effective shear correction factor of the Reddy–Bickford beam theory is $K_s = 5/6$.

2.5 Summary

In this chapter, exact relationships between the bending solutions of the Euler–Bernoulli beam theory and those of the Timoshenko beam theory and the Reddy–Bickford beam theory are presented. For the bending relationships linking Timoshenko and Euler–Bernoulli beam solutions, they are explicit. However, the relationships between Reddy–Bickford beams and Euler–Bernoulli beams require solving an additional second-order differential equation. This arises because the Reddy–Bickford beam theory is a sixth-order beam theory while the Euler–Bernoulli and Timoshenko beam theories are fourth-order ones.

The relationships can be used to generate bending solutions of the Timoshenko and Reddy–Bickford theories whenever the Euler–Bernoulli beam solutions are available. Since solutions of the Euler–Bernoulli beam theory are easily determined or are available in most textbooks on mechanics of materials for a variety of boundary conditions, the correspondence presented herein between the various theories makes it easier to compute the solutions of the Timoshenko beam theory and the Reddy–Bickford beam theory directly from the known Euler–Bernoulli beam solutions. In the next chapter we show how these relationships may be used to develop finite element models of Timoshenko and Reddy–Bickford theories using the finite element model of Euler–Bernoulli beam theory. The stiffness matrix of the shear deformable elements are also 4×4 for the pure bending case, and the finite elements are free from the shear locking phenomenon experienced by the conventional shear deformable finite elements.

The present relationships can be easily extended to symmetrically laminated beams. Indeed the relationships developed herein hold for symmetrically laminated beams in which the Poisson effect is neglected and the transverse deflection is assumed to be only a function of x. The only difference lies in the calculation of the beam stiffnesses, D_{xx}, A_{xz}, and so on, which depend on individual layer stiffnesses and thicknesses.

Further the relationships may be readily modified to link the bending solutions of linear viscoelastic Timoshenko beams and linear viscoelastic Euler–Bernoulli beams under quasi-static loads (see Wang, Yang, and Lam 1997). To do this, one can use the elastic-viscoelastic correspondence principle (Flügge 1975). This principle involves replacing the elastic moduli and the elastic field variables in the elastic solution by the Laplace transformed viscoelastic moduli and viscoelastic field variables. Then, the transformed field variables are converted back to the physical domain.

Problems

2.1 Consider a beam with a rectangular cross section with width b and thickness h. The equilibrium shear stress distribution through the thickness of the beam under a transverse point load Q_0 is given by

$$\sigma_{xz}^c = \frac{3Q_0}{2bh}\left[1 - \left(\frac{2z}{h}\right)^2\right], \quad -\frac{h}{2} \leq z \leq \frac{h}{2} \qquad (i)$$

The transverse shear stress computed using the constitutive equation in the Timoshenko beam theory is constant and is given by $\sigma_{xz}^f = Q_0/bh$. Compute the strain energies due to transverse shear stresses in the two theories and then determine the shear correction factor as the ratio of U_s^f to U_s^c.

2.2 Verify the expressions in Table 2.2.1 for (a) simply supported (SS) beams, (b) clamped-free (CF) beams, (c) clamped-simply supported (CS) beams, and (d) clamped (CC) beams.

2.3 Verify the relations in Eqs. (2.3.3)–(2.3.5).

2.4 Use the deflection relationships to determine (w_0^T, ϕ^T) and (w_0^R, ϕ^R) of a clamped-simply supported beam subjected to uniformly distributed load intensity q_0.

2.5 Starting with a linear distribution of the displacements through the beam thickness in terms of unknown functions (F, G)

$$u(x, z) = zF(x), \quad w(x, z) = w_0(x) + zG(x) \qquad (i)$$

determine the functions F and G such that the following conditions hold:

$$\frac{\partial w}{\partial z} = 0, \quad \frac{\partial u}{\partial z} = \phi \qquad (ii)$$

2.6 Starting with a cubic distribution of the displacements through the beam thickness in terms of unknown functions (F, G, H)

$$u(x, z) = zF(x) + z^2G(x) + z^3H(x), \quad w(x, z) = w_0(x) \qquad (i)$$

determine the functions (F, G, H) in terms of (w_0, ϕ) such that the following conditions are satisfied:

$$\left(\frac{\partial u}{\partial z}\right)_{z=0} = \phi, \quad \sigma_{xz}(x, -\frac{h}{2}) = 0, \quad \sigma_{xz}(x, \frac{h}{2}) = 0 \qquad (ii)$$

2.7 The Levinson Beam Theory.
The Levinson beam theory is based on the same displacement field, Eqs. (2.15a,b), as the Reddy–Bickford beam theory. As opposed to using the variationally-derived equations of equilibrium, Levinson (1981) used the thickness-integrated equations of elasticity, which are exactly the same as those of the Timoshenko beam theory:

$$-\frac{dM_{xx}^L}{dx} + Q_x^L = 0, \quad -\frac{dQ_x^L}{dx} = q \qquad (i)$$

The stress resultant-displacement relations for the Levinson beam theory are the same as those in Reddy-Bickford beam theory and they are

$$M_{xx}^L = \hat{D}_{xx}\frac{d\phi^L}{dx} - \alpha F_{xx}\frac{d^2 w_0^L}{dx^2} \qquad (ii)$$

$$Q_x^L = \hat{A}_{xz}\left(\phi^L + \frac{dw_0^L}{dx}\right) \qquad (iii)$$

where the stiffnesses $\hat{D}_{xx}$, F_{xx}, and $\hat{A}_{xz}$ are defined in Eqs. (2.151a,b) and (2.1.52a,b). Show that [cf. Eqs. (2.2.1)–(2.2.4)]

$$Q_x^L = Q_x^E + C_1 \qquad (v)$$

$$M_{xx}^L = M_{xx}^E + C_1 x + C_2 \qquad (vi)$$

$$\phi^L = -\frac{dw_0^E}{dx} + \frac{\alpha F_{xx}}{D_{xx}\hat{A}_{xz}}\left(Q_x^E + C_1\right)$$

$$+ \frac{1}{D_{xx}}\left(C_1\frac{x^2}{2} + C_2 x + C_3\right) \qquad (vii)$$

$$w_0^L = w_0^E + \frac{\hat{D}_{xx}}{\hat{A}_{xz}D_{xx}}\left(M_x^E + C_1 x\right)$$

$$- \frac{1}{D_{xx}}\left(C_1\frac{x^3}{6} + C_2\frac{x^2}{2} + C_3 x + C_4\right) \qquad (viii)$$

Chapter 3

Shear-Flexural Stiffness Matrix

Presented in this chapter is a unified element stiffness matrix that incorporates the element stiffness matrices of the Euler–Bernoulli, Timoshenko and the simplified Reddy-Bickford third-order beam theories. The beam element has only four degrees of freedom, namely, deflection and rotation at each of its two nodes. Depending on the choice of the element type, the general stiffness matrix can be specialized to any of the three theories by merely assigning proper values to parameters introduced in the development. The element does not experience shear locking, and gives exact generalized nodal displacements for Euler–Bernoulli and Timoshenko beam theories when the beam is uniform and homogeneous. While the Timoshenko beam theory requires a shear correction factor, the third-order beam theory does not require the specification of such a factor.

3.1 Introduction

The finite element models of the Euler–Bernoulli beam theory and the Timoshenko beam theory are now standard (see Reddy 1993). A number of Timoshenko beam finite elements have appeared in the literature. They differ from each other in the choice of interpolation functions used for the transverse deflection w_0 and rotation ϕ. Some are based on equal interpolation and others on unequal interpolation of w_0 and ϕ.

The Timoshenko beam finite element with linear interpolation of both w_0 and ϕ is the simplest element. However, it behaves in a very stiff manner in the thin beam limit, i.e. as the length-to-thickness ratio becomes very large (say, 100). Such behaviour is known as *shear locking* (see Nickell and Secor 1972, Tessler and Dong 1981, Prathap and Bhashyam 1982, and Averill and Reddy 1990). The locking is due to the inconsistency of the interpolation used for w_0 and ϕ. To overcome the locking, one may use equal interpolation for both w_0 and ϕ but use a lower-order polynomial for the shear strain, $\epsilon_{xz} = (dw_0/dx) + \phi$. This is often realized by using selective integration, in which reduced-order integration is used to evaluate the stiffness coefficients associated with the transverse shear strain, and all other coefficients of the stiffness matrix are evaluated using full integration. The selective integration Timoshenko beam element is known to exhibit spurious energy modes (see Prathap and Bhashyam 1982 and Averill and Reddy 1990). Prathap and Bhashyam (1982) used a consistent interpolation of the variables to alleviate locking.

The transverse shear strain in the Timoshenko beam theory (Timoshenko 1921, 1922) is represented as a constant through the beam thickness, and a shear correction factor is thus introduced to calculate the transverse shear force that would be equal in magnitude to the actual shear force. Since the actual shear stress distribution through beam thickness is quadratic, Jemielita (1975), Levinson (1981), Bickford (1982) and Reddy (1984a) developed third-order beam theories to capture the true variation of the shear stress. The displacement field of these third-order theories accommodates a quadratic variation of the transverse shear strain and stresses, and there is no need to use shear correction factors in a third-order theory. The Levinson third-order beam theory has the same equations of equilibrium as the Timoshenko beam theory but the force and moment resultants contain higher-order strain terms. Bickford (1982) used Levinson's displacement field and developed variationally consistent equations of motion of isotropic beams while Reddy (1984) developed a variationally consistent third-order theory of laminated composite plates.

Heyliger and Reddy (1988) used the third-order laminate theory of Reddy to develop a beam finite element and studied bending and vibrations of isotropic beams. The element is based on Lagrange linear interpolation of the rotation ϕ and Hermite cubic interpolation of w_0, as they are the minimum requirements imposed by the weak form of the third-order theory (also see Phan and Reddy 1985 and Reddy 1997a).

In this chapter, we present the development of a unified beam finite element that contains the finite element models of the Euler–Bernoulli, Timoshenko and the refined third-order beam theory. The derivation of the unified element is based on the exact relationships between the various theories presented in Chapter 2. The relationships allow interdependent interpolation of w_0 and ϕ and the rank deficiency is removed, resulting in an efficient and accurate locking-free finite element for the analysis of beams according to classical as well as refined beam theories.

3.2 Summary of Relationships
3.2.1 Relationships Between TBT and EBT

As discussed in Chapter 2, the shear force, bending moment, slope and deflection of Timoshenko beam theory can be expressed in terms of the corresponding quantities of the Euler–Bernoulli beam theory. These relationships are summarized below [see Eqs. (2.2.1)–(2.2.4)]:

$$Q_x^T = Q_x^E + C_1 \tag{3.2.1}$$

$$M_{xx}^T = M_{xx}^E + C_1 x + C_2 \tag{3.2.2}$$

$$D_{xx}\phi^T = -D_{xx}\frac{dw_0^E}{dx} + C_1\frac{x^2}{2} + C_2 x + C_3 \tag{3.2.3}$$

$$D_{xx}w_0^T = D_{xx}w_0^E + \frac{D_{xx}}{K_s A_{xz}}M_{xx}^E - C_1\left(\frac{x^3}{6} - \frac{D_{xx}}{K_s A_{xz}}x\right)$$
$$- C_2\frac{x^2}{2} - C_3 x - C_4 \tag{3.2.4}$$

where A_{xz} and D_{xx} are defined in Eqs. (2.1.52a,b), and C_1, C_2, C_3, C_4 are constants of integration, which are to be determined using the boundary conditions of the particular beam.

3.2.2 Relationships Between RBT and EBT

Equations for the force and moment resultants, and the rotation and deflection of the Reddy–Bickford beam theory in terms of the Euler–Bernoulli beam theory are given by [see Eqs. (2.3.2), (2.3.8), (2.3.10), and (2.3.12)]

$$\hat{V}_x^R(x) = Q_x^R - \beta R_x + \alpha\frac{dP_{xx}}{dx} = \frac{dM_{xx}^R}{dx}$$
$$= Q_x^E(x) + C_1 \tag{3.2.5}$$

$$M_{xx}^R(x) = M_{xx}^E + C_1 x + C_2 \tag{3.2.6}$$

$$D_{xx}\phi^R(x) = -D_{xx}\frac{dw_0^E}{dx} + \frac{\alpha F_{xx}}{\hat{A}_{xz}}Q_x^R + C_1\frac{x^2}{2} + C_2 x + C_3 \tag{3.2.7}$$

$$D_{xx}w_0^R(x) = D_{xx}w_0^E(x) + \frac{\hat{D}_{xx}}{\hat{A}_{xz}}\left(\int Q_x^R(\eta)d\eta\right)$$

$$- C_1\frac{x^3}{6} - C_2\frac{x^2}{2} - C_3 x - C_4 \qquad (3.2.8)$$

where $\alpha = 4/(3h^2)$ and $\beta = 4/h^2$. In addition, a second-order equation must be solved to determine Q_x^R in terms of Q_x^E [see Eq. (2.3.7)]. This solution requires another two constants C_5 and C_6. The six constants C_1 through C_6 are determined using the six boundary conditions available in the third-order theory.

3.2.3 Relationships Between Simplified RBT and EBT

As stated earlier, the Reddy–Bickford third-order theory requires, unlike in the Timoshenko beam theory, the solution of an additional second-order equation to establish the relationships. The reason is that both the Euler–Bernoulli beam theory and the Timoshenko beam theory are fourth-order theories, whereas the Reddy–Bickford beam theory is a sixth-order beam theory. The second-order equation can be in terms of Q_x^R, M_{xx}^R, ϕ^R, and w_0. In this section we develop relationships between a simplified Reddy–Bickford beam theory and the Euler–Bernoulli beam theory. The term *simplified Reddy–Bickford beam theory* refers to the fourth-order Reddy–Bickford beam theory obtained by dropping the second-derivative term in the additional differential equation for w_0^R. While this is an approximation of the original Reddy–Bickford beam theory, it is as simple and as accurate as the Timoshenko beam theory while not requiring a shear correction factor.

For the simplified Reddy–Bickford beam theory, we first derive the second-order equation in terms of w_0^E. Substituting Eqs. (2.1.17) and (2.1.47) into Eq. (3.2.6), we obtain

$$\hat{D}_{xx}\frac{d\phi^R}{dx} - \alpha F_{xx}\frac{d^2 w_0^R}{dx^2} = -D_{xx}\frac{d^2 w_0^E}{dx^2} + C_1 x + C_2 \qquad (3.2.9)$$

Integrating the above equation gives

$$\hat{D}_{xx}\phi^R - \alpha F_{xx}\frac{dw_0^R}{dx} = -D_{xx}\frac{dw_0^E}{dx} + C_1\frac{x^2}{2} + C_2 x + C_3 \qquad (3.2.10)$$

From Eqs. (2.1.43), (2.1.44), (2.1.49), and (2.1.50), we have

$$\frac{d\hat{M}_{xx}^R}{dx} = \frac{dM_{xx}^R}{dx} - \alpha\frac{dP_{xx}^R}{dx} = \bar{A}_{xz}\left(\phi^R + \frac{dw_0^R}{dx}\right) \qquad (3.2.11)$$

so that
$$\phi^R = \frac{1}{\bar{A}_{xz}} \frac{d\hat{M}_{xx}^R}{dx} - \frac{dw_0^R}{dx} \qquad (3.2.12)$$

where $\bar{A}_{xz}$ and so on are defined in Eqs. (2.1.51a,b) and (2.1.55). Substituting Eq. (3.2.12) into Eq. (3.2.10), we obtain

$$\left(\frac{\hat{D}_{xx}}{\bar{A}_{xz}}\right)\frac{d\hat{M}_{xx}^R}{dx} - D_{xx}\frac{dw_0^R}{dx} = -D_{xx}\frac{dw_0^E}{dx} + C_1\frac{x^2}{2} + C_2 x + C_3 \qquad (3.2.13)$$

which on integration yields

$$D_{xx} w_0^R(x) = \left(\frac{\hat{D}_{xx}}{\bar{A}_{xz}}\right)\hat{M}_{xx}^R + D_{xx} w_0^E(x) - C_1\frac{x^3}{6} - C_2\frac{x^2}{2} - C_3 x - C_4 \qquad (3.2.14)$$

From Eq. (2.1.47), we have

$$M_{xx}^R = \hat{D}_{xx}\frac{d\phi^R}{dx} - \alpha F_{xx}\frac{d^2 w_0^R}{dx^2} \qquad (3.2.15)$$

$$\hat{M}_{xx}^R = M_{xx}^R - \alpha P_{xx} = \bar{D}_{xx}\frac{d\phi^R}{dx} - \alpha \hat{F}_{xx}\frac{d^2 w_0^R}{dx^2} \qquad (3.2.16)$$

Eliminating $d\phi^R/dx$ from Eqs. (3.2.15) and (3.2.16), we obtain

$$\bar{D}_{xx} M_{xx}^R - \hat{D}_{xx}\hat{M}_{xx}^R = \alpha\left(-F_{xx}\bar{D}_{xx} + \hat{F}_{xx}\hat{D}_{xx}\right)\frac{d^2 w_0^R}{dx^2} \qquad (3.2.17)$$

and using Eq. (3.2.6), we can write

$$\hat{M}_{xx}^R = \frac{\bar{D}_{xx}}{\hat{D}_{xx}}\left(M_{xx}^E + C_1 x + C_2\right) + \frac{\alpha}{\hat{D}_{xx}}\left(-F_{xx}\bar{D}_{xx} + \hat{F}_{xx}\hat{D}_{xx}\right)\frac{d^2 w_0^R}{dx^2} \qquad (3.2.18)$$

Finally, substituting Eq. (3.2.18) into Eq. (3.2.14), we obtain

$$D_{xx} w_0^R(x) - \frac{\alpha}{\bar{A}_{xz}}\left(F_{xx}\bar{D}_{xx} - \hat{F}_{xx}\hat{D}_{xx}\right)\frac{d^2 w_0^R}{dx^2}$$
$$= D_{xx} w_0^E(x) + \left(\frac{\hat{D}_{xx}}{\bar{A}_{xz}}\right) M_{xx}^E - C_1\left[\frac{x^3}{6} - \left(\frac{\hat{D}_{xx}}{\bar{A}_{xz}}\right)x\right]$$
$$- C_2\left[\frac{x^2}{2} - \left(\frac{\hat{D}_{xx}}{\bar{A}_{xz}}\right)\right] - C_3 x - C_4 \qquad (3.2.19)$$

Now, we wish to simplify the Reddy–Bickford beam theory by neglecting the second-order derivative term in Eq. (3.2.19). This amounts to reducing the order of the theory from six to four. We obtain

$$D_{xx}w_0^R(x) = D_{xx}w_0^E(x) + \left(\frac{\hat{D}_{xx}}{\bar{A}_{xz}}\right)M_{xx}^E - C_1\left[\frac{x^3}{6} - \left(\frac{\hat{D}_{xx}}{\bar{A}_{xz}}\right)x\right]$$
$$- C_2\left[\frac{x^2}{2} - \left(\frac{\hat{D}_{xx}}{\bar{A}_{xz}}\right)\right] - C_3 x - C_4 \qquad (3.2.20)$$

In summary, we have the following relations from Eqs. (3.2.5), (3.2.6), (3.2.10), and (3.2.20):

$$V_x^S(x) = Q_x^E(x) + C_1 \qquad (3.2.21)$$
$$M_{xx}^S(x) = M_{xx}^E(x) + C_1 x + C_2 \qquad (3.2.22)$$
$$D_{xx}\phi^S(x) = -D_{xx}\frac{dw_0^E}{dx} + C_1\frac{x^2}{2} + C_2 x + C_3 \qquad (3.2.23)$$
$$D_{xx}w_0^S(x) = D_{xx}w_0^E(x) + \frac{\hat{D}_{xx}}{\bar{A}_{xz}}M_{xx}^E - C_1\left(\frac{x^3}{6} - \frac{\hat{D}_{xx}}{\bar{A}_{xz}}x\right)$$
$$- C_2\left(\frac{x^2}{2} - \frac{\hat{D}_{xx}}{\bar{A}_{xz}}\right) - C_3 x - C_4 \qquad (3.2.24)$$

where we have introduced the following equivalent slope:

$$D_{xx}\phi^S(x) = \hat{D}_{xx}\phi^R - \alpha F_{xx}\frac{dw_0^R}{dx} \qquad (3.2.25)$$

and the superscript 'S' denotes the quantities in the simplified theory. Note that the relationships for the shear force and bending moment remain unchanged between the original and simplified theories.

3.3 Stiffness Matrix

Next we develop the stiffness matrix of a beam finite element that incorporates the stiffness matrices of all three theories. The development utilizes the relationships between the solutions of the three theories (see Section 3.2).

The relationships (3.2.1)–(3.2.4) between the Timoshenko and Euler–Bernoulli beam theories as well as the relationships (3.2.21)–(3.2.24) between the simplified Reddy–Bickford and Euler–Bernoulli beam theories can now be expressed in one set by introducing tracers A and B:

$$V_x^U(x) = Q_x^E(x) + C_1 \tag{3.3.1}$$

$$M_{xx}^U(x) = M_{xx}^E(x) + C_1 x + C_2 \tag{3.3.2}$$

$$D_{xx}\theta^U(x) = -D_{xx}\frac{dw_0^E}{dx} + C_1\frac{x^2}{2} + C_2 x + C_3 \tag{3.3.3}$$

$$D_{xx}w_0^U(x) = D_{xx}w_0^E(x) + \mathcal{A}M_{xx}^E - C_1\left(\frac{x^3}{6} - \mathcal{A}x\right)$$

$$- C_2\left(\frac{x^2}{2} - \mathcal{B}\right) - C_3 x - C_4 \tag{3.3.4}$$

where the quantities with superscript 'U' belong to either the Timoshenko beam theory ($\alpha = 0$) or the Reddy–Bickford beam theory ($\alpha \neq 0$),

$$\mathcal{A} = \begin{cases} D_{xx}/(A_{xz}K_s) & \text{for Timoshenko beam theory} \\ \hat{D}_{xx}/\bar{A}_{xz} & \text{for simplified Reddy–Bickford beam theory} \end{cases} \tag{3.3.5a}$$

$$\mathcal{B} = \begin{cases} 0 & \text{for Timoshenko beam theory} \\ \hat{D}_{xx}/\bar{A}_{xz} & \text{for simplified Reddy–Bickford beam theory} \end{cases} \tag{3.3.5b}$$

and $\theta^U(\bar{x})$ denotes the slope, which has a different meaning in different theories, as defined below:

$$\theta^U(\bar{x}) = \begin{cases} \phi^T(\bar{x}) & \text{for Timoshenko beam theory} \\ \phi^S(\bar{x}) & \text{for simplified Reddy–Bickford beam theory} \end{cases} \tag{3.3.6}$$

Clearly, $\theta^U = \phi^S = \phi^R$ when $\alpha = 0$. When $C_i = 0$ and $\mathcal{A} = 0$, the relationships in Eqs. (3.3.1)–(3.3.4) degenerate to the trivial statements

$$V_x^E(x) = Q_x^E(x) \tag{3.3.7}$$

$$M_{xx}^E(x) = M_{xx}^E(x)$$

$$\theta^E(x) = -\frac{dw_0^E}{dx} \tag{3.3.8}$$

$$w_0^E(x) = w_0^E(x)$$

46 SHEAR DEFORMABLE BEAMS AND PLATES

Based on the foregoing unified relationships, we now derive the stiffness matrix for the unified beam element (UBE). Consider a (Hermite cubic) beam element of length and element-wise uniform material and geometric properties. Let the generalized displacements at nodes 1 and 2 of a typical element associated with any of the three beam theories be defined as (see Figure 3.3.1a)

$$w_0(0) = \Delta_1, \qquad \theta(0) = \Delta_2$$
$$w_0(L) = \Delta_3, \qquad \theta(L) = \Delta_4 \qquad (3.3.9)$$

where L denotes the length of the element, and $\theta(\bar{x})$ denotes the slope, which has a different meaning in different theories, as defined below:

$$\theta(\bar{x}) = \begin{cases} -\frac{dw_0}{dx} & \text{for Euler–Bernoulli beam theory} \\ \phi^T(\bar{x}) & \text{for Timoshenko beam theory} \\ \phi^R(\bar{x}) & \text{for simplified Reddy–Bickford beam theory} \end{cases} \qquad (3.3.10)$$

where $\bar{x}$ denotes the element coordinate whose origin is located at node 1 of the element. Next, let Q_1 and Q_3 denote the shear forces (i.e. values of V_x^U) at nodes 1 and 2, respectively; similarly, let Q_2 and Q_4 denote the bending moments (i.e. values of M_{xx}^U) at nodes 1 and 2, respectively. Figure 3.3.1 shows the sign convention used for the generalized displacements and forces.

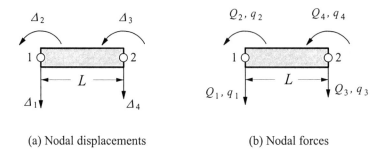

(a) Nodal displacements (b) Nodal forces

Figure 3.3.1. A typical unified beam finite element with the generalized displacements and forces for the derivation of the stiffness matrix.

The stiffness matrix for the unified element is derived using the traditional method to calculate stiffnesses in structural analysis. The method involves imposing a unit generalized displacement, while all other generalized displacements are kept zero, and determining the generalized forces required to keep the beam in equilibrium (i.e., equivalent to using the unit-dummy-displacement method). The formulation utilizes the relationships between the Euler–Bernoulli beam theory, the Timoshenko beam theory and the simplified Reddy–Bickford beam theory. This amounts to using Hermite cubic interpolation for the transverse deflection and a dependent interpolation for the slope. The procedure is outlined briefly here.

To obtain the first column of the element stiffness matrix, we set (see Figure 3.3.1b)

$$\text{at } \bar{x} = 0 : w_0^E = w_0^T = w_0^R \equiv \Delta_1, \quad \frac{dw_0^E}{dx} = \phi^T = \phi^R = 0 \quad (3.3.11a)$$

$$\text{at } \bar{x} = L : w_0^E = w_0^T = w_0^R \equiv 0, \quad \frac{dw_0^E}{dx} = \phi^T = \phi^R = 0 \quad (3.3.11b)$$

and determine the constants C_1 through C_4 from Eqs. (3.3.13)–(3.3.16). We obtain

$$C_1 = -\frac{12 D_{xx}}{L^3} \left[\frac{12\mathcal{A}}{L^2 + 12\left(\mathcal{A} - \frac{1}{2}\mathcal{B}\right)} \right] \Delta_1 \quad (3.3.12a)$$

$$C_2 = \frac{L}{2} C_1 \quad (3.3.12b)$$

$$C_3 = \left(\frac{6 D_{xx}}{L^3}\mathcal{A}\right) \Delta_1 - \left(\frac{L^2}{6} - \mathcal{A}\right) C_1$$

$$\quad - \left(\frac{L}{2} - \frac{\mathcal{B}}{L}\right) C_2 - \frac{1}{L} C_4 \quad (3.3.12c)$$

$$C_4 = -\left(\frac{6 D_{xx}}{L^2}\mathcal{A}\right) \Delta_1 \quad (3.3.12d)$$

The substitution of these constants into Eqs. (3.3.1)–(3.3.4) gives

$$Q_1 \equiv -V_x^U(0) = \left(\frac{12 D_{xx}}{\mu L^3}\right) \Delta_1 = k_{11} \Delta_1 \quad (3.3.13a)$$

$$Q_2 \equiv -M_{xx}^U(0) = \left(\frac{6 D_{xx}}{\mu L^2}\right) \Delta_1 = k_{21} \Delta_1 \quad (3.3.13b)$$

$$Q_3 \equiv -V_x^U(L) = \left(\frac{12D_{xx}}{\mu L^3}\right)\Delta_1 = k_{31}\Delta_1 \qquad (3.3.13c)$$

$$Q_4 \equiv M_{xx}^U(0) = -\left(\frac{6D_{xx}}{\mu L^2}\right)\Delta_1 = k_{41}\Delta_1 \qquad (3.3.13d)$$

$$\mu = 1 + 12\Omega, \qquad \Omega = \frac{\mathcal{A}}{L^2 - 6\mathcal{B}} \qquad (3.3.14)$$

This completes the derivation of the stiffness coefficients of the first column of the stiffness matrix. The same procedure can be repeated, with different generalized displacements set to unity, to obtain the remaining stiffness coefficients. The complete unified beam finite element model is given by

$$\frac{2D_{xx}}{\mu L^3}\begin{bmatrix} 6 & -3L & -6 & -3L \\ -3L & 2L^2\lambda & 3L & L^2\xi \\ -6 & 3L & 6 & 3L \\ -3L & L^2\xi & 3L & 2L^2\lambda \end{bmatrix}\begin{Bmatrix}\Delta_1\\\Delta_2\\\Delta_3\\\Delta_4\end{Bmatrix} = \begin{Bmatrix}q_1\\q_2\\q_3\\q_4\end{Bmatrix} + \begin{Bmatrix}Q_1\\Q_2\\Q_3\\Q_4\end{Bmatrix} \qquad (3.3.15)$$

$$\lambda = 1 + 3\Omega, \qquad \xi = 1 - 6\Omega \qquad (3.3.16)$$

and is the load vector due to the distributed load $q(x)$

$$q_i^e = \int_0^L q(x)\varphi_i(x)\,dx \qquad (3.3.17)$$

Here $\varphi_i(x)$ denote the Hermite interpolation functions implied by Eqs. (3.3.1)–(3.3.4) (see Problem 3.2 at the end of the chapter). The stiffness matrix in Eq. (3.3.15) is also reported by Gere and Weaver (1965), Przemieniecki (1968), and Meek (1971), among others (see Reddy 1999b for additional references).

3.4 Frame Structure - An Example

Consider the two–member frame structure shown in Figure 3.4.1a. The following geometric and material parameters are used in the analysis:

Member 1: $L = 144$ in., $A = 10$ in^2., $I = 10$ in^4., $E = 10^6$ psi., $\nu = 0.3$
Member 2: $L = 180$ in., $A = 10$ in^2., $I = 10$ in^4., $E = 10^6$ psi., $\nu = 0.3$

The shear correction coefficient for the Timoshenko beam element is taken to be $K_s = 5/6$.

The structure is analyzed using the aforementioned stiffness method according to the Euler–Bernoulli theory and the Timoshenko beam theory. The simplified Reddy beam element essentially gives the same results as the Timoshenko beam element, and hence is not included. The exact Timoshenko beam element $[\mathcal{A} = D_{xx}/(K_s A_{xz}) = EI/(GAK_s)$ and $\mathcal{B} = 0]$ is denoted by UBE. The results are also compared with those predicted by two other commonly used Timoshenko beam finite elements, namely the linear equal-interpolation reduced-integration element (RIE) and the consistent interpolation element (CIE) [see Reddy 1993, (1997b,1999b)]. Figures 3.4.1 shows the two, four and eight element meshes of the structure. Note that all these elements are extended to include the axial displacement degrees of freedom (i.e., linear interpolation of the axial displacement is used), and each element stiffness matrix is of the order 6×6.

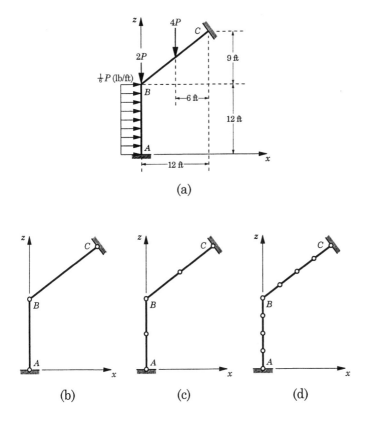

Figure 3.4.1. Analysis of a frame structure. (a) Frame structure analyzed. (b) Meshes of 2, 4, and 8 elements.

50 SHEAR DEFORMABLE BEAMS AND PLATES

Table 3.4.1 contains the displacements at point B obtained using various types of elements. Note that one Euler–Bernoulli element (EBE) or unified beam element (UBE) per member of a structure gives exact displacements, whereas at least two RIE or CIE per member are needed to obtain acceptable results. The forces in each element are included in Table 3.4.2. The forces calculated from the element equations are also exact for EBE and UBE.

Table 3.4.1. Comparison of the generalized displacements $[\bar{v} = (v/P) \times 10^4$ where v is a typical displacement] at point A of the frame structure shown in Figure 3.4.1.

Displ.	RIE(1)*	RIE (2)	RIE (4)	CIE (1)	CIE (2)	CIE (4)	UBE†	EBE†
$\bar{u}_B$	0.2709	0.8477	0.8411	0.2844	0.8415	0.8396	0.8390	0.8390
$\bar{w}_B$	0.4661	0.6806	0.6811	0.4432	0.6808	0.6811	0.6812	0.6812
$\bar{\phi}_B$	-0.0016	0.8665	0.9450	0.0004	0.7703	0.9164	0.9621	0.9610

* Number in the parenthesis denotes the number of elements per member.

† Values independent of the number of elements (and coincide with the exact values predicated by the respective beam theories).

3.5 Concluding Remarks

In this chapter, a unified finite element model of the Euler–Bernoulli, Timoshenko, and simplified Reddy–Bickford third-order beam theories is developed. Bending stiffness coefficients of the unified element are derived. The development is based on the exact relationships between the bending solutions of the Euler–Bernoulli beam theory, Timoshenko beam theory and the simplified Reddy–Bickford third-order beam theory. The relationships provide an interdependent interpolation of the deflection and rotation of the form (for more details, see Problem 3.2 at the end of the chapter)

$$w_0^U(x) = \sum_{j=1}^{4} \Delta_j \varphi_j^{(1)}(x), \quad \theta^U(x) = \sum_{j=1}^{4} \Delta_j \varphi_j^{(2)}(x) \quad (3.5.1)$$

where $\varphi_j^{(2)}$ are quadratic interpolation functions related to $\varphi_j^{(1)}$. Hence,

Table 3.4.2. Comparison of the generalized forces (divided by P) at the nodes of each member of the frame structure shown in Figure 3.4.1.

Element*	F_1	F_2	F_3	F_4	F_5	F_6
RIE(1)	3.237	1.865	-62.24	-3.237	0.136	-62.26
	0.850	0.908	62.26	1.550	2.292	62.28
RIE(2)	4.723	0.671	-0.332	-4.723	1.329	47.70
	2.699	1.384	-47.70	-0.299	1.816	86.67
RIE(4)	4.730	0.713	-8.362	-4.730	1.288	49.76
	2.668	1.411	-49.76	-0.268	1.789	83.74
CIE(1)	3.007	1.575	-65.39	-3.077	0.425	-17.38
	0.987	0.607	17.38	1.413	2.593	161.4
CIE(2)	4.728	0.708	-8.327	-4.728	1.292	50.37
	2.670	1.407	-50.37	-0.270	1.793	85.07
CIE(4)	4.730	0.721	-10.30	-4.730	1.279	50.43
	2.661	1.417	-50.43	-0.261	1.783	83.39
UBE[†]	4.731	0.725	-10.92	-4.731	1.275	50.45
	2.658	1.420	-50.45	-0.258	1.780	82.87
EBE[†]	4.731	0.725	-10.90	-4.731	1.275	50.45
	2.658	1.420	-50.45	-0.258	1.780	82.87

* Number in the parenthesis denotes the number of elements per member, and the two rows correspond to the two members of the structure.

† Values independent of the number of elements (and coincide with the exact values predicated by the respective beam theories).

the element stiffness matrix is of the order 4×4, and it gives exact nodal values of the generalized displacements (i.e., w_0 and ϕ) for Euler–Bernoulli and Timoshenko beams with uniform cross-section and homogeneous material properties. An independent interpolation of the form

$$w_0^U(x) = \sum_{j=1}^{4} \Delta_j \Phi_j(x), \quad \theta^U(x) = \sum_{j=1}^{3} \Theta_j \Psi_j(x) \qquad (3.5.2)$$

would result in a 7×7 stiffness matrix for the same accuracy as the element derived here. This shear deformable finite element based on the Timoshenko and third-order beam theories can be included in any computer program by simply replacing the stiffness matrix of the existing Euler–Bernoulli beam finite element with that given in Eq. (3.3.25). Note that conventional Timoshenko beam elements are not completely shear locking free and one-element discretization per member using such elements in the analysis of a frame structure will yield erroneous displacements as well as member forces as demonstrated by the given two-member frame example.

The element stiffness matrix developed herein can be applied to beams with element-wise constant geometric and material properties. The element can also be extended to symmetrically laminated beams under appropriate assumptions (see Reddy 1997a, Chapter 6). Extension to buckling is also straightforward. However, extension of the unified beam element to dynamic problems is not possible because of the mass inertia terms [see Reddy (1999b)].

Problems

3.1 Verify the relations in Eqs. (3.3.22)–(3.3.24).

3.2 Consider the following equilibrium equations of the Timoshenko beam theory in the absence of distributed load q:

$$-\frac{d}{dx}\left(D_{xx}\frac{d\phi}{dx}\right) + A_{xz}K_s\left(\phi + \frac{dw_0}{dx}\right) = 0 \quad (i)$$

$$-\frac{d}{dx}\left[A_{xz}K_s\left(\phi + \frac{dw_0}{dx}\right)\right] = 0 \quad (ii)$$

The exact solution of Eqs. (i) and (ii) is of the form

$$w(x) = -\frac{1}{D_{xx}}\left(C_1\frac{x^3}{6} + C_2\frac{x^2}{2} + C_3 x + C_4\right) + \frac{1}{A_{xz}K_s}(C_1 x) \quad (iii)$$

$$D_{xx}\phi(x) = C_1\frac{x^2}{2} + C_2 x + C_3 \quad (iv)$$

where C_1 through C_4 are the constants of integration. Note that the constants C_1, C_2, and C_3 appearing in (iv) are the same as those in Eq. (iii). Equations (iii) and (iv) suggest that one may use cubic approximation of w_0 and an interdependent quadratic approximation of ϕ. Use Eqs. (iii) and (iv) to express the constants C_1 through C_4 in terms of the nodal variables

$$\Delta_1 = w_0(0), \quad \Delta_2 = \phi(0), \quad \Delta_3 = w_0(L), \quad \Delta_4 = \phi(L) \qquad (v)$$

where L denotes the length of the beam element and $\bar{x}$ is the element coordinate with its origin at node 1, $0 \leq \bar{x} \leq L$, and express $w_0(\bar{x})$ and $\phi(\bar{x})$ in the form

$$w_0(\bar{x}) \approx \sum_{j=1}^{4} \varphi_j^{(1)} \Delta_j, \qquad \phi(\bar{x}) \approx \sum_{j=1}^{4} \varphi_j^{(2)} \Delta_j \qquad (vi)$$

In particular, show that $\varphi_i^{(1)}$ and $\varphi_i^{(2)}$ are given by

$$\varphi_1^{(1)} = \frac{1}{\mu}\left[\mu - 12\Omega\eta - (3 - 2\eta)\eta^2\right]$$

$$\varphi_2^{(1)} = -\frac{L}{\mu}\left[(1-\eta)^2\eta + 6\Omega(1-\eta)\eta\right]$$

$$\varphi_3^{(1)} = \frac{1}{\mu}\left[(3-2\eta)\eta^2 + 12\Omega\eta\right]$$

$$\varphi_4^{(1)} = \frac{L}{\mu}\left[(1-\eta)\eta^2 + 6\Omega(1-\eta)\eta\right] \qquad (vii)$$

$$\varphi_1^{(2)} = \frac{6}{L\mu}(1-\eta)\eta$$

$$\varphi_2^{(2)} = \frac{1}{\mu}(\mu - 4\eta + 3\eta^2 - 12\Omega\eta)$$

$$\varphi_3^{(2)} = -\frac{6}{L\mu}(1-\eta)\eta$$

$$\varphi_4^{(2)} = \frac{1}{\mu}(3\eta^2 - 2\eta + 12\Omega\eta) \qquad (viii)$$

Here η is the non-dimensional local coordinate

$$\eta = \frac{\bar{x}}{L}, \quad \mu = 1 + 12\Omega, \quad \Omega = \frac{D_{xx}}{A_{xz}L^2} \qquad (ix)$$

3.3 (*Continuation of Problem 3.2*) The displacement finite element model of the Timoshenko beam theory is constructed using the principle of minimum total potential energy, or equivalently, using the weak form

$$0 = \int_0^L \left[D_{xx} \frac{d\delta\phi}{dx}\frac{d\phi}{dx} + A_{xz}K_s \left(\delta\phi + \frac{d\delta w_0}{dx}\right)\left(\phi + \frac{dw_0}{dx}\right)\right] dx$$
$$- \int_0^L q(x)\delta w \, dx - V_1\delta w(0) - V_2\delta w(L) - M_1\delta\phi(0) - M_2\delta\phi(L) \qquad (i)$$

where

$$V_1 \equiv -Q_x(0) = -\left[A_{xz}K_s\left(\frac{dw}{dx}+\phi\right)\right]_{x=0}$$
$$M_1 \equiv -M_{xx}(0) = -\left[D_{xx}\frac{d\phi}{dx}\right]_{x=0}$$
$$V_2 \equiv Q_x(L) = \left[A_{xz}K_s\left(\frac{dw_0}{dx}+\phi\right)\right]_{x=L}$$
$$M_2 \equiv M_{xx}(L) = \left[D_{xx}\frac{d\phi}{dx}\right]_{x=L} \qquad (ii)$$

Substitute the approximation (vi) of Problem 3.2 into the weak form and show that the finite element model is of the form

$$[K]\{\Delta\} = \{q\} + \{Q\} \qquad (iii)$$

where

$$K_{ij} = \int_0^L \left[A_{xz}K_s\left(\varphi_i^{(2)} + \frac{d\varphi_i^{(1)}}{dx}\right)\left(\varphi_j^{(2)} + \frac{d\varphi_j^{(1)}}{dx}\right)\right.$$
$$\left. + D_{xx}\frac{d\varphi_i^{(2)}}{dx}\frac{d\varphi_j^{(2)}}{dx}\right] dx \qquad (iv)$$
$$q_i = \int_0^L \varphi_i^{(1)} q(x) \, dx \qquad (v)$$
$$Q_1 = V_1, \quad Q_2 = M_1, \quad Q_3 = V_2, \quad Q_4 = M_2 \qquad (vi)$$

3.4 (*Continuation of Problem 3.3*) Show that Eq. (iii) of Problem 3.3 has the explicit form given in Eq. (3.3.25).

3.5 Develop the beam finite element based on the Levinson beam theory (see Problem 2.7).

Chapter 4

Buckling of Columns

This chapter presents an approach by means of an analogy for deriving the exact relationship between the elastic buckling loads of columns based on the Euler–Bernoulli beam theory, the Timoshenko beam theory and the Reddy–Bickford beam theory. The exact relationship applies to axially loaded columns with boundary conditions that result in zero lateral force in the members. The resulting Reddy–Bickford buckling solutions are found to be higher than the Timoshenko buckling solutions. For the cases of pinned-pinned and fixed-free columns, the buckling loads are practically the same for the two types of columns. However, in the case of fixed-fixed columns, the buckling loads are somewhat different, especially when the columns have relatively large values of the shear parameter $\Omega = D_{xx}/(K_s A_{xz} L^2)$. It is worth noting that the advantage of the Bickford–Reddy theory over the Timoshenko theory is that the former does not require a shear correction factor.

4.1 Introduction

When the column is stocky, or of a built-up or of a composite-type construction, the application of the Euler–Bernoulli (classical) beam theory will overestimate the buckling loads. This is due to the neglect of transverse shear deformation in the Euler–Bernoulli beam theory. A refined beam theory, known as the first-order shear deformation theory or Timoshenko beam theory, that incorporates the shear deformation effect was proposed by Engesser (1891) and Timoshenko (1921). This first-order shear deformation theory relaxes the normality assumption of the Euler–Bernoulli beam theory but assumes a constant transverse shear strain (and thus constant shear stress when computed using the constitutive equations) through the beam thickness. In order to compensate for the parabolic distribution of the transverse shear stress through the thickness, a shear correction factor is introduced to calculate the effective shear modulus. The usual approaches for estimating the shear correction factor are either by matching the high frequency spectrum of vibrating beams (e.g., Mindlin and Deresiewicz 1954) or by using approximation procedures and simplifying assumptions within the linear theory of elasticity (e.g., Cowper 1966). However, the third-order beam theory derived independently by Bickford (1982) and Heyliger and Reddy (1988) does away with the need for a shear correction factor because the assumed third-order displacement field gives a parabolic distribution of the transverse shear stress and satisfies the zero shear stress condition at the free surfaces.

4.2 Relationship Between Euler–Bernoulli and Timoshenko Columns

4.2.1 General Relationship

Consider a column of flexural rigidity D_{xx}, shear rigidity A_{xz}, length L which is subjected to a compressive axial load N. The stress resultant-displacement relations according to the Euler–Bernoulli beam theory are given by

$$M_{xx}^E = -D_{xx}\frac{d^2 w^E}{dx^2} \qquad (4.2.1)$$

$$Q_x^E = -D_{xx}\frac{d^3 w^E}{dx^3} \qquad (4.2.2)$$

while those according to the Engesser–Timoshenko beam theory are given by

$$M_{xx}^T = D_{xx}\frac{d\phi^T}{dx} \qquad (4.2.3)$$

$$Q_x^T = K_s A_{xz}\left(\phi^T + \frac{dw^T}{dx}\right) \qquad (4.2.4)$$

in which x is the longitudinal coordinate measured from the column base, M_{xx} the bending moment, Q_x the transverse shear force, ϕ the rotation in the Engesser–Timoshenko column and w the transverse deflection, measured from the onset of buckling. The superscripts 'E' and 'T' denote quantities belonging to the Euler–Bernoulli column and the Engesser–Timoshenko column, respectively. The shear correction coefficient K_s in Eq. (4.2.4) is introduced to account for the difference in the constant state of shear stress in the Engesser–Timoshenko column theory and the parabolic variation of the actual shear stress through the depth of the cross-section. The values of K_s for various cross-sections and built-up columns are given in standard texts such as Timoshenko and Gere (1959).

For both kinds of columns, it can be readily shown that the equilibrium equations are

$$\frac{dM_{xx}}{dx} = Q_x \tag{4.2.5}$$

$$\frac{dQ_x}{dx} = N\frac{d^2w}{dx^2} \tag{4.2.6}$$

Substituting Eq. (4.2.2) into Eq. (4.2.6) yields the following equation governing the buckling of Euler–Bernoulli columns:

$$\frac{d^4w^E}{dx^4} + \frac{N^E}{D_{xx}}\frac{d^2w^E}{dx^2} = 0 \tag{4.2.7}$$

By substituting Eqs. (4.2.3) and (4.2.4) into Eqs. (4.2.5) and (4.2.6), the equilibrium equations of the Engesser–Timoshenko column may be written as

$$D_{xx}\frac{d^2\phi^T}{dx^2} = K_s A_{xz}\left(\phi^T + \frac{dw^T}{dx}\right) \tag{4.2.8}$$

$$N^T\frac{d^2w^T}{dx^2} = K_s A_{xz}\left(\frac{d\phi^T}{dx} + \frac{d^2w^T}{dx^2}\right) \tag{4.2.9}$$

By differentiating Eq. (4.2.8) and then using Eq. (4.2.9), we obtain

$$D_{xx}\frac{d^3\phi^T}{dx^3} = N^T\frac{d^2w^T}{dx^2} \tag{4.2.10}$$

Equation (4.2.9) can be solved for $d\phi^T/dx$

$$\frac{d\phi^T}{dx} = -\left(1 - \frac{N^T}{K_s A_{xz}}\right)\frac{d^2w^T}{dx^2} \tag{4.2.11}$$

Substituting Eq. (4.2.11) into Eq. (4.2.10) yields

$$\frac{d^4w^T}{dx^4} + \left(\frac{\frac{N^T}{D_{xx}}}{1 - \frac{N^T}{K_s A_{xz}}}\right)\frac{d^2w^T}{dx^2} = 0 \tag{4.2.12}$$

By differentiating Eq. (4.2.8) and using Eq. (4.2.10), we can also obtain

$$\frac{d^3\phi^T}{dx^4} + \left(\frac{\frac{N^T}{D_{xx}}}{1 - \frac{N^T}{K_s A_{xz}}}\right)\frac{d\phi^T}{dx} = 0 \qquad (4.2.13)$$

In view of the similarity of Eqs. (4.2.7), (4.2.11), (4.2.12) and (4.2.13), and provided the boundary conditions are of the same form, it can be deduced that

$$N^E = \frac{N^T}{1 - \frac{N^T}{K_s A_{xz}}} \quad \text{or} \quad N^T = \frac{N^E}{1 + \frac{N^E}{K_s A_{xz}}} \qquad (4.2.14)$$

and

$$\phi^T = -\left(1 - \frac{N^T}{K_s A_{xz}}\right)\frac{dw^T}{dx} + C_1 \qquad (4.2.15)$$

$$w^T = C_2 w^E + C_3 x + C_4 \qquad (4.2.16)$$

$$\phi^T = -\left(1 - \frac{N^T}{K_s A_{xz}}\right)\left(C_2 \frac{dw^E}{dx} + C_3\right) + C_1 \qquad (4.2.17)$$

where C_1, C_2, C_3, and C_4 are constants.

It is clear that the Euler–Bernoulli buckling load and the Engesser–Timoshenko buckling load are linked together through the relationship in Eq. (4.2.14), provided the boundary conditions of the two theories are also linked together by Eqs. (4.2.15)–(4.2.17). Considering various combinations of free, pinned and fixed end conditions, it will be shown below that the foregoing requirements were met for pinned-ended columns, fixed-ended columns and fixed-free columns but not for fixed-pinned columns. Figure 4.2.1 shows columns with various boundary conditions.

4.2.2 Pinned-Pinned Columns

The boundary conditions for the pin-ended Euler–Bernoulli column are given by

$$w^E = M_{xx}^E = \frac{d^2 w^E}{dx^2} = 0 \quad \text{at } x = 0 \text{ and } x = L \qquad (4.2.18a)$$

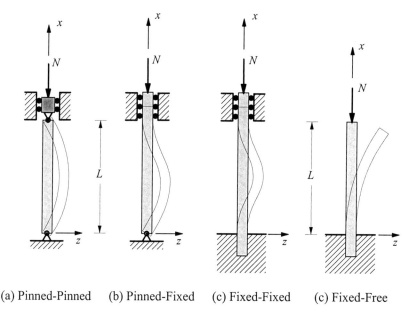

(a) Pinned-Pinned   (b) Pinned-Fixed   (c) Fixed-Fixed   (c) Fixed-Free

Figure 4.2.1. Columns with various boundary conditions.

where L is the column length. In the case of the Engesser–Timoshenko column, the boundary conditions are

$$w^T = M_{xx}^T = \frac{d\phi^T}{dx} = 0 \text{ at } x = 0 \text{ and } x = L \qquad (4.2.18b)$$

From Eqs. (4.2.11) and (4.2.18b), it is clear that

$$\frac{d^2 w^T}{dx^2} = 0, \text{ at } x = 0 \text{ and } x = L \qquad (4.2.18c)$$

Thus the boundary conditions for the Engesser–Timoshenko beam can be written as

$$w^T = M_{xx}^T = \frac{d^2 w^T}{dx^2} = 0 \text{ at } x = 0 \text{ and } x = L \qquad (4.2.18d)$$

In view of the similarity of Eqs. (4.2.18a) and (4.2.18d), it is concluded that the boundary conditions for the pinned-pinned Euler–Bernoulli and

Engesser–Timoshenko columns are of the same form, and that Eqs. (4.2.14)–(4.2.17) are valid.

To determine the constants C_1, C_2, C_3, and C_4, Eq. (4.2.16) is used together with the boundary conditions in Eqs. (4.2.18a) and (4.2.18b) to give

$$C_3 = C_4 = 0 \tag{4.2.19}$$

Next, it is observed that for a pinned-pinned column under a compressive axial load, there is no lateral shear force at the ends so that

$$Q_x^T = \phi^T + \frac{dw^T}{dx} = 0 \text{ at } x = 0 \text{ and } x = L \tag{4.2.20}$$

Combining Eqs. (4.2.17) and (4.2.20), we obtain

$$-\left(\frac{dw^T}{dx}\right)_{x=0} = -\left(1 - \frac{N^T}{K_s A_{xz}}\right) C_2 \left(\frac{dw^E}{dx}\right)_{x=0} + C_1 \tag{4.2.21}$$

Using the above equation with Eq. (4.2.16) results in

$$C_1 = -\frac{N^T}{K_s A_{xz}} C_2 \left(\frac{dw^E}{dx}\right)_{x=0} \tag{4.2.22}$$

Therefore, the relationships between the eigen functions of the pinned-pinned Euler–Bernoulli and Engesser–Timoshenko columns are

$$w^T = C_2 w^E \tag{4.2.23}$$

$$\phi^T = -C_2 \left[\left(1 - \frac{N^T}{K_s A_{xz}}\right) \frac{dw^E}{dx} + \frac{N^T}{K_s A_{xz}} \left(\frac{dw^E}{dx}\right)_{x=0}\right] \tag{4.2.24}$$

where C_2 is an arbitrary constant.

4.2.3 Fixed-Fixed Columns

For fixed-fixed columns, the boundary conditions for the Euler–Bernoulli columns are given by

$$w^E = \frac{dw^E}{dx} = 0 \text{ at } x = 0 \text{ and } x = L \tag{4.2.25a}$$

In the case of the Engesser–Timoshenko column, the boundary conditions are

$$w^T = \phi^T = 0 \quad \text{at} \quad x = 0 \quad \text{and} \quad x = L \tag{4.2.25b}$$

As in the case of the pinned-pinned column, the absence of a lateral shear force at the ends implies [see Eq. (4.2.20)]

$$Q_x^T = \phi^T + \frac{dw^T}{dx} = 0 \quad \text{at} \quad x = 0 \quad \text{and} \quad x = L$$

Due to this condition, Eq. (4.2.25b) can be rewritten as

$$w^T = \frac{dw^T}{dx} = 0 \quad \text{at} \quad x = 0 \quad \text{and} \quad x = L \tag{4.2.25c}$$

By comparing Eqs. (4.2.25a) and (4.2.25c), it is seen that the boundary conditions for both columns are of the same form. Thus, Eqs. (4.2.14)–(4.2.17) are valid.

Substituting Eq. (4.2.25a) and (4.2.25b) into Eqs. (4.2.16) and (4.2.17) yields

$$C_1 = C_3 = C_4 = 0 \tag{4.2.26}$$

Thus, the relationships between the eigenfunctions of the two columns are given by

$$w^T = C_2 w^E \tag{4.2.27}$$

$$\phi^T = -C_2 \left(1 - \frac{N^T}{K_s A_{xz}}\right) \frac{dw^E}{dx} \tag{4.2.28}$$

with C_2 is an arbitrary constant.

4.2.4 Fixed-Free Columns

The boundary conditions for the fixed-free Euler–Bernoulli columns are given by

$$w^E = \frac{dw^E}{dx} = 0 \quad \text{at} \quad x = 0 \tag{4.2.29a}$$

$$M_{xx}^E = \frac{d^2 w^E}{dx^2} = 0 \quad \text{and} \quad Q_x^E = N^E \frac{dw^E}{dx} \quad \text{at} \quad x = L \tag{4.2.29b}$$

In the case of the Engesser–Timoshenko column, the boundary conditions are

$$w^T = \phi^T = 0 \quad \text{at} \quad x = 0 \qquad (4.2.30a)$$

$$M_{xx}^T = \frac{d\phi^T}{dx} = 0, \quad \text{and} \quad Q_x^T = N^T \frac{dw^T}{dx} \quad \text{at} \quad x = L \qquad (4.3.30b)$$

As shown in the previous case, the absence of a lateral shear force at the fixed end allows (4.2.30a) to be expressed as

$$w^T = \frac{dw^T}{dx} = 0 \quad \text{at} \quad x = 0 \qquad (4.2.30c)$$

In order to show that the second of the boundary conditions in Eqs. (4.2.29b) and (4.2.30b) are of the same form, Eq. (4.2.2) is used to rewrite the condition for the Euler–Bernoulli column as

$$-D_{xx}\frac{d^3 w^E}{dx^3} = N^E \frac{dw^E}{dx} = 0 \quad \text{or} \quad \frac{d^3 w^E}{dx^3} + \frac{N^E}{D_{xx}}\frac{dw^E}{dx} = 0 \quad \text{at} \quad x = L \qquad (4.2.31)$$

In the same way, for the Timoshenko column, Eqs. (4.2.3), (4.2.5), and (4.2.11) are used to rewrite the condition as

$$-D_{xx}\left(1 - \frac{N^T}{K_S A_{xz}}\right)\frac{d^3 w^T}{dx^3} = N^T \frac{dw^T}{dx}$$

or

$$\frac{d^3 w^T}{dx^3} + \left(\frac{\frac{N^T}{D_{xx}}}{1 - \frac{N^T}{K_S A_{xz}}}\right)\frac{dw^T}{dx} = 0 \quad \text{at} \quad x = L \qquad (4.2.32)$$

A comparison of Eq. (4.2.29a) with Eq. (4.2.30c), and Eq. (4.2.31) with Eq. (4.2.32) shows clearly that the boundary conditions are of the same form and are consistent with Eq. (4.2.14)–(4.2.17).

Finally, by substituting Eqs. (4.2.29a), (4.2.30a), and (4.2.30c) into Eqs. (4.2.16) and (4.2.17), the constants are found to be

$$C_1 = C_3 = C_4 = 0 \qquad (4.2.33)$$

and the relationships between the eigen functions are the same as those given in Eqs. (4.2.27) and (4.2.28).

It is interesting to note that the deflections of the Euler–Bernoulli and Engesser–Timoshenko columns are proportional to each other for the pinned-pinned, fixed-fixed, and fixed-free columns. Also, the rotation of the Engesser–Timoshenko column is related to the slope of the deflection of the Euler–Bernoulli columns.

It should be noted that the buckling load relationship given by Eq. (4.2.14) does not apply to fixed-pinned columns because the boundary conditions do not match exactly. The Engesser–Timoshenko buckling load for such columns is to be determined from solving the transcendental equation

$$\sqrt{\frac{N^T L^2}{D_{xx}}\left(1 - \frac{N^T}{K_s A_{xz}}\right)} = \tan\left[\sqrt{\frac{\frac{N^T L^2}{D_{xx}}}{\left(1 - \frac{N^T}{K_s A_{xz}}\right)}}\right] \quad (4.2.34)$$

Figure 4.2.2 shows a comparison of the exact Engesser–Timoshenko buckling load from Eq. (4.2.34) for the fixed-pinned column with the approximate solutions predicted by Eq. (4.2.14). It can be seen that as the shear parameter $\Omega = D_{xx}/(K_s A_{xz} L^2)$ increases, the difference in solutions increases. Considering the range of values of the shear parameter Ω between 0 and 0.01, the buckling load relationship in Eq. (4.2.14) can be made more accurate by modifying it to

$$N^T = \frac{N^E}{1 + 1.1 \frac{N^E}{K_s A_{xz}}}, \quad N^E = (4.493)^2 \left(\frac{D_{xx}}{L^2}\right) \quad (4.2.35)$$

Thus for this particular case of fixed-pinned columns, the buckling load relationship (4.2.35) should be used while Eq. (4.2.14) is to be taken for the pinned-pinned columns, fixed-fixed columns, and fixed-free columns.

Ziegler (1982) established that the relationship in Eq. (4.2.14) is also valid for columns with fixed ends and fixed ends with top sway. Moreover, he gave the modified form of the relationship for the effect of pre-buckling shortening. Banerjee and Williams (1994) showed that the buckling relationship applies as well to hinged-hinged columns with rotational springs of equal stiffness added to their ends.

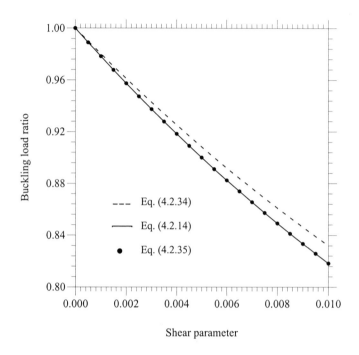

Figure 4.2.2. A comparison of the approximate Engesser–Timoshenko buckling loads with the exact buckling loads of a fixed-pinned column.

It is clear from Eq. (4.2.14) that the effect of transverse shear deformation leads to a reduction in the Euler–Bernoulli buckling load by the factor found in the denominator of the buckling load relationship. This reduction of the Euler–Bernoulli load thus increases with respect to a higher value of Euler–Bernoulli load (especially for columns with highly restrained ends or internal restraints) and also with a lower value of shear rigidity.

4.3 Relationship Between Euler–Bernoulli and Reddy–Bickford Columns

4.3.1 General Relationship

Consider an elastic column of length L, area of cross section A, Young's modulus E, and shear modulus G_{xz}, and subjected to an axial compressive load N^R. The axial displacement u^R of the Reddy–Bickford beam is given by [see Eq. (2.1.5a)]

$$u^R(x,z) = z\phi^R - \alpha z^3 \left(\phi^R + \frac{dw^R}{dx}\right) \quad (4.3.1)$$

where $\alpha = 4/(3h^2)$, x is the longitudinal coordinate measured from the bottom end of the column, w^R the transverse deflection of the centerline of the column at the onset of buckling, ϕ^R the rotation of the cross-section of the column, and h the thickness of the column in the z-direction. Given w^R, it is clear that the axial strain ε_{xx}^R and shear strain γ_{xz}^R can be computed as given in Eqs. (2.1.36a,b). The axial stress σ_{xx}^R and shear stress σ_{xz}^R are given by

$$\sigma_{xx}^R = E_x \varepsilon_{xx}^R = E_x z \frac{d\phi^R}{dx} - E_x \alpha z^3 \left(\frac{d\phi^R}{dx} + \frac{d^2 w^R}{dx^2}\right) \quad (4.3.2a)$$

$$\sigma_{xz}^R = G_{xz} \gamma_{xz}^R = G_{xz}(1 - \beta z^2)\left(\phi^R + \frac{dw^R}{dx}\right) \quad (4.3.2b)$$

where $\beta = 4/h^2$.

Following the procedure presented in section 2.1.4 for the derivation of the equilibrium equations of the Reddy–Bickford theory, we apply the principle of virtual work to the column at the onset of buckling

$$\delta W = 0 = \int_0^L \left[\int_A \left(\sigma_{xx}^R \delta \varepsilon_{xx}^R + \sigma_{xz}^R \delta \gamma_{xz}^R\right) dA - N^R \frac{dw^R}{dx}\frac{d\delta w^R}{dx}\right] dx$$

which, on integrating by parts, gives the governing equations

$$\delta\phi^R : -\frac{d\hat{M}_{xx}^R}{dx} + \hat{Q}_x^R = 0 \quad (4.3.3)$$

$$\delta w^R : -\alpha \frac{d^2 P_{xx}}{dx^2} - \frac{d\hat{Q}_x^R}{dx} + N^R \frac{d^2 w^R}{dx^2} = 0 \quad (4.3.4)$$

where

$$\hat{M}_{xx}^R = M_{xx}^R - \alpha P_{xx}, \quad \hat{Q}_x^R = Q_x^R - \beta R_x \quad (4.3.5)$$

The boundary conditions of the theory are of the form

$$\text{Specify}: \left\{\begin{array}{c} w^R \\ \frac{dw^R}{dx} \\ \phi^R \end{array}\right\} \text{ or } \left\{\begin{array}{c} \alpha \frac{dP_{xx}}{dx} + \hat{Q}_x^R - N^R \frac{dw^R}{dx} \\ \alpha P_{xx} \\ \hat{M}_{xx}^R \end{array}\right\} \quad (4.3.6)$$

The moment and force resultants are given by [see Eqs. (2.1.47)–(2.1.52)]

$$M_{xx}^R = \int_A z\sigma_{xx}\, dA = \hat{D}_{xx}\frac{d\phi^R}{dx} - \alpha F_{xx}\frac{d^2 w^R}{dx^2} \qquad (4.3.7)$$

$$P_{xx} = \int_A z^3 \sigma_{xx}\, dA = \hat{F}_{xx}\frac{d\phi^R}{dx} - \alpha H_{xx}\frac{d^2 w^R}{dx^2} \qquad (4.3.8)$$

$$Q_x^R = \int_A \sigma_{xz}\, dz = \hat{A}_{xz}\left(\phi^R + \frac{dw^R}{dx}\right) \qquad (4.3.9)$$

$$R_x = \int_A z^2 \sigma_{xz}\, dz = \hat{D}_{xz}\left(\phi^R + \frac{dw^R}{dx}\right) \qquad (4.3.10)$$

where [$\alpha = 4/(3h^2)$ and $\beta = 4/h^2$]

$$\hat{D}_{xx} = D_{xx} - \alpha F_{xx}, \quad \hat{F}_{xx} = F_{xx} - \alpha H_{xx} \qquad (4.3.11a)$$

$$\hat{A}_{xz} = A_{xz} - \beta D_{xz}, \quad \hat{D}_{xz} = D_{xz} - \beta F_{xz} \qquad (4.3.11b)$$

$$(A_{xx}, D_{xx}, F_{xx}, H_{xx}) = \int_A (1, z^2, z^4, z^6) E_x\, dA \qquad (4.3.11c)$$

$$(A_{xz}, D_{xz}, F_{xz}) = \int_A (1, z^2, z^4) G_{xz}\, dA \qquad (4.3.11d)$$

By manipulating the expressions in Eqs. (4.3.4) to (4.3.6), it is possible to write the governing equations and boundary conditions as

$$\frac{d^2 M_{xx}^R}{dx^2} - N^R \frac{d^2 w^R}{dx^2} = 0 \qquad (4.3.12)$$

$$\alpha \frac{d^2 P_{xx}}{dx^2} + \frac{d\hat{Q}_x^R}{dx} - N^R \frac{d^2 w^R}{dx^2} = 0 \qquad (4.3.13)$$

$$\text{Specify}:\begin{Bmatrix} w^R \\ \frac{dw^R}{dx} \\ \phi^R \end{Bmatrix} \quad \text{or} \quad \begin{Bmatrix} \frac{dM_{xx}^R}{dx} - N^R \frac{dw^R}{dx} \\ \alpha P_{xx} \\ \hat{M}_{xx}^R \end{Bmatrix} \qquad (4.3.14)$$

In order to obtain differential equations for buckling in terms of a single variable, it is useful to express P_{xx}, $\frac{d Q_x^R}{dx}$ and $\frac{dR}{dx}$ in terms of M_{xx}^R

and $\frac{d^2 w^R}{dx^2}$ by using Eqs. (4.3.7) to (4.3.10). First, from Eq. (4.3.7), it is seen that

$$\frac{d\phi^R}{dx} = \frac{1}{\hat{D}_{xx}}\left(M_{xx}^R + \alpha F_{xx}\frac{d^2 w^R}{dx^2}\right) \quad (4.3.15)$$

which, on substitution into Eqs. (4.3.8) to (4.3.10) yields

$$P_{xx} = \frac{1}{\hat{D}_{xx}}\left[\hat{F}_{xx}M_{xx}^R - \alpha\left(D_{xx}H_{xx} - F_{xx}^2\right)\frac{d^2 w}{dx^2}\right] \quad (4.3.16)$$

$$\frac{dQ_x^R}{dx} = \frac{\hat{A}_{xz}}{\hat{D}_{xz}}\left(M_{xx}^R + \alpha F_{xx}\frac{d^2 w^R}{dx^2}\right) + \hat{A}_{xz}\frac{d^2 w}{dx^2} \quad (4.3.17)$$

$$\frac{dR_x}{dx} = \frac{\hat{D}_{xz}}{\hat{D}_{xx}}\left(M_{xx}^R + \alpha F_{xx}\frac{d^2 w^R}{dx^2}\right) + \hat{D}_{xz}\frac{d^2 w}{dx^2} \quad (4.3.18)$$

By using the foregoing expressions for P_{xx}, $\frac{dQ_x^R}{dx}$ and $\frac{dR_x}{dx}$ in the governing Eqs. (4.3.12) and (4.3.13), the following buckling equation in terms of M_{xx}^R can be derived:

$$\frac{d^4 M_{xx}^R}{dx^4} + \left[\frac{\bar{D}_{xx}N^R - D_{xx}\bar{A}_{xz}}{\alpha^2 \tilde{D}_{xx}}\right]\frac{d^2 M_{xx}^R}{dx^2} - \left[\frac{\bar{A}_{xz}N^R}{\alpha^2 \tilde{D}_{xx}}\right]M_{xx}^R = 0 \quad (4.3.19)$$

where

$$\bar{D}_{xx} = \hat{D}_{xx} - \alpha\hat{F}_{xx} = D_{xx} - 2\alpha F_{xx} + \alpha^2 H_{xx} \quad (4.3.20a)$$

$$\bar{A}_{xz} = \hat{A}_{xz} - \beta\hat{D}_{xz} = A_{xz} - 2\beta D_{xz} + \beta^2 F_{xz} \quad (4.3.20b)$$

$$\tilde{D}_{xx} = D_{xx}H_{xx} - F_{xx}^2 \quad (4.3.20c)$$

Buckling equations in terms of ϕ^R and w^R can also be derived in a similar manner. These are

$$\frac{d^5 \phi^R}{dx^5} + \left[\frac{\bar{D}_{xx}N^R - D_{xx}\bar{A}_{xz}}{\alpha^2 \tilde{D}_{xx}}\right]\frac{d^3 \phi^R}{dx^3}$$
$$- \left[\frac{\bar{A}_{xz}N^R}{\alpha^2 \tilde{D}_{xx}}\right]\frac{d\phi^R}{dx} = 0 \quad (4.3.21)$$

$$\frac{d^6 w^R}{dx^6} + \left[\frac{\bar{D}_{xx}N^R - D_{xx}\bar{A}_{xz}}{\alpha^2 \tilde{D}_{xx}}\right]\frac{d^4 w^R}{dx^4}$$
$$- \left[\frac{\bar{A}_{xz}N^R}{\alpha^2 \tilde{D}_{xx}}\right]\frac{d^2 w^R}{dx^2} = 0 \quad (4.3.22)$$

Equation (4.3.22) shows that the Reddy–Bickford theory is a sixth-order theory in comparison to the fourth-order theories of Euler–Bernoulli and Timoshenko.

Equation (4.3.19) may be factored to give

$$\left(\frac{d^2}{dx^2} + \lambda_1^R\right)\left(\frac{d^2}{dx^2} + \lambda_2^R\right) M_{xx}^R = 0 \qquad (4.3.23)$$

where

$$\lambda_j^R = (-1)^j \sqrt{\left[\frac{\bar{D}_{xx} N^R - D_{xx} \bar{A}_{xz}}{2\alpha^2 \tilde{D}_{xx}}\right]^2 + \frac{\bar{A}_{xz} N^R}{\alpha^2 \tilde{D}_{xx}}} + \left[\frac{\bar{D}_{xx} N^R - D_{xx} \bar{A}_{xz}}{2\alpha^2 \tilde{D}_{xx}}\right] \qquad (4.3.24)$$

with $j = 1, 2$.

By letting

$$\mu_i = \left(\frac{d^2}{dx^2} + \lambda_i^R\right) M_{xx}^R \qquad (4.3.25)$$

Eq. (4.3.23) may be written as

$$\left(\frac{d^2}{dx^2} + \lambda_j^R\right) \mu_i = 0, \quad j = 1 \text{ or } 2, \quad j \neq i \qquad (4.3.26)$$

In the case of the buckling of Euler–Bernoulli columns, it can be shown that Eq. (4.2.7) may be written as (Timoshenko and Gere 1961)

$$\left(\frac{d^2}{dx^2} + \lambda^E\right) M_{xx}^E = 0 \qquad (4.3.27)$$

where M_{xx}^E is the bending moment in the Euler–Bernoulli column and

$$\lambda^E = \frac{N^E}{D_{xx}} \qquad (4.3.28)$$

It can be seen that Eqs. (4.3.26) and (4.3.27) are similar in form. Thus, one may relate the buckling loads between the Reddy–Bickford columns and the Euler–Bernoulli columns based on this analogous second-order governing equation as

$$\lambda_j^R = \lambda^E \qquad (4.3.29)$$

The substitution of Eqs. (4.3.24) and (4.3.28) into Eq. (4.3.29) furnishes the buckling load relationship given by

$$N^R = \frac{N^E \left[1 + \frac{N^E \alpha^2 \tilde{D}_{xx}}{A_{xz} D_{xx}^2}\right]}{1 + \frac{N^E \bar{D}_{xx}}{D_{xx} A_{xz}}} \qquad (4.3.30)$$

However, the foregoing relationship in Eq. (4.3.30) is only valid provided the two boundary conditions, required for solving the second-order differential equations [(4.3.26) or (4.3.27)], for both types of columns must also have the same form. Below, we proved that this is true for the cases of (a) pinned-pinned, (b) fixed-fixed, (c) fixed-free, and (d) pinned-pinned columns with rotational springs of equal stiffness added to their ends.

4.3.2 Pinned-Pinned Columns

Consider the case of pinned-pinned columns. The two boundary conditions for the Euler–Bernoulli column, to be solved with Eq. (4.3.27), are given by

$$M_{xx}^E = 0 \quad \text{at} \quad x = 0 \quad \text{and} \quad x = L \qquad (4.3.31)$$

Thus it must be shown that in the Reddy–Bickford column, $\mu = 0$ at $x = 0$ and $x = L$. Now the boundary conditions of such a column are given by

$$M_{xx}^R - \alpha P_{xx} = 0 \quad \text{and} \quad \alpha P_{xx} = 0 \quad \text{at} \quad x = 0 \quad \text{and} \quad x = L$$
$$\text{or} \quad M_{xx}^R = 0 \quad \text{at} \quad x = 0 \quad \text{and} \quad x = L \qquad (4.3.32)$$

In view of Eqs. (4.3.7), (4.3.8) and (4.3.32), it may be deduced that

$$\frac{d\phi^R}{dx} = \frac{d^2 w^R}{dx} = 0 \quad \text{at} \quad x = 0 \quad \text{and} \quad x = L \qquad (4.3.33)$$

It follows from Eq. (4.3.12) that

$$\frac{d^2 M_{xx}^R}{dx^2} = 0 \quad \text{at} \quad x = 0 \quad \text{and} \quad x = L \qquad (4.3.34)$$

It is clear from Eqs. (4.3.25), (4.3.32), and (4.3.34) that

$$\mu_i = 0 \quad \text{at} \quad x = 0 \quad \text{and} \quad x = L \tag{4.3.35}$$

Equations (4.3.31) and (4.3.35) show an exact matching of form for the boundary conditions and the thus the buckling load relationship is valid for the pinned-pinned case. A comparison of the numerical results will be presented later.

4.3.3 Fixed-Fixed Columns

For a fixed-fixed column, the lateral shearing force is zero along the column length. This means that the two boundary conditions for the Euler–Bernoulli column to be solved with Eq. (4.3.27) are

$$\frac{dM_{xx}^E}{dx} = 0 \quad \text{at} \quad x = 0 \quad \text{and} \quad x = L \tag{4.3.36}$$

This means that one has to prove that $\frac{d\mu}{dx} = 0$ at $x = 0$ and $x = L$ in the fixed-fixed Reddy–Bickford columns. To do this, Eq. (4.3.12) is first integrated with respect to x to give

$$\frac{dM_{xx}^R}{dx} - N^R \frac{dw^R}{dx} = C \tag{4.3.37}$$

where C is a constant. Since the effective shear force is zero for the fixed-fixed Reddy–Bickford column,

$$C = 0 \tag{4.3.38}$$

Moreover, by using the fact that for a fixed end, $\frac{dw^R}{dx} = 0$, it can be deduced from Eqs. (4.3.37) and (4.3.38) that

$$\frac{dM_{xx}^R}{dx} = 0 \quad \text{at} \quad x = 0 \quad \text{and} \quad x = L \tag{4.3.39}$$

The substitution of Eqs. (4.3.9) and (4.3.10) into Eq. (4.3.4) leads to

$$\alpha \frac{dP_{xx}}{dx} + \bar{A}_{xz}\left(\phi^R + \frac{dw^R}{dx}\right) = 0 \quad \text{at} \quad x = 0 \quad \text{and} \quad x = L \tag{4.3.40}$$

Using the fact that $\phi^R = \frac{dw^R}{dx} = 0$ at a fixed end, Eq. (4.3.40) reduces to

$$\frac{dP_{xx}}{dx} = 0 \quad \text{at} \quad x = 0 \quad \text{and} \quad x = L \tag{4.3.41}$$

In view of Eqs. (4.3.39) and (4.3.41), Eqs. (4.3.7) and (4.3.8) furnish

$$\frac{d^2\phi^R}{dx^2} = \frac{d^3w^R}{dx^3} = 0 \quad \text{at} \quad x = 0 \quad \text{and} \quad x = L \tag{4.3.42}$$

and together with Eq. (4.3.12), we have

$$\frac{d^3 M_{xx}^R}{dx^3} = 0 \quad \text{at} \quad x = 0 \quad \text{and} \quad x = L \tag{4.3.43}$$

Thus in view of Eqs. (4.3.39) and (4.3.43), we have

$$\frac{d\mu_i}{dx} = \frac{d^3 M_{xx}^R}{dx^3} + \lambda_i^R \frac{dM_{xx}^R}{dx} = 0 \quad \text{at} \quad x = 0 \quad \text{and} \quad x = L \tag{4.3.44}$$

As in the case of pinned-pinned columns, we have a matching in the form of the boundary conditions [c.f., Eqs. (4.3.36) and (4.3.44)] and therefore the buckling load relationship holds for the fixed-fixed columns.

4.3.4 Fixed-Free Columns

For the fixed end at $x = 0$, the boundary conditions of the Reddy–Bickford and the Euler–Bernoulli columns have already been shown to match in form. Now for the free end at $x = L$, the boundary condition of the Euler–Bernoulli column, to be solved with Eq. (4.3.27), is

$$M_{xx}^E = 0 \quad \text{at} \quad x = L \tag{4.3.45}$$

It is thus necessary to show that in the Reddy–Bickford column, $\mu = 0$ at $x = L$. First it is noted that for the free end, the boundary conditions are given by

$$M_{xx}^R - \alpha P_{xx} = 0 \quad \text{and} \quad \alpha P_{xx} = 0 \quad \text{at} \quad x = L$$
$$\text{or} \quad M_{xx}^R = 0 \quad \text{at} \quad x = L \tag{4.3.46}$$

Next, from Eqs. (4.3.7), (4.3.8), and (4.3.46), it can be shown that

$$\frac{d\phi^R}{dx} = \frac{d^2 w^R}{dx^2} = 0 \quad \text{at} \quad x = L \tag{4.3.47}$$

It follows from Eq. (4.3.12) that

$$\frac{d^2 M_{xx}^R}{dx^2} = 0 \quad \text{at} \quad x = L \qquad (4.3.48)$$

Thus, in view of Eqs. (4.3.25), (4.3.46), and (4.3.48), one obtains

$$\mu = \frac{d^2 M_{xx}^R}{dx^2} + \lambda_i^R = 0 \quad \text{at} \quad x = L \qquad (4.3.49)$$

As before there is a matching in the form of the boundary conditions for the fixed-free columns and thus the buckling load relationship is valid.

4.3.5 Pinned-Pinned Columns with End Rotational Springs of Equal Stiffness

In view of the fact that the buckling load relationship is valid for the pinned-pinned columns and the fixed-fixed columns, it can be readily proved that it also holds for the case of pinned-pinned columns with ends having additional elastic rotational springs of equal stiffness.

Here we simplify the form of the buckling load relationship for columns with square or circular cross-section. Noting the definition of the rigidities in Eqs. (4.3.11) and (4.3.20), it can be readily shown that the buckling load relationships reduce to

$$N^R = \frac{N^E \left(1 + \frac{N^E}{70 A_{xz}}\right)}{1 + \frac{17 N^E}{14 A_{xz}}} \quad \text{for square cross-section} \qquad (4.3.50)$$

$$N^R = \frac{N^E \left(1 + \frac{N^E}{90 A_{xz}}\right)}{1 + \frac{101 N^E}{90 A_{xz}}} \quad \text{for circular cross-section} \qquad (4.3.51)$$

It is worth noting that the relationship obtained for columns of square cross-section has a similar form to the buckling load relationship developed for circular plates under uniform in-plane loading (Wang and Lee 1998).

Tables 4.3.1–4.3.3 show the comparison of the buckling load parameters $\lambda = NL^2/D_{xx}$ between the Engesser–Timoshenko and Reddy–Bickford columns for different end conditions and shear parameter $\Omega = D_{xx}/(A_{xz}L^2)$. Note that the Engesser–Timoshenko buckling load results are computed from Eq. (4.2.14) with $K_s = 5/6$ for a square bar and $K_s = 9/10$ for a circular bar.

Table 4.3.1. Comparison of buckling load parameters of pinned-pinned columns between the Engesser–Timoshenko and Reddy–Bickford column theories.

Pinned-pinned column	Column of square cross section		Column of circular cross section	
Ω	λ^T	λ^R	λ^T	λ^R
0.0	9.8696	9.8696	9.8696	9.8696
0.1	4.5183	4.5526	4.7074	4.7342
0.2	2.9298	2.9874	3.0908	3.1370

Table 4.3.2. Comparison of buckling load parameters of fixed-fixed columns between the Engesser–Timoshenko and Reddy–Bickford column theories.

Fixed-fixed column	Column of square cross section		Column of circular cross section	
Ω	λ^T	λ^R	λ^T	λ^R
0.0	39.4784	39.4784	39.4784	39.4784
0.1	6.8809	7.1982	7.3292	7.5888
0.2	3.7689	4.1498	4.0395	4.3549

Table 4.3.3. Comparison of buckling load parameters of fixed-free columns between Engesser-Timoshenko and Reddy-Bickford column theories.

Fixed-free column	Column of square cross section		Column of circular cross section	
Ω	λ^T	λ^R	λ^T	λ^R
0.0	2.4674	2.4674	2.4674	2.4674
0.1	1.9037	1.9053	1.9365	1.9376
0.2	1.5497	1.5537	1.5936	1.5967

It can be seen that the buckling load parameters predicted by the Reddy–Bickford theory are higher than their Engesser–Timoshenko counterparts due to the factor found in the numerator of the buckling load relationship. The buckling load parameters are, however, in somewhat close agreement but there is a larger difference for the case of fixed-fixed columns as the latter has a higher Euler–Bernoulli buckling load value which magnifies the factor in the numerator. It is also clear that the difference between the Reddy–Bickford and Engesser–Timoshenko buckling solutions increases with higher modes of buckling. These higher modes of buckling become important when dealing with columns with internal restraints. As shown by Rozvany and Mröz (1977) and Olhoff and Akesson (1991), the optimal locations of internal supports for maximizing the buckling load of a column are found at the nodal points of an appropriate higher-order buckling mode.

4.4 Concluding Remarks

This chapter presents exact relationships between the buckling loads of the Engesser–Timoshenko columns, Reddy–Bickford columns, and Euler–Bernoulli columns with the following end conditions:

- Pinned-pinned
- Fixed-fixed
- Fixed-free
- Pinned-pinned with equal rotational stiffnesses at both ends

Using these relationships, buckling solutions of the Engesser–Timoshenko columns and Reddy–Bickford columns can be readily obtained from the corresponding Euler–Bernoulli solutions. It has been found that the Reddy–Bickford column theory predicts a higher buckling load when compared to the corresponding value from the Engesser–Timoshenko column theory. Although the buckling loads are close to those of the Engesser–Timoshenko columns for the cases of pinned-pinned and fixed-free columns, the results are slightly different due to a much higher value of the Euler–Bernoulli buckling load when the column ends are fixed. The differences in the buckling loads of the Engesser–Timoshenko and Reddy–Bickford columns become more significant at higher modes of buckling. The advantage of the Reddy–Bickford beam theory over the Engesser–Timoshenko beam theory is that it does not require shear correction factors.

Problems

4.1 Use the relationship in Eq. (4.2.14) to determine the buckling load $\lambda^T = N^T L^2 / D_{xx}$ of a fixed-free column when the parameter $\Omega = D_{xx}/A_{xz}L^2$ is equal to 0.1 and 0.2.

4.2 Use the relationship in Eq. (4.2.35) to determine the buckling load $\lambda^T = N^T L^2 / D_{xx}$ of a fixed-pinned column when the parameter $\Omega = D_{xx}/A_{xz}L^2$ is equal to 0.1 and 0.2.

4.3 Verify the relation in Eq. (4.3.19).

4.5 Establish the buckling load relationship between the Euler–Bernoulli beam theory and the Levinson beam theory (see Problem 2.7 for the governing equilibrium equations, which must be modified for the buckling case).

Chapter 5

Tapered Beams

The similarity between the equations of the Euler–Bernoulli beam theory and the Timoshenko beam theory for isotropic, variable cross-section, single span beams is used to develop expressions for deflection, slope, bending moment and shear force of the Timoshenko beam theory in terms of the same quantities of the Euler–Bernoulli beam theory. This new set of general relationships derived herein are illustrated for statically determinate and statically indeterminate, single span tapered beams. Extension of the results to laminated beams is also discussed.

5.1 Introduction

The exact relationships for deflections, slopes, shear forces and bending moments between single-span, uniform Timoshenko beams and the corresponding Euler–Bernoulli beams under general transverse loading and end conditions were presented in Chapter 2. These explicit and exact relationships are useful in elucidating the effect of transverse shear deformation on the bending of beams of uniform cross-section. For example, they show clearly that the stress-resultants and the slopes of Timoshenko beams are exactly equal to their Euler–Bernoulli counterparts for statically determinate beams. Further, they allow easy conversion of the well-known Euler–Bernoulli beam solutions to those for Timoshenko beams. They can be used to check numerical solutions to Timoshenko beam problems and even to verify the solutions obtained from other Timoshenko beam formulations such as the $w^b - w^s$ deflection component formulation (see Reddy, 1999b). In fact, Lee and Wang (1997) , while using these relationships, detected erroneous solutions from the common practice of setting $w^b = w^s = 0$ at simply supported and clamped ends for statically indeterminate Timoshenko beams.

Hitherto, these relationships are restricted to beams of uniform cross-section. In practice, it is common to utilize tapered beams due to the more efficient use of materials. It is thus the aim of this chapter to present more general relationships between the bending solutions of Timoshenko beams and Euler–Bernoulli beams of *arbitrary taper*. The extension of the results to laminated beams is also discussed.

5.2 Stress Resultant-Displacement Relations

Consider a tapered beam of length L, area of cross-section $A(x)$, second moment of area $I(x)$, modulus of elasticity E_x, and shear modulus of rigidity G_{xz} under any transverse loading condition $q(x)$. According to the Euler–Bernoulli beam theory, the stress resultant-displacement relations are given by:

$$M_{xx}^E = -D_{xx}\frac{d^2 w_0^E}{dx^2} \tag{5.2.1}$$

$$Q_x^E = \frac{dM_{xx}^E}{dx} = -\frac{d}{dx}\left(D_{xx}\frac{d^2 w_0^E}{dx^2}\right) \tag{5.2.2}$$

where w_0^E, M_{xx}^E and Q_x^E are the deflection, bending moment and shear force, respectively, $D_{xx} = E_x I$ the flexural rigidity, and x the longitudinal coordinate measured along the beam. The superscript E denotes quantities belonging to the Euler–Bernoulli beam theory.

According to the Timoshenko beam theory, the stress-resultant-displacement relations are given by:

$$M_{xx}^T = D_{xx}\frac{d\phi^T}{dx} \tag{5.2.3}$$

$$Q_x^T = A_{xz} K_s \left(\phi^T + \frac{dw_0^T}{dx}\right) \tag{5.2.4}$$

where the superscript T denotes quantities belonging to the Timoshenko beam, ϕ^T and K_s are the rotation of the cross section and the shear correction factor, respectively, and $A_{xz} = G_{xz} A$ the shear stiffness.

5.3 Equilibrium Equations

For both beams, the equilibrium equations are given by:

$$\frac{dM_{xx}}{dx} = Q_x \tag{5.3.1}$$

$$\frac{dQ_x}{dx} = -q \tag{5.3.2}$$

Substituting Eqs. (5.2.2) and (5.3.1) into Eq. (5.3.2) yields the Euler–Bernoulli beam equation:

$$-\frac{d^2}{dx^2}\left(D_{xx}\frac{d^2 w_0^E}{dx^2}\right) = \frac{d^2 M_{xx}^E}{dx^2} = \frac{dQ_x^E}{dx} = -q \quad (5.3.3)$$

By substituting Eqs. (5.2.3) and (5.2.4) into Eqs. (5.3.1) and (5.3.2), one obtains the following two equilibrium equations for the Timoshenko beam:

$$A_{xz}K_s\left(\phi^T + \frac{dw_0^T}{dx}\right) = \frac{d}{dx}\left(D_{xx}\frac{d\phi^T}{dx}\right) \quad (5.3.4)$$

and

$$\frac{d}{dx}\left[A_{xz}K_s\left(\phi^T + \frac{dw_0^T}{dx}\right)\right] = -q \quad (5.3.5)$$

Differentiating Eq. (5.3.4) with respect to x, and using Eqs. (5.2.3) and (5.3.5) lead to

$$\frac{d^2}{dx^2}\left(D_{xx}\frac{d\phi^T}{dx}\right) = \frac{d^2 M_{xx}^T}{dx^2} = \frac{dQ_x^T}{dx} = -q \quad (5.3.6)$$

5.4 Deflection and Force Relationships
5.4.1 General Relationships

From the mathematical similarity of Eqs. (5.3.3) and (5.3.6), it can be deduced that

$$D_{xx}\frac{d\phi^T}{dx} = -D_{xx}\frac{d^2 w_0^E}{dx^2} + C_1 x + C_2$$

or

$$M_{xx}^T = M_{xx}^E + C_1 x + C_2 \quad (5.4.1)$$

and

$$Q_x^T = Q_x^E + C_1 \quad (5.4.2)$$

Assuming that $D_{xx} \neq 0$ anywhere along the length of the beam, the integration of Eq.(5.4.1) yields

$$\phi^T = -\frac{dw_0^E}{dx} + \int_0^x \frac{C_1 x + C_2}{D_{xx}}\,dx + C_3 \quad (5.4.3)$$

The substitution of Eq. (5.4.3) into Eq. (5.3.4) followed by integration with respect to x yield

$$w_0^T = w_0^E + \int_0^x \frac{Q_x^E + C_1}{A_{xz}K_s}dx - \int_0^x \int_0^x \frac{C_1 x + C_2}{D_{xx}}dx\,dx - C_3 x - C_4 \quad (5.4.4)$$

The integration constants C_1, C_2, C_3, and C_4 can be evaluated from the boundary conditions given by:

Free(F): $M_{xx}^E = M_{xx}^T = Q_x^E = Q_x^T = 0$ (5.4.5)

Simple support(S): $w_0^E = w_0^T = M_{xx}^E = M_{xx}^T = 0$ (5.4.6)

Clamped(C): $w_0^E = w_0^T = \dfrac{dw_0^E}{dx} = \phi^T = 0$ (5.4.7)

Below, the constants C_i are evaluated for single-span beams of various combinations of end conditions.

5.4.2 Simply Supported (SS) Beams

The boundary conditions for simply supported beams are given by Eq. (5.4.6) for $x = 0$ and $x = L$. The substitution of these boundary conditions into Eqs. (5.4.1) and (5.4.4) gives the following:

$$C_1 = C_2 = C_4 = 0, \quad C_3 = \frac{1}{L}\int_0^L \frac{Q_x^E}{A_{xz}K_s}dx \quad (5.4.8)$$

In view of these constants, Eqs. (5.4.1) to (5.4.4) for simply supported beams read as:

$$M_{xx}^T = M_{xx}^E \quad (5.4.9a)$$

$$Q_x^T = Q_x^E \quad (5.4.9b)$$

$$\phi^T = -\frac{dw_0^E}{dx} + \frac{1}{L}\int_0^L \frac{Q_x^E}{A_{xz}K_s}dx \quad (5.4.9c)$$

$$w_0^T = w_0^E + \int_0^x \frac{Q_x^E}{A_{xz}K_s}dx - \frac{x}{L}\int_0^L \frac{Q_x^E}{A_{xz}K_s}dx \quad (5.4.9d)$$

5.4.3 Clamped-Free (CF) Beams

The boundary conditions for clamped-free beams are given by Eq. (5.4.7) for $x = 0$ and by Eq. (5.4.5) for $x = L$. The substitution of these boundary conditions into Eqs. (5.4.1) to (5.4.4) gives the following:

$$C_1 = C_2 = C_3 = C_4 = 0 \qquad (5.4.10)$$

In view of these constant values, the relationships given by Eqs. (5.4.1) to (5.4.4) for clamped free beams read as:

$$M_{xx}^T = M_{xx}^E \qquad (5.4.11a)$$
$$Q_x^T = Q_x^E \qquad (5.4.11b)$$
$$\phi^T = -\frac{dw_0^E}{dx} \qquad (5.4.11c)$$
$$w_0^T = w_0^E + \int_0^L \frac{Q_x^E}{A_{xz}K_s} dx \qquad (5.4.11d)$$

5.4.4 Free-Clamped (FC) Beams

The boundary conditions for free-clamped beams are given by Eq. (5.4.5) for $x = 0$ and by Eq. (5.4.7) for $x = L$. The substitution of these boundary conditions into Eqs. (5.4.1) to (5.4.4) gives the following:

$$C_1 = C_2 = C_3 = 0; \quad C_4 = \int_0^L \frac{Q_x^E}{A_{xz}K_s} dx \qquad (5.4.12)$$

In view of these constant values, the relationships given by Eqs. (5.4.1) to (5.4.4) for clamped free beams read as:

$$M_{xx}^T = M_{xx}^E \qquad (5.4.13a)$$
$$Q_x^T = Q_x^E \qquad (5.4.13b)$$
$$\phi^T = -\frac{dw_0^E}{dx} \qquad (5.4.13c)$$
$$w_0^T = w_0^E - \int_0^L \frac{Q_x^E}{A_{xz}K_s} dx \qquad (5.4.13d)$$

5.4.5 Clamped (CC) Beams

The boundary conditions for clamped beams are given by Eq. (5.4.7) for $x = 0$ and $x = L$. The substitution of these boundary conditions into Eqs. (5.4.3) and (5.4.4) gives the following:

$$C_1 = -\frac{\int_0^L \frac{Q_x^E}{A_{xz}K_s} dx}{\int_0^L \left(\frac{1}{A_{xz}K_s} - \int_0^x \frac{x}{D_{xx}} dx \right) dx + \left(\frac{\int_0^L \frac{x}{D_{xx}} dx}{\int_0^L \frac{1}{D_{xx}} dx} \right) \left(\int_0^L \int_0^x \frac{1}{D_{xx}} dx \, dx \right)}$$

$$C_2 = -C_1 \frac{\int_0^L \frac{x}{D_{xx}} dx}{\int_0^L \frac{1}{D_{xx}} dx}$$

$$C_3 = C_4 = 0 \qquad (5.4.14)$$

In view of these constants, the relationships given by Eqs. (5.4.1) to (5.4.4) for clamped beams read as

$$M_{xx}^T = M_{xx}^E + C_1 x + C_2 \qquad (5.4.15a)$$

$$Q_x^T = Q_x^E + C_1 \qquad (5.4.15b)$$

$$\phi^T = -\frac{dw_0^E}{dx} + \int_0^x \frac{C_1 x + C_2}{D_{xx}} dx \qquad (5.4.15c)$$

$$w_0^T = w_0^E + \int_0^x \frac{Q_x^E + C_1}{A_{xz}K_s} dx - \int_0^x \int_0^x \frac{C_1 x + C_2}{D_{xx}} dx \qquad (5.4.15d)$$

5.4.6 Clamped-Simply Supported (CS) Beams

The boundary conditions for clamped-simply supported beams are given by Eq. (5.4.7) for $x = 0$ and by Eq. (5.4.6) for $x = L$. The substitution of these boundary conditions into Eqs. (5.4.1), (5.4.3) and (5.4.4) gives the following:

$$C_1 = -\frac{\int_0^L \frac{Q_x^E}{A_{xz}K_s} dx}{\int_0^L \left\{ \frac{1}{A_{xz}K_s} - \int_0^x \frac{x-L}{D_{xx}} dx \right\} dx}$$

$$C_2 = -C_1 L, \quad C_3 = C_4 = 0 \qquad (5.4.16)$$

In view of these constants, the relationships given by Eqs. (5.4.1) to (5.4.4) for clamped-simply supported beams read as:

$$M_{xx}^T = M_{xx}^E - C_1(L - x) \qquad (5.4.17a)$$

$$Q_x^T = Q_x^E + C_1 \tag{5.4.17b}$$

$$\phi^T = -\frac{dw_0^E}{dx} + C_1 \int_0^x \frac{x-L}{D_{xx}} dx \tag{5.4.17c}$$

$$w_0^T = w_0^E + \int_0^x \frac{Q_x^E + C_1}{A_{xz} K_s} dx - C_1 \int_0^x \int_0^x \frac{x-L}{D_{xx}} dx dx \tag{5.4.17d}$$

5.4.7 Simply Supported-Clamped (SC) Beams

The boundary conditions for clamped-simply supported beams are given by Eq. (5.4.6) for $x = 0$ and by Eq. (5.4.7) for $x = L$. The substitution of these boundary conditions into Eqs. (5.4.1), (5.4.3) and (5.4.4) gives the following:

$$C_1 = -\frac{\int_0^L \frac{Q_x^E}{A_{xz} K_s} dx}{L \int_0^L \frac{x}{D_{xx}} dx + \int_0^L \left\{ \frac{1}{A_{xz} K_s} - \int_0^L \frac{x}{D_{xx}} dx \right\} dx}$$

$$C_2 = C_4 = 0, \quad C_3 = -C_1 \int_0^L \frac{x}{D_{xx}} dx \tag{5.4.18}$$

In view of these constants, the relationships given by Eqs. (5.4.1) to (5.4.4) for simply supported-clamped beams read as:

$$M_{xx}^T = M_{xx}^E + C_1 x \tag{5.4.19a}$$

$$Q_x^T = Q_x^E + C_1 \tag{5.4.19b}$$

$$\phi^T = -\frac{dw_0^E}{dx} + C_1 \int_x^L \frac{x}{D_{xx}} dx \tag{5.4.19c}$$

$$w_0^T = w_0^E + \int_0^x \frac{Q_x^E + C_1}{A_{xz} K_s} dx - C_1 \int_0^x \int_0^x \frac{x}{D_{xx}} dx dx$$

$$+ C_1 x \int_0^L \frac{x}{D_{xx}} dx \tag{5.4.19d}$$

All the foregoing relationships reduce to the same ones as derived in Chapter 2 when the beam has a uniform cross-section, i.e. A and I are constants. Wang, Chen, and Kitipornchai (1998) have also developed relationships for non-uniform beams with elastic end restraints.

5.4.8 An Example

Consider the use of the relationships in determining the deflections of a simply supported, tapered Timoshenko beam under a uniformly distributed load q_0. The width b and depth h of the tapered beam are assumed to vary linearly along the x-direction

$$b = b_0(1 + \bar{x}), \quad h = h_0(1 + \bar{x}), \quad \bar{x} = \frac{x}{L} \qquad (5.4.20)$$

where b_0 and h_0 are the width and depth, respectively, of the beam cross section at $x = 0$, and L is the length of the beam.

By solving directly the governing Euler–Bernoulli beam equations (5.3.3) together with the boundary conditions in Eq. (5.4.6), and noting $D_{xx}(0) = EI_0$, the deflection is found to be

$$\begin{aligned} w_0^E = \frac{q_0 L^4}{12 E I_0 (1+\bar{x})^2} &\Big[2\bar{x} + \bar{x}^2 - 3\bar{x}^3 + 6\ln(L)\left(1 + \bar{x} - \bar{x}^2\right) \\ &+ 6\ln(2L)\left(\bar{x} + 2\bar{x}^2 + \bar{x}^3\right) \\ &- 6[\ln(L) + \ln(1+\bar{x})]\left(1 + 2\bar{x} + \bar{x}^2\right) \Big] \end{aligned} \qquad (5.4.21)$$

where $I_0 = b_0 h_0^3 / 12$. From Eq. (5.3.3), the transverse shear force of the Euler–Bernoulli beam is given by

$$Q_x^E = \frac{q_0 L}{2}(1 - 2\bar{x}), \quad \bar{x} = \frac{x}{L} \qquad (5.4.22)$$

To determine the Timoshenko beam solutions using the relationships, we substitute the width and depth expressions from Eq. (5.4.20), the Euler–Bernoulli deflection (5.4.21) and the shear force (5.4.22) into Eq. (5.4.9d). The deflection corresponding to the Timoshenko tapered beam, noting $A_{xz}(0) = GA_0$, is given by

$$\begin{aligned} w_0^T = \frac{q_0 L^2}{12 K_s G A_0} &\Big[-\frac{18}{(1+\bar{x})} + 6(3 + 2\ln(L)) + \Lambda(7 + 6\ln(L)) \\ &- [\ln(L) + \ln(1+\bar{x})](12 + 6\Lambda) \Big] \\ + \frac{q_0 L^4}{12 E I_0} &\Big[\frac{2}{(1+\bar{x})^2} - \frac{9}{(1+\bar{x})} \\ &- 6\bar{x}\Big(1 + 3\Omega - 2\ln(2)(1 + 2\Omega)\Big) \Big] \end{aligned} \qquad (5.4.23)$$

where $A_0 = b_0 h_0$ and

$$\Omega = \frac{EI_0}{K_s G A_0 L^2}, \quad \Lambda = \frac{1}{\Omega} \tag{5.4.24}$$

5.5 Symmetrically Laminated Beams

The governing equations of symmetrically laminated beams have the same form as those given in Eqs. (5.2.1) to (5.2.4) for isotropic beams. Therefore, the results presented in this paper are also valid for symmetrically laminated beams. The difference between symmetrically laminated beams and isotropic beams is reflected in the flexural and shear stiffnesses, D_{xx} and $K_s A_{xz}$. For a laminated beam, they are computed from the following equations (Reddy 1997a)

$$E^c I = \frac{b}{D_{11}^*}, \quad G^c A K_s = \frac{1}{h A_{55}^*} \tag{5.5.1}$$

where the superscript C denotes quantities of the composite beam, h is the total thickness and b is the width of the laminate, and

$$D_{11}^* = \frac{D_{22} D_{66} - D_{26} D_{26}}{D_{11} \tilde{D}_{11} + D_{12} \tilde{D}_{12} + D_{16} \tilde{D}_{16}}, \quad A_{55}^* = \frac{A_{44}}{\tilde{A}_{44}} \tag{5.5.2}$$

$$\tilde{D}_{11} = D_{22} D_{66} - D_{26} D_{26}, \quad \tilde{D}_{12} = D_{16} D_{26} - D_{12} D_{66} \tag{5.5.3a}$$

$$\tilde{D}_{16} = D_{12} D_{26} - D_{22} D_{16}, \quad \tilde{A}_{44} = A_{44} A_{55} - A_{45} A_{45} \tag{5.5.3b}$$

Here A_{ij} and D_{ij} denote the extensional and bending stiffnesses, respectively, of a laminate

$$(A_{ij}, D_{ij}) = \int_{-\frac{h}{2}}^{\frac{h}{2}} Q_{ij} \left(1, z^2\right) dz \tag{5.5.4}$$

and Q_{ij} denote the plane stress-reduced elastic stiffnesses of a layer referred to laminate coordinates.

It is rare that laminated beams have variable cross sections due to variable thicknesses of individual layers because these layers are generally made of uniform thickness. Variable cross-section beams, either piece-wise constant (but multiple segments) or linearly varying cross-section laminates are constructed by ply drop-off and chopping-off a constant thickness laminate. The present results hold for the latter case directly. In the former case, Eqs. (5.4.1)–(5.4.4) hold within each segment, and the constants of integration C_i in each segment are evaluated using the continuity of deflection, slope, bending moment, and shear force between segments and the boundary conditions of the beam.

5.6 Concluding Remarks

The deflection, slope, and stress-resultant relationships for single-span Timoshenko and Euler–Bernoulli tapered beams under any transverse loads and end conditions are derived. These relationships may be used to readily convert tapered Euler–Bernoulli beam bending solutions to those for Timoshenko beams, thereby bypassing the need for a more complicated shear deformable beam analysis. The exact relationships should also prove useful when checks are needed for numerical results of tapered Timoshenko beams. The results also hold for symmetrically laminated beams with the stiffnesses modified accordingly, as pointed out earlier.

An interesting point to note is that in uniform and statically determinate beams, the slope of the Timoshenko beam is equal to that of the corresponding Euler–Bernoulli beam. This study shows that in the case of tapered beams, while this slope condition holds for cantilevered beams, the condition is not valid for simply supported beams in general. In the latter simply supported case, the equal slope condition only applies to tapered beams when $\int_0^L \frac{Q_x^E}{A_{xz}K_s}dx = 0$; an example being the beam which is tapered and loaded symmetrically about its mid-span.

Problems

5.1 Determine the deflection of the tapered Timoshenko beam defined in section 5.4.8 by solving directly the governing equations (5.3.4) and (5.3.5) together with the boundary conditions given by Eq. (5.4.6). Check the answer against that given in Eq. (5.4.23).

5.2 Derive the bending relationships for tapered beams with ends that are simply supported and constrained by elastic springs with rotational stiffness constants k_0 and k_1 at ends $x = 0$ and $x = L$, respectively.

PART 2
PLATES

Chapter 6

Theories of Plate Bending

Presented in this chapter are the various plate theories, progressing from the Kirchhoff (classical thin) plate theory to the first-order shear deformation plate theory of Mindlin, and finally, to the third-order plate theory of Reddy. The latter two plate theories allow for the effect of transverse shear deformation which has been neglected in the Kirchhoff plate theory. Using the principle of virtual displacements, the governing equations and boundary conditions are derived for uniform thickness plates on the basis of the kinematic assumptions of the aforementioned plate theories.

6.1 Overview of Plate Theories

The two-dimensional plate theories can be classified into two types: (1) classical plate theory, in which the transverse shear deformation effects are neglected, and (2) shear deformation plate theories. The Kirchhoff (classical) plate theory (CPT) for the pure bending case is based on the displacement field

$$u(x, y, z) = -z \frac{\partial w_0}{\partial x} \qquad (6.1.1a)$$

$$v(x, y, z) = -z \frac{\partial w_0}{\partial y} \qquad (6.1.1b)$$

$$w(x, y, z) = w_0(x, y) \qquad (6.1.1c)$$

where (u, v, w) are the displacement components along the (x, y, z) coordinate directions, respectively, and w_0 is the transverse deflection of a point on the mid-plane (i.e., $z = 0$). The displacement field (6.1.1) implies that straight lines normal to the xy-plane before deformation remain straight and normal to the mid-surface after deformation. The

Kirchhoff assumption amounts to neglecting both transverse shear and transverse normal effects, i.e. deformation is due entirely to bending and in-plane stretching.

There are a number of shear deformation plate theories. The simplest is the first-order shear deformation plate theory (or FSDT), also known as the Mindlin plate theory (Mindlin 1951), and it is based on the displacement field

$$u(x, y, z) = z\phi_x(x, y) \quad (6.1.2a)$$
$$v(x, y, z) = z\phi_y(x, y) \quad (6.1.2b)$$
$$w(x, y, z) = w_0(x, y) \quad (6.1.2c)$$

where ϕ_x and $-\phi_y$ denote rotations about the y and x axes, respectively. The FSDT extends the kinematics of the CPT by including a gross transverse shear deformation in its kinematic assumptions, i.e. the transverse shear strain is assumed to be constant with respect to the thickness coordinate. In the FSDT, shear correction factors are introduced to correct the discrepancy between the actual transverse shear force distributions and those computed using the kinematic relations of the FSDT. The shear correction factors depend not only on the geometric parameters, but also on the loading and boundary conditions of the plate.

In both the CPT and FSDT, the plane-stress state assumption is used and the plane-stress-reduced form of the constitutive law is used. In both theories, the inextensibility of a transverse normal can be removed by assuming that the transverse deflection also varies through the thickness. This allows the use of full three-dimensional constitutive equations.

Second- and higher-order shear deformation plate theories use higher-order polynomials in the expansion of the displacement components through the thickness of the plate. The higher-order theories introduce additional unknowns that are often difficult to interpret in physical terms. The second-order theory with transverse inextensibility is based on the displacement field

$$u(x, y, z) = z\phi_x(x, y) + z^2\psi_x(x, y) \quad (6.1.3a)$$
$$v(x, y, z) = z\phi_y(x, y) + z^2\psi_y(x, y) \quad (6.1.3b)$$
$$w(x, y, z) = w_0(x, y) \quad (6.1.3c)$$

There are a number of third-order theories in the literature, and a review of these theories is given by Reddy (1984a). The third-order shear deformation plate theory (TSDT) of Reddy (1984a, 1997a, 1999a) with transverse inextensibility is based on the displacement field

$$u(x,y,z) = z\phi_x(x,y) - \alpha z^3 \left(\phi_x + \frac{\partial w_0}{\partial x}\right) \quad (6.1.4a)$$

$$v(x,y,z) = z\phi_y(x,y) - \alpha z^3 \left(\phi_y + \frac{\partial w_0}{\partial y}\right) \quad (6.1.4b)$$

$$w(x,y,z) = w_0(x,y) \quad (6.1.4c)$$

where $\alpha = 4/(3h^2)$. Note that by setting $\alpha = 0$, we recover the displacement field of the FSDT. The displacement field accommodates a quadratic variation of the transverse shear strains (and hence shear stresses) through the thickness and the vanishing of transverse shear stresses on the top and bottom surfaces of the plate (see Figure 6.1.1). Unlike the FSDT, the TSDT requires no shear correction factors.

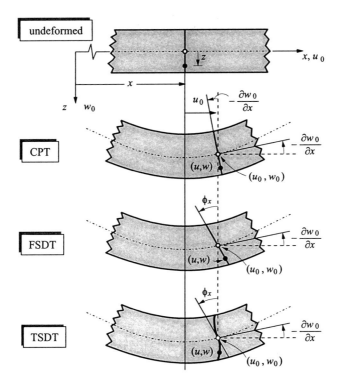

Figure 6.1.1. Undeformed and deformed geometries of an edge of a plate in various plate theories. Here u_0 denotes the in-plane displacement, which is not included in the present study.

In addition to its inherent simplicity and low computational cost, the first-order plate theory often provides a sufficiently accurate description of the global response (e.g., deflections, buckling loads, and natural vibration frequencies) for thin to moderately thick plates. Therefore, it is of interest to determine the global response using FSDT. Third-order theories provide a small increase in accuracy relative to the FSDT solution, at the expense of a significant increase in computational effort.

6.2 Classical (Kirchhoff) Plate Theory (CPT)

6.2.1 Equations of Equilibrium

The principle of virtual displacements has been adopted in the derivation of the equilibrium equations and the boundary conditions for the various plate theories. In the subsections that follow, the equations of equilibrium for the CPT, FSDT, and TSDT are derived.

The non-zero linear strains associated with the displacement field in Eqs. (6.1.1a-c) are

$$\varepsilon_{xx} = \frac{\partial u}{\partial x} = -z\frac{\partial^2 w_0}{\partial x^2} \tag{6.2.1a}$$

$$\varepsilon_{yy} = \frac{\partial u}{\partial y} = -z\frac{\partial^2 w_0}{\partial y^2} \tag{6.2.1b}$$

$$\gamma_{xy} = \left(\frac{\partial u}{\partial y} + \frac{\partial v}{\partial x}\right) = -2z\frac{\partial^2 w_0}{\partial x \partial y} \tag{6.2.1c}$$

where $(\varepsilon_{xx}, \varepsilon_{yy})$ are the normal strains and γ_{xy} is the shear strain.

The virtual strain energy U of the Kirchhoff plate theory is given by

$$\delta U = \int_{\Omega_0} \left[\int_{-\frac{h}{2}}^{\frac{h}{2}} (\sigma_{xx}\delta\varepsilon_{xx} + \sigma_{yy}\delta\varepsilon_{yy} + \sigma_{xy}\delta\gamma_{xy})\,dz \right] dxdy$$

$$= -\int_{\Omega_0} \left(M_{xx}\frac{\partial^2 \delta w_0}{\partial x^2} + M_{yy}\frac{\partial^2 \delta w_0}{\partial y^2} + 2M_{xy}\frac{\partial^2 \delta w_0}{\partial x \partial y} \right) dxdy \tag{6.2.2}$$

where Ω_0 denotes the domain occupied by the mid-plane of the plate, h the thickness of the plate, $(\sigma_{xx}, \sigma_{yy})$ the normal stresses, σ_{xy} the shear stress, and (M_{xx}, M_{yy}, M_{xy}) the moments per unit length (see Figure 6.2.1)

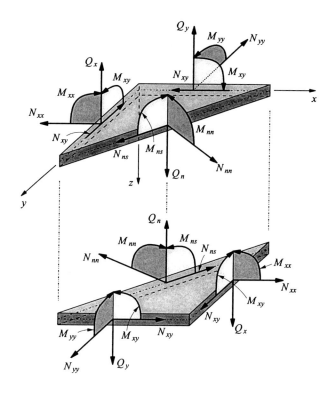

Figure 6.2.1. Forces and moments on a plate element. The in-plane force resultants N_{xx}, N_{yy}, and N_{xy} do not enter the equations in the pure bending case, and they are the specified forces in a buckling problem.

$$\left\{\begin{array}{c} M_{xx} \\ M_{yy} \\ M_{xy} \end{array}\right\} = \int_{-\frac{h}{2}}^{\frac{h}{2}} \left\{\begin{array}{c} \sigma_{xx} \\ \sigma_{yy} \\ \sigma_{xy} \end{array}\right\} z \, dz \qquad (6.2.3)$$

Note that the virtual strain energy associated with the transverse shear strains is zero as $\gamma_{xz} = \gamma_{yz} = 0$ in the Kirchhoff plate theory.

The virtual potential energy δV due to the transverse load $q(x, y)$ is given by

$$\delta V = -\int_{\Omega_0} q(x,y) \delta w_0 \, dxdy \qquad (6.2.4)$$

If there are any nonzero edge forces and moments, the virtual potential energy V should be added to the expression in Eq. (6.2.4).

The principle of virtual displacements requires that $\delta W = \delta U + \delta V = 0$. We obtain, using the divergence theorem,

$$
\begin{aligned}
0 &= -\int_{\Omega_0} \left(M_{xx} \frac{\partial^2 \delta w_0}{\partial x^2} + M_{yy} \frac{\partial^2 \delta w_0}{\partial y^2} + 2 M_{xy} \frac{\partial^2 \delta w_0}{\partial x \partial y} + q \delta w_0 \right) dx\,dy \\
&= -\int_{\Omega_0} (M_{xx,xx} + 2 M_{xy,xy} + M_{yy,yy} + q)\, \delta w_0\, dx\,dy \\
&\quad - \oint_\Gamma \left[(M_{xx} n_x + M_{xy} n_y) \frac{\partial \delta w_0}{\partial x} + (M_{xy} n_x + M_{yy} n_y) \frac{\partial \delta w_0}{\partial y} \right] ds \\
&\quad + \oint_\Gamma \left[(M_{xx,x} + M_{xy,y}) n_x + (M_{yy,y} + M_{xy,x}) n_y \right] \delta w_0\, ds \qquad (6.2.5)
\end{aligned}
$$

where a comma followed by subscripts denotes differentiation with respect to the subscripts, i.e., $M_{xx,x} = \partial M_{xx}/\partial x$, and so on, (n_x, n_y) denote the direction cosines of the unit normal $\hat{\mathbf{n}}$ on the boundary Γ, and a circle on the integral sign signifies integration over the total boundary. Also, s is the coordinate measured along Γ. If the unit normal vector $\hat{\mathbf{n}}$ is oriented at an angle θ from the positive x-axis, then $n_x = \cos\theta$ and $n_y = \sin\theta$. Since δw_0 is arbitrary in Ω_0, and it is independent of $\frac{\partial \delta w_0}{\partial x}$, and $\frac{\partial \delta w_0}{\partial y}$ on the boundary Γ, it follows that

$$
\frac{\partial^2 M_{xx}}{\partial x^2} + 2 \frac{\partial^2 M_{xy}}{\partial x \partial y} + \frac{\partial^2 M_{yy}}{\partial y^2} + q = 0 \quad \text{in } \Omega_0 \qquad (6.2.6)
$$

which represents the equilibrium equation of the Kirchhoff plate theory.

6.2.2 Boundary Conditions

To determine the form of the boundary conditions, we consider the boundary integrals in Eq. (6.2.5). On an edge parallel to the x or y coordinate, the boundary expression in (6.2.5) implies that

$$
\begin{aligned}
&\text{either} \quad \delta w_0 = 0 \quad \text{or} \quad Q_x n_x + Q_y n_y = 0 &(6.2.7a) \\
&\text{either} \quad \frac{\partial \delta w_0}{\partial x} = 0 \quad \text{or} \quad M_{xx} n_x + M_{xy} n_y = 0 &(6.2.7b) \\
&\text{either} \quad \frac{\partial \delta w_0}{\partial y} = 0 \quad \text{or} \quad M_{xy} n_x + M_{yy} n_y = 0 &(6.2.7c)
\end{aligned}
$$

where
$$Q_x \equiv M_{xx,x} + M_{xy,y}, \quad Q_y \equiv M_{yy,y} + M_{xy,x} \qquad (6.2.8)$$
are the shear forces (see Figure 6.2.1). Note that Q_x and Q_y are *defined* by Eq. (6.2.8). Equations (6.2.7a-c) indicate that $(w_0, \frac{\partial w_0}{\partial x}, \frac{\partial w_0}{\partial y})$ are the primary variables and specification of any of them constitutes an essential (or geometric) boundary condition. The associated secondary variables are

$$Q_x n_x + Q_y n_y, \quad M_{xx} n_x + M_{xy} n_y, \quad M_{xy} n_x + M_{yy} n_y$$

The specification of any of the secondary variables constitutes a natural (or force) boundary condition.

In general, not every edge of a plate will be parallel to a coordinate axis. Therefore, it is useful to express the boundary conditions in terms of slopes and moments that are referred to the normal and tangential coordinates (n, s) of an edge (see Figure 6.2.2). The slopes $(\frac{\partial w_0}{\partial x}, \frac{\partial w_0}{\partial y})$ in the (x, y) coordinate system can be expressed in terms of the slopes $(\frac{\partial w_0}{\partial n}, \frac{\partial w_0}{\partial s})$ in the (n, s) system by the relations

$$\frac{\partial w_0}{\partial x} = n_x \frac{\partial w_0}{\partial n} - n_y \frac{\partial w_0}{\partial s} \qquad (6.2.9a)$$
$$\frac{\partial w_0}{\partial y} = n_y \frac{\partial w_0}{\partial n} + n_x \frac{\partial w_0}{\partial s} \qquad (6.2.9b)$$

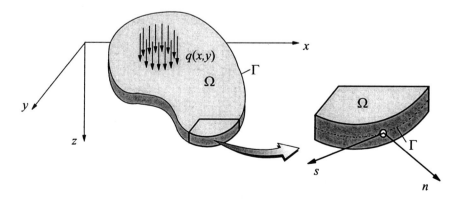

Figure 6.2.2. Plate element with a curved boundary and coordinate system (n, s, z).

The boundary expression in Eq. (6.2.5) can be expressed in terms of the normal and tangential derivatives of w_0:

$$\oint_\Gamma \left[(Q_x n_x + Q_y n_y) \delta w_0 \right.$$
$$- (M_{xx} n_x + M_{xy} n_y) \left(\frac{\partial \delta w_0}{\partial n} n_x - \frac{\partial \delta w_0}{\partial s} n_y \right)$$
$$\left. - (M_{xy} n_x + M_{yy} n_y) \left(\frac{\partial \delta w_0}{\partial n} n_y + \frac{\partial \delta w_0}{\partial s} n_x \right) \right] ds$$
$$= \oint_\Gamma \left[(Q_x n_x + Q_y n_y) \delta w_0 \right.$$
$$- \left(M_{xx} n_x^2 + 2 M_{xy} n_x n_y + M_{yy} n_y^2 \right) \frac{\partial \delta w_0}{\partial n}$$
$$\left. - \left[(M_{yy} - M_{xx}) n_x n_y + M_{xy}(n_x^2 - n_y^2) \right] \frac{\partial \delta w_0}{\partial s} \right] ds \quad (6.2.10)$$

The secondary variables are the coefficients of δw_0, $\frac{\partial \delta w_0}{\partial n}$ and $\frac{\partial \delta w_0}{\partial s}$ on Γ. From Eq. (6.2.10), it is clear that the primary variables (i.e., generalized displacements) and secondary variables (i.e., generalized forces) of the theory are:

$$\text{primary variables:} \quad w_0, \; \frac{\partial w_0}{\partial n}, \; \frac{\partial w_0}{\partial s} \quad (6.2.11a)$$
$$\text{secondary variables:} \quad Q_n, \; M_{nn}, \; M_{ns} \quad (6.2.11b)$$

where

$$Q_n \equiv Q_x n_x + Q_y n_y \quad (6.2.12a)$$
$$M_{nn} \equiv M_{xx} n_x^2 + 2 M_{xy} n_x n_y + M_{yy} n_y^2 \quad (6.2.12b)$$
$$M_{ns} \equiv (M_{yy} - M_{xx}) n_x n_y + M_{xy}(n_x^2 - n_y^2) \quad (6.2.12c)$$

We note that the equation of equilibrium (6.2.6), when expressed in terms of w_0, as will be shown shortly, has a total spatial differential order of four. This implies that there should be only four (two geometric and two force) boundary conditions at a boundary point. However, Eq. (6.2.11) shows three geometric and three force boundary conditions, giving a total of six boundary conditions. To eliminate this discrepancy, one may integrate the tangential derivative term in Eq. (6.2.10) by parts to obtain

$$-\oint_\Gamma M_{ns} \frac{\partial \delta w_0}{\partial s} ds = \oint_\Gamma \frac{\partial M_{ns}}{\partial s} \delta w_0 \, ds - [M_{ns} \delta w_0]_\Gamma \quad (6.2.13)$$

The term $[M_{ns}\delta w_0]_\Gamma$ is zero when the end points of a closed smooth curve coincide or when $M_{ns} = 0$. If $M_{ns} = 0$ is not specified at corners of the boundary Γ of a polygonal plate, concentrated forces of magnitude

$$F_c = -2M_{ns} \tag{6.2.14}$$

will be produced at the corners of rectangular plates. The factor of 2 appears because M_{ns} from two sides of the corner are added there.

The remaining boundary term in Eq. (6.2.10), being the coefficient of δw_0 on Γ, is added to the shear force Q_n, to obtain the effective shear force

$$V_n \equiv Q_n + \frac{\partial M_{ns}}{\partial s} \tag{6.2.15}$$

The specification of this effective shear force V_n is known as the *Kirchhoff free-edge condition*. Finally, the correct boundary conditions of the Kirchhoff plate theory involve specifying the following quantities:

$$\text{generalized displacements: } w_0, \frac{\partial w_0}{\partial n}$$
$$\text{generalized forces: } V_n, M_{nn} \tag{6.2.16}$$

Thus, at every boundary point one must know w_0 or V_n and $\partial w_0/\partial n$ or M_{nn}. On an edge parallel to the x-axis at $y = 0$ (i.e., $n = -y$ and $s = x$), for example, the above boundary conditions involve specifying one quantity in each pair

$$(w_0, V_y) \quad \text{and} \quad \left(\frac{\partial w_0}{\partial n} = -\frac{\partial w_0}{\partial y}, M_{nn} = M_{yy}\right) \tag{6.2.17}$$

Next we discuss some common types of boundary conditions for the bending of rectangular plates with edges parallel to the x and y coordinates. Here we use the edge at $y = 0$ ($n_x = 0$ and $n_y = -1$) to discuss the boundary conditions. It should be noted that only one element of each of the three pairs may (and should) be specified on an edge of a plate. The force boundary conditions may be expressed in terms of the generalized displacements using the plate constitutive equations discussed in the sequel (see Section 6.2.3).

Free edge, $y = 0$: A free edge is one which is geometrically not restrained in any way. Hence, we have

$$w_0 \neq 0, \quad \frac{\partial w_0}{\partial y} \neq 0 \tag{6.2.18a}$$

However, the edge may have applied forces and/or moments

$$V_y \equiv Q_y + \frac{\partial M_{xy}}{\partial x} = \hat{V}_y, \quad M_{yy} = \hat{M}_{yy} \qquad (6.2.18b)$$

where quantities with a hat are specified forces/moments. For free rectangular plates, $M_{xy} = 0$; hence no corner forces are developed.

Fixed (or clamped) edge, $y = 0$: A fixed edge is one that is geometrically fully restrained so that

$$w_0 = 0, \quad \frac{\partial w_0}{\partial y} = 0 \qquad (6.2.19)$$

Therefore, the forces and moments on a fixed edge are not known a priori (i.e., they are reactions to be determined as a part of the analysis). For clamped rectangular plates, $M_{xy} = 0$, hence no corner forces are developed.

Simply supported edge $y = 0$: Here we *define* simply supported boundary conditions as specifying

$$w_0 = 0, \quad M_{yy} = \hat{M}_{yy} \qquad (6.2.20)$$

where $\hat{M}_{yy}$ is the applied normal bending moment on the edge. For simply supported rectangular plates, a reacting force of $2M_{xy}$ is developed at each corner of the plate.

6.2.3 Governing Equations in Terms of the Deflection

Suppose that the material of the plate is isotropic and obeys Hooke's law. Then the stress-strain relations are given by

$$\sigma_{xx} = \frac{E}{1-\nu^2}(\varepsilon_{xx} + \nu\varepsilon_{yy}) \qquad (6.2.21a)$$

$$\sigma_{yy} = \frac{E}{1-\nu^2}(\varepsilon_{yy} + \nu\varepsilon_{xx}) \qquad (6.2.21b)$$

$$\sigma_{xy} = G\gamma_{xy} = \frac{E}{2(1+\nu)}\gamma_{xy} \qquad (6.2.21c)$$

where E denotes Young's modulus, G shear modulus, and ν Poisson's ratio. Using Eqs. (6.2.21a-c) in Eq. (6.2.3) and carrying out the indicated integration over the plate thickness, we arrive at

$$M_{xx} = \int_{-\frac{h}{2}}^{\frac{h}{2}} \sigma_{xx} z \, dz = \frac{E}{(1-\nu^2)} \int_{-\frac{h}{2}}^{\frac{h}{2}} (\varepsilon_{xx} + \nu \varepsilon_{yy}) z \, dz$$

$$= -D \left(\frac{\partial^2 w_0}{\partial x^2} + \nu \frac{\partial^2 w_0}{\partial y^2} \right) \quad (6.2.22a)$$

$$M_{yy} = \int_{-\frac{h}{2}}^{\frac{h}{2}} \sigma_{yy} z \, dz = \frac{E}{(1-\nu^2)} \int_{-\frac{h}{2}}^{\frac{h}{2}} (\varepsilon_{yy} + \nu \varepsilon_{xx}) z \, dz$$

$$= -D \left(\nu \frac{\partial^2 w_0}{\partial x^2} + \frac{\partial^2 w_0}{\partial y^2} \right) \quad (6.2.22b)$$

$$M_{xy} = \int_{-\frac{h}{2}}^{\frac{h}{2}} \sigma_{xy} z \, dz = G \int_{-\frac{h}{2}}^{\frac{h}{2}} \gamma_{xy} z \, dz$$

$$= -(1-\nu) D \frac{\partial^2 w_0}{\partial x \partial y} \quad (6.2.22c)$$

where D is the flexural rigidity

$$D = \frac{Eh^3}{12(1-\nu^2)} \quad (6.2.23)$$

Substituting the expressions for (M_{xx}, M_{yy}, M_{xy}) into Eq. (6.2.6), we obtain the biharmonic equation governing plate bending:

$$D \left(\frac{\partial^4 w_0}{\partial x^4} + 2 \frac{\partial^4 w_0}{\partial x^2 \partial y^2} + \frac{\partial^4 w_0}{\partial x^4} \right) = q \quad (6.2.24)$$

In terms of the Laplace operator ∇^2, we have

$$D \nabla^2 \nabla^2 w_0 = q \quad \text{or} \quad D \nabla^4 w_0 = q \quad (6.2.25)$$

The boundary conditions involve specifying

$$\left\{ \begin{array}{c} w_0 \\ \frac{\partial w_0}{\partial n} \end{array} \right\} \quad \text{or} \quad \left\{ \begin{array}{c} V_n \equiv Q_n + \frac{\partial M_{ns}}{\partial s} \\ M_{nn} \end{array} \right\} \quad (6.2.26)$$

This completes the development of the governing equations of the Kirchhoff plate theory. We shall make use of these equations in subsequent chapters.

6.3 First-Order Shear Deformation Plate Theory (FSDT)

6.3.1 Equations of Equilibrium

In view of the displacement field given in Eqs. (6.1.2a-c), the components of the linear strains are given by

$$\varepsilon_{xx} = z\frac{\partial \phi_x}{\partial x} \tag{6.3.1a}$$

$$\varepsilon_{yy} = z\frac{\partial \phi_y}{\partial y} \tag{6.3.1b}$$

$$\gamma_{xy} = z\left(\frac{\partial \phi_x}{\partial y} + \frac{\partial \phi_y}{\partial x}\right) \tag{6.3.1c}$$

$$\gamma_{xz} = \phi_x + \frac{\partial w_0}{\partial x} \tag{6.3.1d}$$

$$\gamma_{yz} = \phi_y + \frac{\partial w_0}{\partial y} \tag{6.3.1e}$$

Note that the strains $(\varepsilon_{xx}, \varepsilon_{yy}, \gamma_{xy})$ are linear through the plate thickness, while the transverse shear strains $(\gamma_{xz}, \gamma_{yz})$ are constant.

The equations of equilibrium of the first-order plate theory are derived, once again, using the principle of virtual displacements

$$\delta W = \delta U + \delta V = 0 \tag{6.3.2}$$

where the virtual strain energy δU, and virtual potential energy δV due to the transverse load $q(x,y)$ are given by

$$\delta U = \int_{\Omega_0}\left[\int_{-\frac{h}{2}}^{\frac{h}{2}} (\sigma_{xx}\delta\varepsilon_{xx} + \sigma_{yy}\delta\varepsilon_{yy} + \sigma_{xy}\delta\gamma_{xy}\right.$$
$$\left. + \sigma_{xz}\delta\gamma_{xz} + \sigma_{yz}\delta\gamma_{yz}\ ) dz\right] dxdy \tag{6.3.3a}$$

$$\delta V = -\int_{\Omega_0} q(x,y)\delta w_0\ dxdy \tag{6.3.3b}$$

Substituting for δU and δV from Eqs. (6.3.3a,b) into the virtual work statement in Eq. (6.3.2), expressing the virtual strains in terms of the virtual displacements $(\delta w_0, \delta\phi_x, \delta\phi_y)$ using Eqs. (6.3.1a-e), and integrating through the thickness of the plate, we obtain

$$0 = \int_{\Omega_0} \left[M_{xx}\frac{\partial\delta\phi_x}{\partial x} + M_{yy}\frac{\partial\delta\phi_y}{\partial y} + M_{xy}\left(\frac{\partial\delta\phi_x}{\partial y} + \frac{\partial\delta\phi_y}{\partial x}\right) - q\delta w_0 \right.$$
$$\left. + Q_x\left(\frac{\partial\delta w_0}{\partial x} + \delta\phi_x\right) + Q_y\left(\frac{\partial\delta w_0}{\partial y} + \delta\phi_y\right) \right] dxdy \quad (6.3.4)$$

where (M_{xx}, M_{yy}, M_{xy}) are the moments defined in Eq. (6.2.3), and the transverse shear forces per unit length (Q_x, Q_y) are defined by

$$\left\{ \begin{array}{c} Q_x \\ Q_y \end{array} \right\} = \int_{-\frac{h}{2}}^{\frac{h}{2}} \left\{ \begin{array}{c} \sigma_{xz} \\ \sigma_{yz} \end{array} \right\} dz \quad (6.3.5)$$

Since the transverse shear strains are represented as constant through the plate thickness, the transverse shear stresses will also be constant through the thickness. This contradicts the well known fact that the transverse shear stresses are parabolic (i.e. quadratic) through the plate thickness. While this discrepancy between the parabolic variation of transverse shear stresses and the constant state of shear stresses predicted by the first-order plate theory cannot be corrected within the limitations of the kinematics of FSDT, the shear forces (Q_x, Q_y) may be corrected by multiplying the integrals in Eq. (6.3.5) with a parameter K_s, called the shear correction factor:

$$\left\{ \begin{array}{c} Q_x \\ Q_y \end{array} \right\} = K_s \int_{-\frac{h}{2}}^{\frac{h}{2}} \left\{ \begin{array}{c} \sigma_{xz} \\ \sigma_{yz} \end{array} \right\} dz \quad (6.3.6)$$

This amounts to modifying the transverse shear stiffnesses of the plate. The factor K_s is computed such that the strain energy due to the transverse shear stresses of the FSDT equals the strain energy due to the transverse shear stresses predicted by the three-dimensional elasticity theory or its equivalent.

Returning to the virtual work statement in Eq. (6.3.4), we integrate by parts to relieve the virtual generalized displacements $(\delta w_0, \delta\phi_x, \delta\phi_y)$ in Ω_0 of any differentiation. We obtain, using the divergence theorem,

$$0 = \int_{\Omega_0} \left[-(M_{xx,x} + M_{xy,y} - Q_x)\delta\phi_x - (M_{xy,x} + M_{yy,y} - Q_y)\delta\phi_y \right.$$
$$\left. -(Q_{x,x} + Q_{y,y} + q)\delta w_0 \right] dxdy$$
$$+ \oint_\Gamma (Q_n \delta w_0 + M_{nn}\delta\phi_n + M_{ns}\delta\phi_s)\, ds \quad (6.3.7)$$

where the boundary expressions were arrived by expressing ϕ_x and ϕ_y in terms of the normal and tangential rotations, (ϕ_n, ϕ_s):

$$\phi_x = n_x\phi_n - n_y\phi_s , \quad \phi_y = n_y\phi_n + n_x\phi_s \tag{6.3.8}$$

The equations of equilibrium are

$$\delta w_0: \quad -\left(\frac{\partial Q_x}{\partial x} + \frac{\partial Q_y}{\partial y}\right) = q \tag{6.3.9a}$$

$$\delta\phi_x: \quad -\left(\frac{\partial M_{xx}}{\partial x} + \frac{\partial M_{xy}}{\partial y}\right) + Q_x = 0 \tag{6.3.9b}$$

$$\delta\phi_y: \quad -\left(\frac{\partial M_{xy}}{\partial x} + \frac{\partial M_{yy}}{\partial y}\right) + Q_y = 0 \tag{6.3.9c}$$

The primary and secondary variables of the theory are

$$\begin{aligned}\text{primary variables:} \quad & w_0,\ \phi_n,\ \phi_s \\ \text{secondary variables:} \quad & Q_n,\ M_{nn},\ M_{ns}\end{aligned} \tag{6.3.10}$$

where

$$Q_n \equiv Q_x n_x + Q_y n_y \tag{6.3.11}$$

The boundary conditions involve specifying one element of each of the following pairs:

$$(w_0, Q_n),\ (\phi_n, M_{nn}),\ (\phi_s, M_{ns})$$

6.3.2 Plate Constitutive Equations

Assuming the plate material is isotropic and obeys Hooke's law

$$\begin{Bmatrix}\sigma_{xx}\\ \sigma_{yy}\\ \sigma_{xy}\\ \sigma_{xz}\\ \sigma_{yz}\end{Bmatrix} = \begin{bmatrix}\frac{E}{1-\nu^2} & \frac{\nu E}{1-\nu^2} & 0 & 0 & 0\\ \frac{\nu E}{1-\nu^2} & \frac{E}{1-\nu^2} & 0 & 0 & 0\\ 0 & 0 & G & 0 & 0\\ 0 & 0 & 0 & G & 0\\ 0 & 0 & 0 & 0 & G\end{bmatrix}\begin{Bmatrix}\varepsilon_{xx}\\ \varepsilon_{yy}\\ \gamma_{xy}\\ \gamma_{xz}\\ \gamma_{yz}\end{Bmatrix} \tag{6.3.12}$$

where E is the Young's modulus and ν Poisson's ratio. The shear modulus G is related to E and ν by $G = E/2(1+\nu)$.

The plate constitutive equations are given by

$$M_{xx} = \int_{-\frac{h}{2}}^{\frac{h}{2}} \sigma_{xx} z \, dz = D \left(\frac{\partial \phi_x}{\partial x} + \nu \frac{\partial \phi_y}{\partial y} \right) \tag{6.3.13a}$$

$$M_{yy} = \int_{-\frac{h}{2}}^{\frac{h}{2}} \sigma_{yy} z \, dz = D \left(\nu \frac{\partial \phi_x}{\partial x} + \frac{\partial \phi_y}{\partial y} \right) \tag{6.3.13b}$$

$$M_{xy} = \int_{-\frac{h}{2}}^{\frac{h}{2}} \sigma_{xy} z \, dz = \frac{D(1-\nu)}{2} \left(\frac{\partial \phi_x}{\partial y} + \frac{\partial \phi_y}{\partial x} \right) \tag{6.3.13c}$$

$$Q_x = K_s \int_{-\frac{h}{2}}^{\frac{h}{2}} \sigma_{xz} \, dz = \frac{K_s E h}{2(1+\nu)} \left(\phi_x + \frac{\partial w_0}{\partial x} \right) \tag{6.3.13d}$$

$$Q_y = K_s \int_{-\frac{h}{2}}^{\frac{h}{2}} \sigma_{yz} \, dz = \frac{K_s E h}{2(1+\nu)} \left(\phi_y + \frac{\partial w_0}{\partial y} \right) \tag{6.3.13e}$$

6.3.3 Governing Equations in Terms of Displacements

The equations of equilibrium (6.3.9a-c) can be expressed in terms of displacements (w_0, ϕ_x, ϕ_y) by substituting for the force and moment resultants from Eqs. (6.3.13a-e). We have

$$-\frac{K_s E h}{2(1+\nu)} \left(\frac{\partial^2 w_0}{\partial x^2} + \frac{\partial^2 w_0}{\partial y^2} + \frac{\partial \phi_x}{\partial x} + \frac{\partial \phi_y}{\partial y} \right) = q(x,y) \tag{6.3.14}$$

$$-\frac{D(1-\nu)}{2} \left(\frac{\partial^2 \phi_x}{\partial x^2} + \frac{\partial^2 \phi_x}{\partial y^2} \right) - \frac{D(1+\nu)}{2} \frac{\partial}{\partial x} \left(\frac{\partial \phi_x}{\partial x} + \frac{\partial \phi_y}{\partial y} \right)$$
$$+ \frac{K_s E h}{2(1+\nu)} \left(\frac{\partial w_0}{\partial x} + \phi_x \right) = 0 \tag{6.3.15}$$

$$-\frac{D(1-\nu)}{2} \left(\frac{\partial^2 \phi_y}{\partial x^2} + \frac{\partial^2 \phi_y}{\partial y^2} \right) - \frac{D(1+\nu)}{2} \frac{\partial}{\partial y} \left(\frac{\partial \phi_x}{\partial x} + \frac{\partial \phi_y}{\partial y} \right)$$
$$+ \frac{K_s E h}{2(1+\nu)} \left(\frac{\partial w_0}{\partial y} + \phi_y \right) = 0 \tag{6.3.16}$$

Introducing the moment sum

$$\mathcal{M} \equiv \frac{M_{xx} + M_{yy}}{1+\nu} = D \left(\frac{\partial \phi_x}{\partial x} + \frac{\partial \phi_y}{\partial y} \right) \tag{6.3.17}$$

and using the Laplace operator, the equilibrium equations (6.3.14)–(6.3.16) can be expressed in the form

$$-\frac{K_s E h}{2(1+\nu)}\left(\nabla^2 w_0 + \frac{M}{D}\right) = q(x,y) \tag{6.3.18}$$

$$-D(1-\nu)\nabla^2 \phi_x - (1+\nu)\frac{\partial M}{\partial x} + \frac{K_s E h}{(1+\nu)}\left(\frac{\partial w_0}{\partial x} + \phi_x\right) = 0 \tag{6.3.19}$$

$$-D(1-\nu)\nabla^2 \phi_y - (1+\nu)\frac{\partial M}{\partial y} + \frac{K_s E h}{(1+\nu)}\left(\frac{\partial w_0}{\partial y} + \phi_y\right) = 0 \tag{6.3.20}$$

The common edge conditions for the Mindlin plate theory are given below.

Free edge (F): For this type of edge condition

$$Q_n = K_s G h \left(\phi_n + \frac{\partial w_0}{\partial n}\right) = 0 \tag{6.3.21a}$$

$$M_{nn} = D\left(\frac{\partial \phi_n}{\partial n} + \nu \frac{\partial \phi_s}{\partial s}\right) = 0 \tag{6.3.21b}$$

$$M_{ns} = \frac{D(1-\nu)}{2}\left(\frac{\partial \phi_n}{\partial s} + \frac{\partial \phi_s}{\partial n}\right) = 0 \tag{6.3.21c}$$

Simply supported edge (S and S*) There are two kinds of simply supported edges. The first kind (S), which is referred to as the hard type simple support, requires

$$w_0 = 0, \quad M_{nn} = 0, \quad \phi_s = 0 \tag{6.3.22}$$

The second kind (S*), commonly referred to as the soft type simple support, requires

$$w_0 = 0, \quad M_{nn} = 0, \quad M_{ns} = 0 \tag{6.3.23}$$

Clamped edge (C) This type of edge condition requires

$$w_0 = 0, \quad \phi_n = 0, \quad \phi_s = 0 \tag{6.3.24}$$

6.4 Third-Order Shear Deformation Plate Theory (TSDT)

6.4.1 Equations of Equilibrium

The displacement field in Eq. (6.1.4a-c) results in the following linear strains:

$$\varepsilon_{xx} = z\frac{\partial \phi_x}{\partial x} - \alpha z^3 \left(\frac{\partial \phi_x}{\partial x} + \frac{\partial^2 w_0}{\partial x^2}\right) \quad (6.4.1a)$$

$$\varepsilon_{yy} = z\frac{\partial \phi_y}{\partial y} - \alpha z^3 \left(\frac{\partial \phi_y}{\partial y} + \frac{\partial^2 w_0}{\partial y^2}\right) \quad (6.4.1b)$$

$$\varepsilon_{xy} = z\left(\frac{\partial \phi_x}{\partial y} + \frac{\partial \phi_y}{\partial x}\right) - \alpha z^3 \left(\frac{\partial \phi_x}{\partial y} + \frac{\partial \phi_y}{\partial x} + 2\frac{\partial^2 w_0}{\partial x \partial y}\right) \quad (6.4.1c)$$

$$\gamma_{xz} = \left(1 - \beta z^2\right)\left(\phi_x + \frac{\partial w_0}{\partial x}\right) \quad (6.4.1d)$$

$$\gamma_{yz} = \left(1 - \beta z^2\right)\left(\phi_y + \frac{\partial w_0}{\partial y}\right) \quad (6.4.1e)$$

where $\alpha = 4/(3h^2)$ and $\beta = 4/h^2$.

The substitution of the virtual strains associated with the strains in Eqs. (6.4.1a-e) into Eq. (6.3.3a) and the result, along with δV from Eq. (6.3.3b), into the statement of the principle of virtual displacements, Eq. (6.3.2) gives

$$\int_{\Omega_0} \left[M_{xx}\frac{\partial \delta \phi_x}{\delta x} + M_{yy}\frac{\partial \delta \phi_y}{\partial y} + M_{xy}\left(\frac{\partial \delta \phi_x}{\partial y} + \frac{\partial \delta \phi_y}{\partial x}\right) \right.$$

$$- \alpha P_{xx}\left(\frac{\partial \delta \phi_x}{\partial x} + \frac{\partial^2 \delta w_0}{\partial x^2}\right) - \alpha P_{yy}\left(\frac{\partial \delta \phi_y}{\partial y} + \frac{\partial^2 \delta w_0}{\partial y^2}\right)$$

$$- \alpha P_{xy}\left(\frac{\partial \delta \phi_x}{\partial y} + \frac{\partial \delta \phi_y}{\partial x} + 2\frac{\partial^2 \delta w_0}{\partial x \partial y}\right)$$

$$+ (Q_x - \beta R_x)\left(\delta \phi_x + \frac{\partial \delta w_0}{\partial x}\right)$$

$$\left. + (Q_x - \beta R_x)\left(\delta \phi_y + \frac{\partial \delta w_0}{\partial y}\right) - q\delta w_0 \right] dA = 0 \quad (6.4.2)$$

where the moments (M_{xx}, M_{yy}, M_{xy}) and transverse shear forces (Q_x, Q_y) are the same as defined in Eqs. (6.2.3) and (6.3.5), respectively, and the higher-order stress resultants (P_{xx}, P_{yy}, P_{xy}) and (R_x, R_y) are defined by

$$\begin{Bmatrix} P_{xx} \\ P_{yy} \\ P_{xy} \end{Bmatrix} = \int_{-\frac{h}{2}}^{\frac{h}{2}} \begin{Bmatrix} \sigma_{xx} \\ \sigma_{yy} \\ \sigma_{xy} \end{Bmatrix} z^3 \, dz \qquad (6.4.3a)$$

$$\begin{Bmatrix} R_x \\ R_y \end{Bmatrix} = \int_{-\frac{h}{2}}^{\frac{h}{2}} \begin{Bmatrix} \sigma_{xz} \\ \sigma_{yz} \end{Bmatrix} z^2 \, dz \qquad (6.4.3b)$$

Integrating the expressions in Eq. (6.4.2) by parts, and collecting the coefficients of $\delta\phi_x$, $\delta\phi_y$, δw_0, one obtains the following equilibrium equations of the third-order plate theory:

$$-\left(\frac{\partial \hat{M}_{xx}}{\partial x} + \frac{\partial \hat{M}_{xy}}{\partial y} \right) + \hat{Q}_x = 0 \qquad (6.4.4)$$

$$-\left(\frac{\partial \hat{M}_{xy}}{\partial x} + \frac{\partial \hat{M}_{yy}}{\partial y} \right) + \hat{Q}_y = 0 \qquad (6.4.5)$$

$$-\left(\frac{\partial \hat{Q}_x}{\partial x} + \frac{\partial \hat{Q}_y}{\partial y} \right) - \alpha \left(\frac{\partial^2 P_{xx}}{\partial x^2} + 2\frac{\partial^2 P_{xy}}{\partial x \partial y} + \frac{\partial^2 P_{yy}}{\partial y^2} \right) = q \qquad (6.4.6)$$

where

$$\hat{M}_{\xi\eta} = M_{\xi\eta} - \alpha P_{\xi\eta} \qquad (6.4.7a)$$
$$\hat{Q}_\xi = Q_\xi - \beta R_\xi \qquad (6.4.7b)$$

and $\xi, \eta = x, y$. The boundary conditions involve specifying

$$w_0 \quad \text{or} \quad Q_n \qquad (6.4.8a)$$
$$\frac{\partial w_0}{\partial n} \quad \text{or} \quad P_{nn} \qquad (6.4.8b)$$
$$\phi_n \quad \text{or} \quad M_{nn} \qquad (6.4.9a)$$
$$\phi_{ns} \quad \text{or} \quad M_{ns} \qquad (6.4.9b)$$

where

$$M_{nn} = \hat{M}_{xx} \cos^2\theta - 2\hat{M}_{xy} \sin\theta \cos\theta + \hat{M}_{yy} \sin^2\theta \qquad (6.4.10a)$$
$$M_{ns} = -\left(\hat{M}_{xx} - \hat{M}_{yy} \right) \sin\theta \cos\theta + \hat{M}_{xy} \left(\cos^2\theta - \sin^2\theta \right) \qquad (6.4.10b)$$
$$Q_n = \hat{Q}_x \sin\theta - \hat{Q}_y \cos\theta - \frac{4}{3h^2} \frac{\partial P_{ns}}{\partial s} \qquad (6.4.10c)$$

Like in the CPT, corner forces exist in this theory as well.

6.4.2 Plate Constitutive Equations

Using the stress-strain relations (6.3.12) in Eqs. (6.2.3), (6.3.5), and (6.4.3a,b), and using the strain-displacement relations (6.4.1a-e), we obtain

$$M_{xx} = \frac{4D}{5}\left(\frac{\partial \phi_x}{\partial x} + \nu \frac{\partial \phi_y}{\partial y}\right) - \frac{D}{5}\left(\frac{\partial^2 w_0}{\partial x^2} + \nu \frac{\partial^2 w_0}{\partial y^2}\right) \quad (6.4.11a)$$

$$P_{xx} = \frac{4h^2 D}{35}\left(\frac{\partial \phi_x}{\partial x} + \nu \frac{\partial \phi_y}{\partial y}\right) - \frac{h^2 D}{28}\left(\frac{\partial^2 w_0}{\partial x^2} + \nu \frac{\partial^2 w_0}{\partial y^2}\right) \quad (6.4.11b)$$

$$M_{yy} = \frac{4D}{5}\left(\nu \frac{\partial \phi_x}{\partial x} + \frac{\partial \phi_y}{\partial y}\right) - \frac{D}{5}\left(\nu \frac{\partial^2 w_0}{\partial x^2} + \frac{\partial^2 w_0}{\partial y^2}\right) \quad (6.4.11c)$$

$$P_{yy} = \frac{4h^2 D}{35}\left(\nu \frac{\partial \phi_x}{\partial x} + \frac{\partial \phi_y}{\partial y}\right) - \frac{h^2 D}{28}\left(\nu \frac{\partial^2 w_0}{\partial y^2} + \frac{\partial^2 w_0}{\partial y^2}\right) \quad (6.4.11d)$$

$$M_{xy} = \left(\frac{1-\nu}{2}\right)\left[\frac{4D}{5}\left(\frac{\partial \phi_x}{\partial y} + \frac{\partial \phi_y}{\partial x}\right) - \frac{D}{5}\left(2\frac{\partial^2 w_0}{\partial x \partial y}\right)\right] \quad (6.4.11e)$$

$$P_{xy} = \left(\frac{1-\nu}{2}\right)\left[\frac{4h^2 D}{35}\left(\frac{\partial \phi_x}{\partial y} + \frac{\partial \phi_y}{\partial x}\right) - \frac{h^2 D}{28}\left(2\frac{\partial^2 w_0}{\partial x \partial y}\right)\right] \quad (6.4.11f)$$

$$Q_x = \frac{2hG}{3}\left(\phi_x + \frac{\partial w_0}{\partial x}\right) \quad (6.4.11g)$$

$$R_x = \frac{h^3 G}{30}\left(\phi_x + \frac{\partial w_0}{\partial x}\right) \quad (6.4.11h)$$

$$Q_y = \frac{2hG}{3}\left(\phi_y + \frac{\partial w_0}{\partial y}\right) \quad (6.4.11i)$$

$$R_y = \frac{h^3 G}{30}\left(\phi_y + \frac{\partial w_0}{\partial y}\right) \quad (6.4.11j)$$

These moment/force-deflection relationships can be substituted into Eqs. (6.4.4)–(6.4.6) to express the equilibrium equations in terms of the generalized displacements (w_0, ϕ_x, ϕ_y). We will return to these equations in the following chapters.

Problems

6.1 Starting with a linear distribution of the displacements through the plate thickness in terms of unknown functions (f_1, f_2, f_3)

$$u(x,y,z) = zf_1(x,y), \quad v(x,y,z) = zf_2(x,y)$$
$$w(x,y,z) = w_0(x,y) + zf_3(x,y) \tag{i}$$

determine the functions (f_1, f_2, f_3) such that the Kirchhoff hypothesis holds:

$$\frac{\partial w}{\partial z} = 0, \quad \frac{\partial u}{\partial z} = -\frac{\partial w}{\partial x}, \quad \frac{\partial v}{\partial z} = -\frac{\partial w}{\partial y} \tag{ii}$$

6.2 Repeat Problem 6.1 for the Mindlin plate theory and determine the functions (f_1, f_2, f_3) such that the following conditions hold:

$$\frac{\partial w}{\partial z} = 0, \quad \frac{\partial u}{\partial z} = \phi_x, \quad \frac{\partial v}{\partial z} = \phi_y \tag{i}$$

6.3 Starting with a cubic distribution of the displacements through the plate thickness in terms of unknown functions $(f_1, f_2, g_1, g_2, h_1, h_2)$

$$u(x,y,z) = zf_1(x,y) + z^2 g_1(x,y) + z^3 h_1(x,y)$$
$$v(x,y,z) = zf_2(x,y) + z^2 g_2(x,y) + z^3 h_2(x,y)$$
$$w(x,y,z) = w_0(x,y) \tag{i}$$

determine the functions (f_i, g_i, h_i) in terms of (w_0, ϕ_x, ϕ_y) such that the following conditions are satisfied:

$$\left(\frac{\partial u}{\partial z}\right)_{z=0} = \phi_x, \quad \left(\frac{\partial v}{\partial z}\right)_{z=0} = \phi_y \tag{ii}$$

$$\sigma_{xz}(x,y,-\frac{h}{2}) = 0, \quad \sigma_{xz}(x,y,\frac{h}{2}) = 0 \tag{iii}$$

$$\sigma_{yz}(x,y,-\frac{h}{2}) = 0, \quad \sigma_{yz}(x,y,\frac{h}{2}) = 0 \tag{iv}$$

6.4 Consider the following equations of equilibrium of 3-D elasticity in the absence of body forces:

$$\frac{\partial \sigma_{xx}}{\partial x} + \frac{\partial \sigma_{xy}}{\partial y} + \frac{\partial \sigma_{xz}}{\partial z} = 0 \tag{i}$$

$$\frac{\partial \sigma_{xy}}{\partial x} + \frac{\partial \sigma_{yy}}{\partial y} + \frac{\partial \sigma_{yz}}{\partial z} = 0 \tag{ii}$$

$$\frac{\partial \sigma_{xz}}{\partial x} + \frac{\partial \sigma_{yz}}{\partial y} + \frac{\partial \sigma_{zz}}{\partial z} = 0 \tag{iii}$$

subject to the following boundary conditions:

$$\sigma_{xz}(x,y,-\tfrac{h}{2}) = 0, \quad \sigma_{xz}(x,y,\tfrac{h}{2}) = 0, \quad \sigma_{yz}(x,y,-\tfrac{h}{2}) = 0 \quad (iv)$$

$$\sigma_{yz}(x,y,\tfrac{h}{2}) = 0, \quad \sigma_{zz}(x,y,-\tfrac{h}{2}) = -q_b, \quad \sigma_{zz}(x,y,\tfrac{h}{2}) = q_t \quad (v)$$

Integrate equations (i)–(iii) with respect to z over the interval $(-h/2, h/2)$ and express the results in terms of the stress resultants (N_{xx}, N_{yy}, N_{xy}). Next, multiply the equations of motion with z, integrate with respect to z over the interval $(-h/2, h/2)$, and express the results in terms of the moments (M_{xx}, M_{yy}, M_{xy}). Be sure that the boundary conditions (iv) and (v) on the stresses are satisfied.

6.5 Use Eqs. (6.3.13a-c) in Eqs. (6.3.9b,c) to establish the shear force-rotation relations

$$Q_x = D\left[\frac{\partial}{\partial x}\left(\frac{\partial \phi_x}{\partial x} + \frac{\partial \phi_y}{\partial y}\right) + \frac{1-\nu}{2}\frac{\partial}{\partial y}\left(\frac{\partial \phi_x}{\partial y} - \frac{\partial \phi_y}{\partial x}\right)\right] \quad (i)$$

$$Q_y = D\left[\frac{\partial}{\partial y}\left(\frac{\partial \phi_x}{\partial x} + \frac{\partial \phi_y}{\partial y}\right) - \frac{1-\nu}{2}\frac{\partial}{\partial x}\left(\frac{\partial \phi_x}{\partial y} - \frac{\partial \phi_y}{\partial x}\right)\right] \quad (ii)$$

and in terms of the Marcus moment defined in Eq. (6.3.17)

$$Q_x = \frac{\partial M}{\partial x} + \frac{(1-\nu)D}{2}\frac{\partial}{\partial y}\left(\frac{\partial \phi_x}{\partial y} - \frac{\partial \phi_y}{\partial x}\right) \quad (iii)$$

$$Q_y = \frac{\partial M}{\partial y} - \frac{(1-\nu)D}{2}\frac{\partial}{\partial x}\left(\frac{\partial \phi_x}{\partial y} - \frac{\partial \phi_y}{\partial x}\right) \quad (iv)$$

6.6 Use Eqs. (6.3.13d,e) and Eqs. (iii) and (iv) of Problem 6.5 to establish the following relationship:

$$\nabla^2 \Omega = c^2 \Omega, \quad c^2 = \frac{2K_s G h}{(1-\nu)D} = \frac{12 K_s}{h^2} \quad (i)$$

where

$$\Omega = \left(\frac{\partial \phi_x}{\partial y} - \frac{\partial \phi_y}{\partial x}\right) \quad (ii)$$

Chapter 7

Bending Relationships for Simply Supported Plates

In this chapter the differential relationships between the deflections of the classical Kirchhoff plate theory and first- and third-order plate theories are developed for simply supported, polygonal plates. As examples, the deflections of simply supported triangular and rectangular plates are obtained using these relationships.

7.1 Introduction

The subject of plate bending based on the Kirchhoff and Mindlin plate theories for a variety of transverse loading and boundary conditions has been studied by numerous investigators. The works have been compiled in standard texts on plates such as the ones by Timoshenko and Woinowsky-Krieger (1970), Szilard (1974), Roark and Young (1975), Reismann (1988), Huang (1988), and Reddy (1999a). Closed-form solutions for the stress resultants and deflections have been derived for some plate cases. Where these exact solutions cannot be obtained, the analysts can draw on very general finite element software, such as ABAQUS and COSMOS, to solve their plate bending problems. These software packages for plate analysis usually provide classical (or Kirchhoff) plate theory (CPT) elements and first-order shear deformation (or Mindlin) plate theory (FSDT) elements. The latter type of elements allows for the effect of transverse shear deformation. In this chapter, we present exact relationships linking the stress-resultants and deflections of the first-order shear deformation theory to those of the classical plate theory for simply supported polygonal plates.

7.2 Relationships Between CPT and FSDT

The governing equations of static equilibrium of plates according to the Kirchhoff [Eq. (6.2.25)] and Mindlin [Eqs. (6.3.18)–(6.3.20)] plate theories can be expressed in terms of the deflection w_0 and the moment sum (or Marcus moment) $\mathcal{M}$ as (see Problem 7.9)

$$\nabla^2 \mathcal{M}^K = -q, \qquad \nabla^2 w_0^K = -\frac{\mathcal{M}^K}{D} \qquad (7.2.1\text{a, b})$$

$$\nabla^2 \mathcal{M}^M = -q, \quad \nabla^2 \left(w_0^M - \frac{\mathcal{M}^M}{K_s G h} \right) = -\frac{\mathcal{M}^M}{D} \qquad (7.2.2\text{a, b})$$

where the superscripts K and M refer to quantities of the Kirchhoff and Mindlin plate theories, respectively, D is the flexural rigidity, and the moment sum is related to the generalized displacements by the relations

$$\mathcal{M}^K = \frac{M_{xx}^K + M_{yy}^K}{1+\nu} = -D \left(\frac{\partial^2 w_0^K}{\partial x^2} + \frac{\partial^2 w_0^K}{\partial y^2} \right) = -D\nabla^2 w_0^K \quad (7.2.3\text{a})$$

$$\mathcal{M}^M = \frac{M_{xx}^M + M_{yy}^M}{1+\nu} = D \left(\frac{\partial \phi_x}{\partial x} + \frac{\partial \phi_y}{\partial y} \right) \qquad (7.2.3\text{b})$$

From Eqs. (7.2.1a) and (7.2.2a), in view of the load equivalence, it follows that

$$\mathcal{M}^M = \mathcal{M}^K + D\nabla^2 \Phi \qquad (7.2.4)$$

where Φ is a function such that it satisfies the biharmonic equation

$$\nabla^4 \Phi = 0 \qquad (7.2.5)$$

Using this result in Eqs. (7.2.1b) and (7.2.2b), one may arrive at the relationship

$$w_0^M = w_0^K + \frac{\mathcal{M}^K}{K_s G h} + \Psi - \Phi \qquad (7.2.6\text{a})$$

$$= w_0^K + \frac{h^2}{6K_s(1-\nu)} \nabla^2 w_0^K + \Psi - \Phi \qquad (7.2.6\text{b})$$

where Ψ is a harmonic function that satisfies the Laplace equation

$$\nabla^2 \Psi = 0 \qquad (7.2.7)$$

Note that the relationship (7.2.6b) is valid for all plates with arbitrary boundary conditions and transverse load. One must determine Φ and Ψ from Eqs. (7.2.5) and (7.2.7), respectively, subject to the boundary conditions of the plate. It is worth noting that Barrett and Ellis (1988) also obtained a form similar to Eq. (7.2.6a,b) but they have assumed Φ to be a constant.

In cases where $w_0^M = w_0^K$ on the boundaries and $\mathcal{M}^K$ is either zero or equal to a constant $\mathcal{M}^{*K}$ (which can be zero) over the boundaries, $\Psi - \Phi$ simply takes on the value of $-\mathcal{M}^{*K}/(K_s G h)$. However, if $\mathcal{M}^K$ varies over the boundaries, the functions Ψ and Φ must be determined separately. Restricting the analysis to the former case allows Eq. (7.2.6a) to be written as

$$w_0^M = w_0^K + \frac{\mathcal{M}^K - \mathcal{M}^{*K}}{K_s G h} \qquad (7.2.8)$$

Using Eq. (7.2.8), it can be readily shown that the relationships between deflection gradients, bending moments, twisting moment and shear forces of the Kirchhoff and Mindlin plate theories are given by

$$\frac{\partial w_0^M}{\partial x} = \frac{\partial w_0^K}{\partial x} + \frac{Q_x^K}{K_s G h} \qquad (7.2.9a)$$

$$\frac{\partial w_0^M}{\partial y} = \frac{\partial w_0^K}{\partial y} + \frac{Q_y^K}{K_s G h} \qquad (7.2.9b)$$

$$M_{xx}^M = M_{xx}^K + \frac{D(1-\nu)}{2K_s G h} \left[\frac{\partial}{\partial x} \left(Q_x^M - Q_x^K \right) - \frac{\partial}{\partial y} \left(Q_y^M - Q_y^K \right) \right]$$

$$= M_{xx}^K + \frac{D}{K_s G h} \left[\frac{\partial}{\partial x} \left(Q_x^M - Q_x^K \right) + \nu \frac{\partial}{\partial y} \left(Q_y^M - Q_y^K \right) \right]$$

$$= M_{xx}^K + \frac{D(1-\nu)}{K_s G h} \frac{\partial}{\partial x} \left(Q_x^M - Q_x^K \right) \qquad (7.2.10a)$$

$$M_{yy}^M = M_{yy}^K + \frac{D(1-\nu)}{2K_s G h} \left[\frac{\partial}{\partial y} \left(Q_y^M - Q_y^K \right) - \frac{\partial}{\partial y} \left(Q_x^M - Q_x^K \right) \right]$$

$$= M_{yy}^K + \frac{D}{K_s G h} \left[\frac{\partial}{\partial y} \left(Q_y^M - Q_y^K \right) + \nu \frac{\partial}{\partial x} \left(Q_x^M - Q_x^K \right) \right]$$

$$= M_{yy}^K + \frac{D(1-\nu)}{K_s G h} \frac{\partial}{\partial y} \left(Q_y^M - Q_y^K \right) \qquad (7.2.10b)$$

$$M_{xy}^M = M_{xy}^K + \frac{D(1-\nu)}{2K_s G h} \left[\frac{\partial}{\partial y} \left(Q_x^M - Q_x^K \right) - \frac{\partial}{\partial x} \left(Q_y^M - Q_y^K \right) \right]$$

$$(7.2.10c)$$

Note that there are a few variations in presenting the above relationships of the bending moments because of the shear force relationship given by

$$\frac{\partial Q_x^M}{\partial x} + \frac{\partial Q_y^M}{\partial y} = \frac{\partial Q_x^K}{\partial x} + \frac{\partial Q_y^K}{\partial y} = -q \qquad (7.2.11)$$

Also, it is interesting to note that

$$Q_x^M = Q_x^K + \frac{D(1-\nu)}{2K_sGh}\nabla^2\left(Q_x^M - Q_x^K\right) \qquad (7.2.12a)$$

$$Q_y^M = Q_y^K + \frac{D(1-\nu)}{2K_sGh}\nabla^2\left(Q_y^M - Q_y^K\right) \qquad (7.2.12b)$$

The foregoing relationships are exact if $w_0^M = w_0^K$ at the boundaries and the Marcus moments at the boundaries are equal to the same constant in the Kirchhoff and Mindlin plate theories.

Consider the case of simply supported, polygonal plates with straight edges. In the Kirchhoff plate theory, it is well known that in addition to the deflection being zero along the simply supported edges, the Marcus moment is also zero (see Timoshenko and Woinowsky-Krieger 1970). That is,

$$w_0^K = \mathcal{M}^K = 0 \text{ along the straight simply supported edges} \qquad (7.2.13)$$

In the Mindlin plate theory, the simply supported boundary condition considered is of the "hard" type such that $w_0^M = 0$, $M_{nn}^M = 0$ and $\phi_s = 0$ where n is the direction normal to the simply supported edge and s the direction tangential to the edge. Owing to the latter two conditions, $\partial \phi_s/\partial s = 0$ and the Marcus moment is thus equal to zero. The boundary conditions of the FSDT for the simply supported plate are therefore

$$w_0^M = \mathcal{M}^M = 0 \text{ along the straight simply supported edges} \qquad (7.2.14)$$

Since the Marcus moments at the boundaries of plates with any polygonal shape and simply supported edges are equal to zero, Eqs. (7.2.8) to (7.2.12) apply to such plates. As the Marcus moments are zero at the boundaries, $\mathcal{M}^{*K} = 0$ and the deflection relationship is simply given by

$$w_0^M = w_0^K + \frac{\mathcal{M}^K}{K_sGh} \qquad (7.2.15)$$

Equation (7.2.15) furnishes an important relationship between the deflections of a simply supported polygonal Mindlin plate and the corresponding simply supported polygonal Kirchhoff plate. This means that the deflection of simply supported Mindlin plate can be readily calculated from Eq. (7.2.15) upon supplying the deflection of the Kirchhoff plate and the Marcus moment, thus bypassing the necessity for a shear-deformable plate bending analysis.

Using the same reasoning, one may readily deduce that Eq. (7.2.) holds for simply supported plates under a constant distributed moment $\mathcal{M}^{*K}$ along their edges.

Remarks:

- For "soft" simply supported condition, the Marcus moment is nonzero and thus the foregoing derived relationship does not apply. There are important differences between "soft" and "hard" simply supported conditions which are discussed in Arnold and Falk (1990) and Haggblad and Bathe (1990).

- Donnell (1976) has derived an apparently similar result to that of Eq. (7.2.15) [see Eq. (5.84e) in Donnell (1976)]. However, Donnell's result is misleading. A reader would be led to believe that Donnell's result is independent of boundary conditions and plate shape. It can be readily shown that this is not the case. Donnell's deflection component w_0^s due to the transverse shear deformation is given by

$$w_0^s = -\frac{\alpha h^2}{6(1-\nu)}\nabla^2 w_0^K \qquad (7.2.16)$$

  where α is a shear correction factor. On a simply supported boundary, $\nabla^2 w_0^K$ would not be equal to zero in general, thus violating the null displacement requirement for $w_0^s + w_0^K$ on the boundary. As mentioned above, $\nabla^2 w_0^K$ is zero for the special case of simply supported polygonal-shaped plates. The reason for the incompleteness of Donnell's derivation is that he has developed the relevant relations on an augmentation basis without a comprehensive treatment of the governing conditions. Consequently, a harmonic function has been left out from the right hand side of Eq. (7.2.16).

- Equation (7.2.15) has been derived by disregarding the stress singularity at the corners of simply supported edges. The stress singularity is significant in the case of obtuse corners and hence the relationship will not yield good results for simply supported plates with large obtuse corners.

- Interestingly, a conjugate plate analogy can be observed from Eqs. (7.2.1a), (7.2.1b) and (7.2.13). The equilibrium problem expressed by these equations is analogous to the compatibility problem expressed by Eqs. (7.2.1b) and (7.2.13). The kinematics of the plate can be determined by solving the equilibrium equation of a conjugate plate loaded by the Marcus curvature $\mathcal{M}^K/D$ and interpreting the conjugate Marcus moment as the actual deflection w_0^K. In the case of the Mindlin plate theory, the additional shear-deflection component can also be determined as in the case of the conjugate beam/frame analogy. As with a simply supported beam, the conjugate plate for shear deflection is the original plate itself with the original load/reactions factored by $1/(K_s Gh)$. The conjugate Marcus moment would then become the shear-deflection component. Note that because the load/reactions are factored, the shear deflection component is given by $w_0^s = \mathcal{M}/(K_s Gh)$.

7.3 Examples

7.3.1 Simply Supported, Uniformly Loaded, Equilateral Triangular Plate

Consider a simply supported, equilateral triangular plate of side length $2L/\sqrt{3}$ as shown in Figure 7.3.1. The plate is subjected to a uniformly distributed load q_0. The deflection of this Kirchhoff plate is given by [see Woinowsky-Krieger (1933) and Reddy (1999a)]

$$w_0^K = \frac{q_0 L^4}{64 D} \left[\bar{x}^3 - 3\bar{y}^2 \bar{x} - \left(\bar{x}^2 + \bar{y}^2 \right) + \frac{4}{27} \right] \left(\frac{4}{9} - \bar{x}^2 - \bar{y}^2 \right) \quad (7.3.1)$$

where $\bar{x} = x/L$ and $\bar{y} = y/L$. In view of Eq. (7.2.3a), the Marcus moment is given by

$$\mathcal{M}^K = -D\nabla^2 w_0^K = \frac{q_0 L^2}{4} \left[\bar{x}^3 - 3\bar{x}\bar{y}^2 - \left(\bar{x}^2 - \bar{y}^2 \right) + \frac{4}{27} \right] \quad (7.3.2)$$

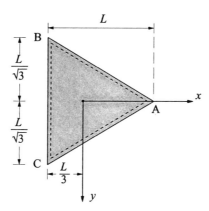

Figure 7.3.1. Simply supported equilateral triangular plate.

Using Eq. (7.2.15), the deflection w_0^M of the uniformly distributed loaded equilateral triangular plate according to the FSDT is given by

$$w_0^M = \frac{q_0 L^4}{4D}\left[\bar{x}^3 - 3\bar{y}^2 - \left(\bar{x}^2 + \bar{y}^2\right) + \frac{4}{27}\right]\left[\frac{\frac{4}{9} - \bar{x}^2 - \bar{y}^2}{16} + \frac{D}{K_s G L^2}\right] \quad (7.3.3)$$

7.3.2 Simply Supported, Uniformly Loaded, Rectangular Plate

Consider a simply supported, rectangular plate of side lengths $a \times b$ as shown in Figure 7.3.2. The plate is subjected to a distributed load $q(x,y)$. The deflection of this Kirchhoff plate is given by (see Timoshenko and Woinowsky-Krieger 1970 and Reddy 1999a)

$$w_0^K = \sum_{n=1}^{\infty}\sum_{m=1}^{\infty}\left[\frac{q_{mn}}{\pi^4 D\left(\frac{m^2}{a^2} + \frac{n^2}{b^2}\right)^2}\right]\sin\frac{m\pi x}{a}\sin\frac{n\pi y}{b} \quad (7.3.4)$$

where

$$q_{mn} = \frac{4}{ab}\int_0^b\int_0^a q(x,y)\sin\frac{m\pi x}{a}\sin\frac{n\pi y}{b}\,dxdy \quad (7.3.5)$$

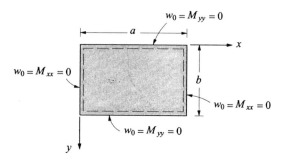

Figure 7.3.2. Simply supported rectangular plate.

In view of Eq. (7.2.3a), the Marcus moment is given by

$$\mathcal{M}^K = -D\nabla^2 w_0^K = -\sum_{n=1}^{\infty}\sum_{m=1}^{\infty}\left[\frac{q_{mn}}{\pi^2\left(\frac{m^2}{a^2}+\frac{n^2}{b^2}\right)}\right]\sin\frac{m\pi x}{a}\sin\frac{n\pi y}{b} \quad (7.3.6)$$

Using Eq. (7.2.15), the deflection w_0^M of the simply supported rectangular plate under distributed load is given by the Navier solution (see Reddy 1999a)

$$w_0^M = \sum_{n=1}^{\infty}\sum_{m=1}^{\infty}\left[\frac{Q_{mn}}{\pi^4\left(\frac{m^2}{a^2}+\frac{n^2}{b^2}\right)^2}\right]\sin\frac{m\pi x}{a}\sin\frac{n\pi y}{b} \quad (7.3.7)$$

where

$$Q_{mn} = q_{mn}\left[1+\frac{\pi^2 h\left(\frac{m^2}{a^2}+\frac{n^2}{b^2}\right)}{6K_s Gh}\right] \quad (7.3.8)$$

7.4 Relationships Between CPT and TSDT

7.4.1 Introduction

In this section, we develop the differential deflection relationship between the CPT and third-order shear deformation theory (TSDT) of

Reddy (1984a) for polygonal plates. Based on this general differential relationship, the relation between the deflections of simply supported polygonal plates is derived. Since the FSDT can be deduced as a special case of the TSDT, the relationship derived herein includes that of Section 7.2. As the TSDT is algebraically complicated, analytical solutions are more difficult to obtain than for the CPT. The relationship between the deflections of the CPT and the TSDT for simply supported polygonal plates enables one to obtain the deflection, and hence the moments and stresses in the TSDT in terms of the deflection of the CPT.

7.4.2 Governing Equations

The equations governing the static bending of an isotropic elastic plate under a transverse load $q(x,y)$ are given by

$$\frac{\partial \bar{M}_{xx}}{\partial x} + \frac{\partial \bar{M}_{xy}}{\partial y} - \bar{Q}_x = 0 \quad (7.4.1)$$

$$\frac{\partial \bar{M}_{xy}}{\partial x} + \frac{\partial \bar{M}_{yy}}{\partial y} - \bar{Q}_y = 0 \quad (7.4.2)$$

$$\frac{\partial \bar{Q}_x}{\partial x} + \frac{\partial \bar{Q}_y}{\partial y} + \alpha \left(\frac{\partial^2 P_{xx}}{\partial x^2} + 2 \frac{\partial^2 P_{xy}}{\partial x \partial y} + \frac{\partial^2 P_{yy}}{\partial y^2} \right) + q = 0 \quad (7.4.3)$$

where

$$\bar{M}_{\xi\eta} = M_{\xi\eta} - \alpha P_{\xi\eta}, \quad \bar{Q}_\xi = Q_\xi - \beta R_\xi \quad (7.4.4)$$

$$\xi, \eta = x, y, \quad \alpha = \frac{4}{3h^2}, \quad \beta = \frac{4}{h^2} \quad (7.4.5)$$

(M_{xx}, M_{yy}, M_{xy}) are the moments, (Q_x, Q_y) the transverse shear forces, and (P_{xx}, P_{yy}, P_{xy}) and (R_x, R_y) denote the higher-order stress resultants as defined below:

$$\begin{Bmatrix} M_{xx} \\ M_{yy} \\ M_{xy} \end{Bmatrix} = \int_{-h/2}^{h/2} \begin{Bmatrix} \sigma_{xx} \\ \sigma_{yy} \\ \sigma_{xy} \end{Bmatrix} z \, dz, \quad \begin{Bmatrix} Q_x \\ Q_y \end{Bmatrix} = \int_{-h/2}^{h/2} \begin{Bmatrix} \sigma_{yz} \\ \sigma_{xz} \end{Bmatrix} dz \quad (7.4.6)$$

$$\begin{Bmatrix} P_{xx} \\ P_{yy} \\ P_{xy} \end{Bmatrix} = \int_{-h/2}^{h/2} \begin{Bmatrix} \sigma_{xx} \\ \sigma_{yy} \\ \sigma_{xy} \end{Bmatrix} z^3 \, dz, \quad \begin{Bmatrix} R_x \\ R_y \end{Bmatrix} = \int_{-h/2}^{h/2} \begin{Bmatrix} \sigma_{yz} \\ \sigma_{xz} \end{Bmatrix} z^2 \, dz \quad (7.4.7)$$

and h is the thickness of the plate. We note again that we recover the governing equations of the first-order shear deformation theory (FSDT) from equations (7.4.1)–(7.4.3) by setting $\alpha = 0$.

The primary (i.e., geometric) and secondary (i.e., force) variables of the theory are

$$\text{Primary Variables}: \quad w_0, \frac{\partial w_0}{\partial n}, \phi_n, \phi_s \qquad (7.4.8)$$
$$\text{Secondary Variables}: \quad \bar{V}_n, \bar{P}_{nn}, \bar{M}_{nn}, \bar{M}_{ns} \qquad (7.4.9)$$

where

$$\bar{V}_n \equiv \alpha\left[\left(\frac{\partial P_{xx}}{\partial x} + \frac{\partial P_{xy}}{\partial y}\right)n_x + \left(\frac{\partial P_{xy}}{\partial x} + \frac{\partial P_{yy}}{\partial y}\right)n_y\right]$$
$$+ (\bar{Q}_x n_x + \bar{Q}_y n_y) + \alpha\frac{\partial P_{ns}}{\partial s} \qquad (7.4.10)$$

and the relations between the components in the (x,y) and (n,s) coordinate systems are given by

$$\phi_x = n_x\phi_n - n_y\phi_s, \quad \phi_y = n_x\phi_s + n_y\phi_n \qquad (7.4.11a)$$

$$\frac{\partial w_0}{\partial x} = n_x\frac{\partial w_0}{\partial n} - n_y\frac{\partial w_0}{\partial s}, \quad \frac{\partial w_0}{\partial y} = n_y\frac{\partial w_0}{\partial n} + n_x\frac{\partial w_0}{\partial s} \qquad (7.4.11b)$$

In the above equations, (n,s) are the local coordinates on the boundary with n being the in-plane normal and s being the in-plane tangential coordinate, and (n_x, n_y) are the direction cosines of the unit normal $\hat{n}$. The normal and tangential components of M's, Q's, P's, and R's are related to the corresponding quantities in the plate coordinates (x,y) by the tensor transformations

$$\begin{Bmatrix} M_{nn} \\ M_{ss} \\ M_{ns} \end{Bmatrix} = \begin{bmatrix} n_x^2 & n_y^2 & 2n_xn_y \\ n_y^2 & n_x^2 & -2n_xn_y \\ -n_xn_y & n_xn_y & n_x^2 - n_y^2 \end{bmatrix} \begin{Bmatrix} M_{xx} \\ M_{yy} \\ M_{xy} \end{Bmatrix} \qquad (7.4.12a)$$

$$\begin{Bmatrix} P_{nn} \\ P_{ss} \\ P_{ns} \end{Bmatrix} = \begin{bmatrix} n_x^2 & n_y^2 & 2n_xn_y \\ n_y^2 & n_x^2 & -2n_xn_y \\ -n_xn_y & n_xn_y & n_x^2 - n_y^2 \end{bmatrix} \begin{Bmatrix} P_{xx} \\ P_{yy} \\ P_{xy} \end{Bmatrix} \qquad (7.4.12b)$$

The plate constitutive equations (i.e., relations between the resultants and the displacement gradients) are given by

$$M_{xx} = D\left[(1-\alpha_0)\left(\frac{\partial \phi_x}{\partial x} + \nu\frac{\partial \phi_y}{\partial y}\right) - \alpha_0\left(\frac{\partial^2 w_0}{\partial x^2} + \nu\frac{\partial^2 w_0}{\partial y^2}\right)\right]$$

$$M_{yy} = D\left[(1-\alpha_0)\left(\frac{\partial \phi_y}{\partial y} + \nu\frac{\partial \phi_x}{\partial x}\right) - \alpha_0\left(\frac{\partial^2 w_0}{\partial y^2} + \nu\frac{\partial^2 w_0}{\partial x^2}\right)\right]$$

$$M_{xy} = \frac{(1-\nu)D}{2}\left[(1-\alpha_0)\left(\frac{\partial \phi_x}{\partial y} + \frac{\partial \phi_y}{\partial x}\right) - 2\alpha_0\frac{\partial^2 w_0}{\partial x \partial y}\right]$$

$$P_{xx} = D\left[\beta_0\left(\frac{\partial \phi_x}{\partial x} + \nu\frac{\partial \phi_y}{\partial y}\right) - \gamma\left(\frac{\partial^2 w_0}{\partial x^2} + \nu\frac{\partial^2 w_0}{\partial y^2}\right)\right]$$

$$P_{yy} = D\left[\beta_0\left(\frac{\partial \phi_y}{\partial y} + \nu\frac{\partial \phi_x}{\partial x}\right) - \gamma\left(\frac{\partial^2 w_0}{\partial y^2} + \nu\frac{\partial^2 w_0}{\partial x^2}\right)\right]$$

$$P_{xy} = \frac{(1-\nu)D}{2}\left[\beta_0\left(\frac{\partial \phi_x}{\partial y} + \frac{\partial \phi_y}{\partial x}\right) - 2\gamma\frac{\partial^2 w_0}{\partial x \partial y}\right]$$

$$Q_x = (1-\mu)Gh\left(\phi_x + \frac{\partial w_0}{\partial x}\right), \quad Q_y = (1-\mu)Gh\left(\phi_y + \frac{\partial w_0}{\partial y}\right)$$

$$R_x = \lambda Gh\left(\phi_x + \frac{\partial w_0}{\partial x}\right), \quad R_y = \lambda Gh\left(\phi_y + \frac{\partial w_0}{\partial y}\right) \qquad (7.4.13)$$

where

$$\alpha_0 = \frac{3}{20}\alpha h^2 = \frac{1}{5}, \quad \beta_0 = \frac{4h^2}{35}, \quad \gamma = \frac{h^2}{28}, \quad \lambda = \frac{h^2}{30}, \quad \mu = \frac{\alpha h^2}{4} = \frac{1}{3} \qquad (7.4.14)$$

D is the flexural rigidity, G the shear modulus

$$D = \frac{Eh^3}{12(1-\nu^2)}, \quad G = \frac{E}{2(1+\nu)} \qquad (7.4.15)$$

ν Poisson's ratio, and E the Young's modulus of the plate.

Eliminating Q_x and Q_y from equation (7.4.3) by using equations (7.4.1) and (7.4.2), one obtains

$$\frac{\partial^2 M_{xx}}{\partial x^2} + 2\frac{\partial^2 M_{xy}}{\partial x \partial y} + \frac{\partial^2 M_{yy}}{\partial y^2} = -q \qquad (7.4.16)$$

Next, we introduce the moment sum $\mathcal{M}$ and the higher-order moment sum $\mathcal{P}$ as

$$\mathcal{M} = \frac{M_{xx} + M_{yy}}{(1+\nu)} = D\left[(1-\alpha_0)\left(\frac{\partial \phi_x}{\partial x} + \frac{\partial \phi_y}{\partial y}\right) - \alpha_0 \nabla^2 w_0\right] \qquad (7.4.17)$$

$$\mathcal{P} = \frac{P_{xx} + P_{yy}}{(1+\nu)} = D\left[\beta_0\left(\frac{\partial \phi_x}{\partial x} + \frac{\partial \phi_y}{\partial y}\right) - \gamma \nabla^2 w_0\right] \qquad (7.4.18)$$

where ∇^2 is the Laplace operator, $\nabla^2 = \partial^2/\partial x^2 + \partial^2/\partial y^2$. In view of the definition of the moment sum and the expressions for M_{xx}, M_{yy}, and M_{xy} in equation (7.4.13), equation (7.4.16) takes the form

$$\nabla^2 \mathcal{M}^R = -q \tag{7.4.19}$$

where the superscript 'R' indicates that the variable belongs to the Reddy third-order shear deformation theory (TSDT).

Using expressions for the moment and shear force resultants from equation (7.4.13), equation (7.4.3) can be expressed as

$$(1 - \mu - 3\alpha\lambda)Gh\left(\frac{\partial \phi_x}{\partial x} + \frac{\partial \phi_y}{\partial y}\right) + \alpha\beta_0 D\nabla^2\left(\frac{\partial \phi_x}{\partial x} + \frac{\partial \phi_y}{\partial y}\right)$$
$$= -(1 - \mu - 3\alpha\lambda)Gh\nabla^2 w_0 + \alpha\gamma D\nabla^4 w_0 - q \tag{7.4.20}$$

Next, we use equation (7.4.17) to write

$$\frac{\partial \phi_x}{\partial x} + \frac{\partial \phi_y}{\partial y} = \frac{\alpha_0}{(1-\alpha_0)}\nabla^2 w_0 + \frac{1}{D(1-\alpha_0)}\mathcal{M}^R \tag{7.4.21}$$

and substitute the result into equation (7.4.20) and obtain

$$\left[(1-\mu-\beta\lambda)\left(\frac{\alpha_0}{1-\alpha_0}\right)\right]Gh\nabla^2 w_0 + \left(\frac{1-\mu-\beta\lambda}{1-\alpha_0}\right)\frac{Gh}{D}\mathcal{M}^R$$
$$+ \alpha\left[\left(\frac{\alpha_0\beta_0}{1-\alpha_0}\right)D\nabla^4 w_0 + \left(\frac{\beta_0}{1-\alpha_0}\right)\nabla^2 \mathcal{M}^R\right]$$
$$= -(1-\mu-\beta\lambda)Gh\nabla^2 w_0 + \alpha\gamma D\nabla^4 w_0 - q \tag{7.4.22}$$

or

$$\mathcal{M}^R = -D\nabla^2 w_0 + \frac{D}{Gh}\left[\alpha\left(C_1 D\nabla^4 w_0 - C_2\nabla^2 \mathcal{M}^R\right) - C_3 q\right] \tag{7.4.23a}$$
$$= -D\nabla^2 w_0 + \frac{D}{Gh}\left(\alpha C_1 D\nabla^4 w_0 + C_4\nabla^2 \mathcal{M}^R\right) \tag{7.4.23b}$$

where

$$C_1 = \frac{\gamma - \alpha_0(\beta_0+\gamma)}{1-\mu-\beta\lambda} = \frac{3h^2}{280}, \quad C_2 = \frac{\beta_0}{1-\mu-\beta\lambda} = \frac{3h^2}{14} \tag{7.4.23c}$$

$$C_3 = \frac{1-\alpha_0}{1-\mu-\beta\lambda} = \frac{3}{2}, \quad C_4 = C_3 - \alpha C_2 = \frac{17}{14} \tag{7.4.23d}$$

and equation (7.4.19) was used in arriving at equation (7.4.23b). The substitution of equation (7.4.23b) into equation (7.4.19) furnishes the following governing differential equation ($\alpha C_1 = 1/70$):

$$D\nabla^4 \left[w_0^R - \frac{1}{Gh} \left(\alpha C_1 D\nabla^2 w_0^R + C_4 \mathcal{M}^R \right) \right] = q \qquad (7.4.24)$$

7.4.3. The Kirchhoff Plate Theory (CPT)

The equation governing static bending of isotropic plates according to the CPT is

$$\frac{\partial^2 M_{xx}^K}{\partial x^2} + 2\frac{\partial^2 M_{xy}^K}{\partial x \partial y} + \frac{\partial^2 M_{yy}^K}{\partial y^2} = -q \qquad (7.4.25)$$

where the superscript 'K' indicates that the variable belongs to the Kirchhoff plate theory. The primary (i.e., geometric) and secondary (i.e., force) variables of the theory are

$$\text{Primary Variables}: \ w_0^K, \ \frac{\partial w_0^K}{\partial n} \qquad (7.4.26)$$

$$\text{Secondary Variables}: \ \bar{V}_n^K, \ \bar{M}_{nn}^K \qquad (7.4.27)$$

where

$$\bar{V}_n^K \equiv \left[\left(\frac{\partial M_{xx}^K}{\partial x} + \frac{\partial M_{xy}^K}{\partial y} \right) n_x + \left(\frac{\partial M_{xy}^K}{\partial x} + \frac{\partial M_{yy}^K}{\partial y} \right) n_y \right] + \frac{\partial M_{ns}^K}{\partial s} \qquad (7.4.28)$$

The moment-deflection relationships for the Kirchhoff plate theory are given by

$$M_{xx}^K = -D \left(\frac{\partial^2 w_0^K}{\partial x^2} + \nu \frac{\partial^2 w_0^K}{\partial y^2} \right)$$

$$M_{yy}^K = -D \left(\frac{\partial^2 w_0^K}{\partial y^2} + \nu \frac{\partial^2 w_0^K}{\partial x^2} \right)$$

$$M_{xy}^K = (1-\nu) D \frac{\partial^2 w_0^K}{\partial x \partial y} \qquad (7.4.29)$$

Using the definition of the moment sum

$$\mathcal{M}^K = \frac{M_{xx}^K + M_{yy}^K}{(1+\nu)} = -D\nabla^2 w_0^K \qquad (7.4.30)$$

we can rewrite equation (7.4.25) in the form [see Eqs. (7.2.1a,b)]

$$\nabla^2 \mathcal{M}^K = -q \qquad (7.4.31)$$

Also, the substitution of equations (7.4.29) in (7.4.25) yields

$$D\nabla^4 w_0^K = q \qquad (7.4.32)$$

7.4.4 Relationships Between the Theories

The simply supported boundary conditions for the two theories are given by

$$w_0^R = 0, \quad M_{nn}^R = 0, \quad P_{nn} = 0, \quad \phi_s = 0 \qquad (7.4.33)$$

and

$$w_0^K = 0, \quad M_{nn}^K = 0 \qquad (7.4.34)$$

The condition $w_0^R = 0$ implies that $\partial^2 w_0^R / \partial s^2 = 0$ and $\phi_s = 0$ implies that $\partial \phi_s / \partial s = 0$. Then together with the conditions $M_{nn}^R = 0$ and $P_{nn} = 0$, we have

$$\frac{\partial^2 w_0^R}{\partial n^2} = 0, \quad \frac{\partial \phi_n}{\partial n} = 0, \quad M_{ss}^R = 0, \quad P_{ss} = 0 \qquad (7.4.35)$$

Hence we have

$$w_0^R = 0, \quad \mathcal{M}^R = 0, \quad \nabla^2 w_0^R = 0 \qquad (7.4.36)$$

on a simply supported edge. For simply supported Kirchhoff plates, the boundary conditions reduce to

$$w_0^K = 0, \quad \mathcal{M}^K = 0, \quad \nabla^2 w_0^K = 0 \qquad (7.4.37)$$

on the boundary.

From equations (7.4.24) and (7.4.32) it follows that

$$w_0^K = w_0^R - \frac{1}{Gh}\left(\alpha C_1 D \nabla^2 w_0^R + C_4 \mathcal{M}^R\right) \qquad (7.4.38)$$

Comparing equations (7.4.19) and (7.4.31), one may conclude that

$$\nabla^2 \mathcal{M}^R = \nabla^2 \mathcal{M}^K \qquad (7.4.39)$$

which does not necessarily imply, in general, that $\mathcal{M}^R = \mathcal{M}^K$. However, for simply supported boundary conditions, we have shown the equality. Thus equation (7.4.38) can be written as

$$w_0^K = w_0^R - \frac{1}{Gh}\left(\alpha C_1 D \nabla^2 w_0^R + C_4 \mathcal{M}^K\right) \tag{7.4.40a}$$

or

$$\nabla^2 w_0^R - \lambda_0^2 w_0^R = -\lambda_0^2\left(w_0^K + \frac{C_4}{Gh}\mathcal{M}^K\right) \tag{7.4.40b}$$

where

$$\lambda_0^2 = \frac{Gh}{\alpha C_1 D} = \frac{70 Gh}{D} \tag{7.4.40c}$$

For a given CPT solution, the right hand side of equation (7.4.40b) is known. Thus from equation (7.4.40b), one needs to solve only a second-order partial differential equation for the deflection of the corresponding Reddy plate theory. This means that one may bypass solving a sixth-order differential equation in terms of w_0^R. The second-order differential equation (7.4.40b), together with the boundary condition along the simply supported edges, may be solved using the finite difference method or the finite element method. In Section 7.4.6, we illustrate this with a simply supported rectangular plate.

Equation (7.4.40a) can be specialized to the Mindlin plate theory by setting $\alpha = 0$ and replacing Gh with $K_s Gh$, where K_s is the shear correction factor. We have ($\alpha_0 = 0, \mu = 0, C_4 = C_3 = 1$)

$$w_0^K = w_0^M - \frac{1}{K_s Gh}\mathcal{M}^K \tag{7.4.41}$$

or

$$w_0^M = w_0^K + \frac{1}{K_s Gh}\mathcal{M}^K \tag{7.4.42}$$

as shown in Section 7.2.

7.4.5 An Accurate Simplified Relationship

If one were to avoid solving the second order equation (7.4.40b) for w_0^R, the term involving $\nabla^2 w_0^R$ may be dropped. This approximation may be justified as discussed below.

Eliminating $(\partial\phi_x/\partial x+\partial\phi_y/\partial y)$ from equations (7.4.17) and (7.4.18), we obtain

$$D\nabla^2 w_0^R = \frac{1-\alpha_0}{\gamma-\alpha_0(\beta_0+\gamma)}\left(\frac{\beta_0}{1-\alpha_0}\mathcal{M}^R-\mathcal{P}\right)=\frac{C_3}{C_1}\left(\frac{\beta_0}{1-\alpha_0}\mathcal{M}^R-\mathcal{P}\right) \quad (7.4.43)$$

Substituting equation (7.4.43) into (7.4.40b) and noting that $\mathcal{M}^R = \mathcal{M}^K$, we obtain

$$w_0^R = w_0^K + \frac{1}{Gh}\left[\left(C_4+\frac{\beta_0\alpha C_3}{1-\alpha_0}\right)\mathcal{M}^K - \alpha C_3\mathcal{P}\right] \quad (7.4.44a)$$

$$= w_0^K + \frac{3}{2Gh}\mathcal{M}^K - \frac{2}{Gh^3}\mathcal{P} \quad (7.4.44b)$$

Examining equations (7.4.17) and (7.4.18), we note that $\mathcal{P}$ is a higher-order function compared to $\mathcal{M}$. From the first expression of equations (7.4.17) and (7.4.18), we see that

$$\mathcal{P} = \frac{\beta_0}{1-\alpha_0}\mathcal{M}^R = \frac{h^2}{7}\mathcal{M}^R = 0.14286h^2\mathcal{M}^R \quad (7.4.45)$$

From the second expression of the same two equations, we have

$$\mathcal{P} = \frac{\gamma}{\alpha_0}\mathcal{M}^R = \frac{5h^2}{28}\mathcal{M}^R = 0.17857h^2\mathcal{M}^R \quad (7.4.46)$$

On the other hand, if we assume $(\partial\phi_x/\partial x+\partial\phi_y/\partial y) = -\nabla^2 w_0$, then we have

$$\mathcal{P} = (\beta_0+\gamma)\mathcal{M}^R = \frac{3h^2}{20}\mathcal{M}^R = 0.15h^2\mathcal{M}^R \quad (7.4.47)$$

Thus, we obtain different expressions from equation (7.4.44b) for different choices of $\mathcal{P}$ in terms of $\mathcal{M}^R = \mathcal{M}^K$. For the choice of $\mathcal{P}$ given by equation (7.4.45), we have $\nabla^2 w_0^R = 0$, and the deflection relationship in equation (7.4.40a) becomes

$$\bar{w}_0^R = w_0^K + \frac{C_4}{Gh}\mathcal{M}^K = w_0^K + \frac{1}{K_sGh}\mathcal{M}^K \quad (7.4.48)$$

Note that equation (7.4.48) is similar to equation (7.4.42) with the shear correction factor $K_s = 1/C_4 = 14/17$. If we choose the relationship in equation (7.4.47), we obtain again equation (7.4.48) with the shear correction factor $K_s = 5/6$.

7.4.6 An Example

Here we illustrate the use of the exact relationship in equation (7.4.40b) and the simplified relationship (7.4.48) in obtaining the deflection of a simply supported rectangular plate under sinusoidally distributed transverse load

$$q(x,y) = q_0 \sin \frac{\pi x}{a} \sin \frac{\pi y}{b} \qquad (7.4.49)$$

The CPT plate solution is given by

$$w_0^K(x,y) = \frac{q_0}{D\Omega_0^2} \sin \frac{\pi x}{a} \sin \frac{\pi y}{b} \qquad (7.4.50a)$$

where

$$\Omega_0 = \frac{\pi^2}{a^2} + \frac{\pi^2}{b^2} \qquad (7.4.50b)$$

and a and b are the dimensions of the plate along the x and y directions. The moment sum $\mathcal{M}^K$ is given by

$$\mathcal{M}^K = -D\nabla^2 w_0^K = D\Omega_0 w_0^K \qquad (7.4.51)$$

The solution of the simplified Reddy plate theory is given by equation (7.4.48)

$$\bar{w}_0^R(x,y) = \left(1 + \frac{D\Omega_0}{K_s G h}\right) \frac{q_0}{D\Omega_0^2} \sin \frac{\pi x}{a} \sin \frac{\pi y}{b} \qquad (7.4.52)$$

where the shear correction factor is $K_s = 14/17$. On the other hand, the exact Reddy plate solution is obtained by solving equation (7.4.40b)

$$\nabla^2 w_0^R - \lambda_0^2 w_0^R = -\lambda_0^2 \left(1 + \frac{17 D \Omega_0}{14 G h}\right) \frac{q_0}{D \Omega_0^2} \sin \frac{\pi x}{a} \sin \frac{\pi y}{b} \qquad (7.4.53)$$

The solution is of the form

$$w_0^R = A \sin \frac{\pi x}{a} \sin \frac{\pi y}{b} \qquad (7.4.54)$$

where the amplitude A is obtained by substituting equation (54) into (53). We obtain

$$-A\Omega_0 - \lambda_0^2 A = -\lambda_0^2 \left(1 + \frac{17D\Omega_0}{14Gh}\right) \frac{q_0}{D\Omega_0^2}$$

or

$$A = \frac{\lambda_0^2 \left(1 + \frac{17D\Omega_0}{14Gh}\right) \frac{q_0}{D\Omega_0^2}}{\Omega_0 + \lambda_0^2} \qquad (7.4.55)$$

Thus the exact Reddy plate solution for a simply supported rectangular plate under sinusoidally distributed transverse load is

$$w_0^R(x,y) = \left(\frac{1}{1 + \frac{\Omega_0}{\lambda_0^2}}\right)\left(1 + \frac{17D\Omega_0}{14Gh}\right) \frac{q_0}{D\Omega_0^2} \sin\frac{\pi x}{a} \sin\frac{\pi y}{b} \qquad (7.4.56)$$

Comparing the simplified solution (7.4.52) and exact solution (7.4.56) of the Reddy plate theory, we note the following relationship

$$\bar{w}_0^R(x,y) = \left(1 + \frac{\Omega_0}{\lambda_0^2}\right) w_0^R(x,y) \qquad (7.4.57)$$

To see the error in the simplified solution, consider a square plate ($a = b$). We have

$$\lambda_0^2 = \frac{70Gh}{D} = \frac{420(1-\nu)}{h^2}, \quad \frac{\Omega_0}{\lambda_0^2} = \frac{\pi^2}{210(1-\nu)}\left(\frac{h}{a}\right)^2 \qquad (7.4.58)$$

For $\nu = 0.3$ and $a/h = 10$, we have $\bar{w}_0^R = 1.0007 w_0^R$. Thus the simplified solution is very close to the exact solution. Even for $a/h = 5$, the simplified solution is in error by only 0.6% !

7.5 Closure

In this chapter a differential relationship between the Kirchhoff and Mindlin plate theories and Kirchhoff and Reddy plate theories for isotropic elastic plates is developed. In the case of the third-order plate theory, the relationship is a second-order differential equation for the deflection and requires the moment sum to be known. To avoid solving the second order equation, a simplified relationship that is very accurate is presented. The accuracy of the simplified solution is illustrated through the Navier solution of a simply supported rectangular plate under sinusoidally distributed transverse load, although the procedure is valid for a general transverse load. The simplified relationship for deflection is valid for any plate for which one has the knowledge of the Marcus moment of the Reddy plate theory in terms of the same for the Kirchhoff plate theory.

Problems

7.1 The solution of Eqs.(7.2.1a,b) [or Eq.(6.2.24)] for rectangular plates with simply supported boundary conditions can be obtained using Navier's method. In Navier's method, the displacement w_0 is expanded in trigonometric series such that the boundary conditions of the problem are satisfied. Substitution of the expansion for the deflection into the governing equation (6.2.24) will dictate the choice of the expansion used for the load. The simply supported boundary conditions are met by the expansion (see Reddy 1999a)

$$w_0(x,y) = \sum_{n=1}^{\infty} \sum_{m=1}^{\infty} W_{mn} \sin \frac{m\pi x}{a} \sin \frac{n\pi y}{b} \qquad (i)$$

where W_{mn} are coefficients to be determined such that Eq. (6.2.24) is satisfied everywhere in the domain of the plate and $a \times b$ denote the plate dimensions along the x and y axes. The coordinates (x, y, z) are taken at the upper left corner of the plate with the z axis downward positive (see Figure P7.1). Assuming that the load can also be expanded in double sine series

$$q(x,y) = \sum_{n=1}^{\infty} \sum_{m=1}^{\infty} q_{mn} \sin \frac{m\pi x}{a} \sin \frac{n\pi y}{b} \qquad (ii)$$

determine the coefficients W_{mn} in terms of q_{mn} and flexural rigidity D.

7.2 Use the constitutive equations to compute the stresses $(\sigma_{xx}, \sigma_{yy}, \sigma_{xy})$ in an isotropic plate based on the CPT for the pure bending. Then use the equilibrium equations of the three-dimensional elasticity theory to determine the transverse stresses $(\sigma_{xz}, \sigma_{yz}, \sigma_{zz})$ as a function of the thickness coordinate.

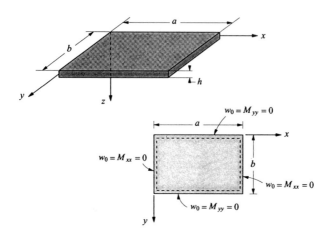

Figure P7.1

7.3 Repeat Problem 7.1 for Eqs. (6.3.14)–(6.3.16). Assume the following expansion for ϕ_x and ϕ_y (see Figure P7.3 for the boundary conditions):

$$\phi_x(x,y) = \sum_{n=1}^{\infty}\sum_{m=1}^{\infty} X_{mn} \cos\frac{m\pi x}{a} \sin\frac{n\pi y}{b} \qquad (i)$$

$$\phi_y(x,y) = \sum_{n=1}^{\infty}\sum_{m=1}^{\infty} Y_{mn} \sin\frac{m\pi x}{a} \cos\frac{n\pi y}{b} \qquad (ii)$$

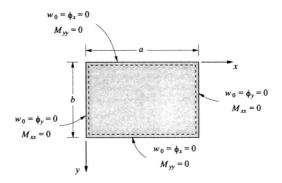

Figure P7.3

7.4 Repeat Problem 7.2 for the Mindlin plate theory.

7.5 Determine the deflection of a simply supported plate when the plate is subjected to a load of the type

$$q(x, y) = q_0 \sin \frac{\pi x}{a}$$

where q_0 is a constant.

7.6 Determine the deflection of a simply supported plate when the plate is subjected to a line load of the type

$$q(x, y) = q_1 \sin \frac{\pi x}{a} \delta(y - y_0)$$

where q_1 is a constant and $\delta(y)$ is the Dirac delta function.

7.7 Verify that w_0^K of Eq. (7.3.1) and w_0^M of Eq. (7.3.3) the solutions of the respective plates.

7.8 Consider a clamped, isotropic elliptic plate with major and minor axes $2a$ and $2b$, respectively. Assume a solution of the form

$$w_0^K = c\left(1 - \frac{x^2}{a^2} - \frac{y^2}{b^2}\right)^2 \qquad (i)$$

and determine the constant c such the w_0^K satisfies the governing equation.

7.9 Use Eqs. (6.3.18)–(6.3.20) to show that

$$\nabla^2 \mathcal{M}^M = -q, \quad \nabla^2 w_0^M + \frac{\mathcal{M}^M}{D} = \frac{\nabla^2 \mathcal{M}^M}{K_s G h} \qquad (i)$$

Then establish the equation

$$D\nabla^4 w_0^M = \left(1 - \frac{D}{K_s G h}\nabla^2\right) q \qquad (iii)$$

Hint: Differentiate Eq. (6.3.19) with respect to x and Eq. (6.3.20) with respect to y and add the resulting equations.

7.10 Use the results of Problem 6.6 and Problem 7.9 to show that

$$\nabla^4 \phi_x^M = -\frac{1}{D}\frac{\partial q}{\partial x} + c^2 \frac{\partial}{\partial y}\left(\frac{\partial \phi_x^M}{\partial y} - \frac{\partial \phi_y^M}{\partial x}\right) \qquad (i)$$

$$\nabla^4 \phi_y^M = -\frac{1}{D}\frac{\partial q}{\partial y} - c^2 \frac{\partial}{\partial x}\left(\frac{\partial \phi_x^M}{\partial y} - \frac{\partial \phi_y^M}{\partial x}\right) \qquad (ii)$$

where $c^2 = 12K_s/h^2$.

7.11 Show that the governing equations of static equilibrium of plates according to the Levinson plate theory can be expressed in terms of the moment sum (or Marcus moment) $\mathcal{M}^L$

$$\mathcal{M}^L = \frac{M_{xx}^L + M_{yy}^L}{1+\nu} \qquad (i)$$

as

$$\nabla^2 \mathcal{M}^L = -q, \qquad \frac{\partial \phi_x^L}{\partial x} + \frac{\partial \phi_y^L}{\partial y} = \frac{1}{4}\left(\nabla^2 w_0^L + \frac{5\mathcal{M}^L}{D}\right) \qquad (ii)$$

7.12 Establish the following relationships between the Levinson plate theory and Kirchhoff plate theory:

$$\mathcal{M}^L = \mathcal{M}^K + D\nabla^2 \Phi, \quad w_0^L = w_0^K + \frac{\mathcal{M}^K}{\frac{5}{6}Gh} - \Phi + \Psi \qquad (i)$$

$$\phi_x^L = -\frac{\partial w_0^K}{\partial x} + \frac{3}{10Gh}\frac{\partial \mathcal{M}^K}{\partial x} + \frac{\partial \Theta}{\partial x} + \frac{h^2}{10}\frac{\partial \Omega}{\partial y} \qquad (ii)$$

$$\phi_y^L = -\frac{\partial w_0^K}{\partial y} + \frac{3}{10Gh}\frac{\partial \mathcal{M}^K}{\partial x} + \frac{\partial \Theta}{\partial y} - \frac{h^2}{10}\frac{\partial \Omega}{\partial x} \qquad (iii)$$

$$M_{xx}^L = M_{xx}^K - D(1-\nu)\frac{\partial}{\partial y}\left(\frac{\partial \Lambda}{\partial y} - \frac{2h^2}{25}\frac{\partial \Omega}{\partial x}\right) + D\nabla^2 \Phi \qquad (iv)$$

$$M_{yy}^L = M_{yy}^K - D(1-\nu)\frac{\partial}{\partial x}\left(\frac{\partial \Lambda}{\partial x} + \frac{2h^2}{25}\frac{\partial \Omega}{\partial y}\right) + D\nabla^2 \Phi \qquad (v)$$

$$M_{xy}^L = M_{xy}^K + D(1-\nu)\left[\frac{\partial^2 \Lambda}{\partial x \partial y} + \frac{h^2}{25}\left(\frac{\partial^2}{\partial y^2} - \frac{\partial^2}{\partial x^2}\right)\Omega\right] \qquad (vi)$$

$$Q_x^L = Q_x^K + D\frac{\partial}{\partial x}\nabla^2 \Phi + \frac{2}{5}D(1-\nu)\frac{\partial \Omega}{\partial y} \qquad (vii)$$

$$Q_y^L = Q_y^K + D\frac{\partial}{\partial y}\nabla^2 \Phi - \frac{2}{5}D(1-\nu)\frac{\partial \Omega}{\partial x} \qquad (viii)$$

where

$$\Theta = \frac{3D}{2Gh}\nabla^2 \Phi + \Phi - \Psi, \quad \Lambda = \frac{D}{\frac{5}{6}Gh}\nabla^2 \Phi + \Phi - \Psi \qquad (ix)$$

and Φ, Ψ, and Ω are functions such that

$$\nabla^4 \Phi = 0, \quad \nabla^2 \Psi = 0, \quad -\nabla^2 \Omega + \frac{10}{h^2}\Omega = 0 \qquad (x)$$

Chapter 8

Bending Relationships for Lévy Solutions

In this chapter, the exact relationships are obtained between the Mindlin and Kirchhoff solutions for the bending of rectangular plates with two opposite edges simply supported and the other two edges under general boundary conditions. These relationships enable the deflections, rotations, and stress-resultants of the Mindlin plate theory to be determined readily from the corresponding solutions of the Kirchhoff plate theory for any combination of boundary conditions on the remaining two edges.

8.1 Introduction

In Chapter 7, an exact deflection relationship between the Kirchhoff (CPT) and Mindlin (FSDT) polygonal plates was presented. All the straight edges of the plates must, however, be simply supported but the transverse loading can be of arbitrary distribution. The derivation of the relationship was based on an analogy approach and the assumption that the moment sum vanishes along the edges including the corner points. When using the relationship, Mindlin solutions obtained for plates with obtuse and re-entrant corners are somewhat less accurate due to the moment singularities at such corner points. Nevertheless, the relationship allows easy and exact determination of the more complicated Mindlin plate solutions from the simpler Kirchhoff plate solutions for scalene triangular plates and rectangular plates or near rectangular shaped plates. Such Kirchhoff plate solutions abound in the open literature for use in the relationship.

In the present chapter, we treat the bending problem of rectangular plates with two opposite edges simply supported while the other two edges are supported in an arbitrary manner. In using Lévy's method of analysis, the load distribution is restricted to be constant with respect to the coordinate parallel to the direction of the two simply supported edges. In this chapter, the exact relationships between the Lévy solutions of Kirchhoff plate theory (or CPT) and the Mindlin plate theory (or FSDT) are derived. These relationships, hitherto not available, enable engineers to determine the Lévy solutions of the Mindlin plate theory from the corresponding solutions of the Kirchhoff plate theory. The solutions of the Kirchhoff plate theory are available in standard textbooks on plates (see, for example, Timoshenko and Woinowsky-Krieger 1970, Mansfield 1989, and Reddy 1984b, 1997a, and 1999a). Using these relationships, it was discovered that the Lévy solutions of the FSDT developed by Cooke and Levinson (1983) are erroneous. Furthermore, the exact FSDT solutions furnished by the relationships (and the corresponding exact CPT solutions) should be useful to researchers for checking the validity, convergence and accuracy of their numerical methods for the bending analysis of plates based on FSDT (see Khdeir and his colleagues, 1987). Some examples of these numerical methods include the segmentation method proposed by Kant and Hinton (1983), and the finite element method by Reddy and Chao (1981), Huang and Hinton (1984), and Hinton and Huang (1986).

8.2 Governing Equations
8.2.1 Introduction

Consider an isotropic plate with uniform thickness h, length a, width b, Young's modulus E, Poisson's ratio ν, and shear modulus $G = E/[2(1+\nu)]$. Adopting the rectangular Cartesian coordinate system as shown in Figure 8.2.1 with its origin at the mid-left side of the plate, the plate is simply supported along the edges $x = 0$ and $x = a$ while the other two edges $y = b/2$ and $y = -b/2$ may be clamped, simply supported, or free. The transverse loading on the plate is characterized by

$$q(x,y) = \sum_{m=1}^{\infty} q_m(y) \sin \frac{m\pi x}{a} \qquad (8.2.1a)$$

where the coefficients q_m are determined from

$$q_m(y) = \frac{2}{a} \int_0^a q(x,y) \sin \frac{m\pi x}{a} \, dx \qquad (8.2.1b)$$

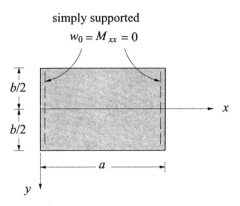

Figure 8.2.1. A rectangular plate with two opposite edges simply supported (for the Lévy solution).

8.2.2 Stress Resultant-Displacement Relations

Based on the Kirchhoff plate theory, the transverse deflection w_0 of the Lévy solution may be written as

$$w_0^K(x,y) = \sum_{m=1}^{\infty} W_m^K(y) \sin \frac{m\pi x}{a} \quad (8.2.2)$$

where the superscript 'K' denotes quantities in the Kirchhoff plate theory. In view of Eq. (8.2.2), the stress resultant-displacement relations are given by [see Eqs. (6.2.22a-c) and (6.2.8)]

$$M_{xx}^K = -D \sum_{m=1}^{\infty} \left[\nu \frac{d^2 W_m^K}{dy^2} - \left(\frac{m\pi}{a}\right)^2 W_m^K \right] \sin \frac{m\pi x}{a} \quad (8.2.3a)$$

$$M_{yy}^K = -D \sum_{m=1}^{\infty} \left[\frac{d^2 W_m^K}{dy^2} - \nu \left(\frac{m\pi}{a}\right)^2 W_m^K \right] \sin \frac{m\pi x}{a} \quad (8.2.3b)$$

$$M_{xy}^K = -(1-\nu)D \sum_{m=1}^{\infty} \left[\left(\frac{m\pi}{a}\right) \frac{dW_m^K}{dy} \right] \cos \frac{m\pi x}{a} \quad (8.2.3c)$$

$$Q_x^K = -D \sum_{m=1}^{\infty} \left[\frac{m\pi}{a} \frac{d^2 W_m^K}{dy^2} - \left(\frac{m\pi}{a}\right)^3 W_m^K \right] \cos \frac{m\pi x}{a} \quad (8.2.3d)$$

$$Q_y^K = -D \sum_{m=1}^{\infty} \left[\frac{d^3 W_m^K}{dy^3} - \left(\frac{m\pi}{a}\right)^2 \frac{dW_m^K}{dy} \right] \sin \frac{m\pi x}{a} \quad (8.2.3e)$$

where (M_{xx}, M_{yy}, M_{xy}) are the bending moments per unit length, (Q_x, Q_y), the shear forces per unit length, $D = Eh^3/[12(1-\nu^2)]$ the flexural rigidity, and h is the thickness of the plate.

For the Mindlin plate theory, the transverse deflection w_0^M and rotations (ϕ_x, ϕ_y) are assumed to be of the form

$$w_0^M(x,y) = \sum_{m=1}^{\infty} W_m^M(y) \sin \frac{m\pi x}{a} \qquad (8.2.4)$$

$$\phi_x(x,y) = \sum_{m=1}^{\infty} \Phi_{xm}(y) \cos \frac{m\pi x}{a} \qquad (8.2.5)$$

$$\phi_y(x,y) = \sum_{m=1}^{\infty} \Phi_{ym}(y) \sin \frac{m\pi x}{a} \qquad (8.2.6)$$

where the superscript 'M' denotes quantities in the Mindlin plate theory. In view of Eqs. (8.2.4)–(8.2.6), the stress resultant-displacement relations (6.3.13a-e) take the form (here K_s denotes the shear correction factor)

$$M_{xx}^M = D \sum_{m=1}^{\infty} \left[\nu \frac{d\Phi_{ym}}{dy} - \left(\frac{m\pi}{a}\right) \Phi_{xm} \right] \sin \frac{m\pi x}{a}$$

$$\equiv M_{mxx}^M \sin \frac{m\pi x}{a} \qquad (8.2.7a)$$

$$M_{yy}^M = D \sum_{m=1}^{\infty} \left[\frac{d\Phi_{xm}}{dy} - \nu \left(\frac{m\pi}{a}\right) \Phi_{xm} \right] \sin \frac{m\pi x}{a}$$

$$\equiv M_{myy}^M \sin \frac{m\pi x}{a} \qquad (8.2.7b)$$

$$M_{xy}^M = \frac{1}{2}(1-\nu) D \sum_{m=1}^{\infty} \left[\frac{d\Phi_{xm}}{dy} + \left(\frac{m\pi}{a}\right) \Phi_{ym} \right] \cos \frac{m\pi x}{a}$$

$$\equiv M_{mxy}^M \cos \frac{m\pi x}{a} \qquad (8.2.7c)$$

$$Q_x^M = K_s Gh \sum_{m=1}^{\infty} \left[\Phi_{xm} + \left(\frac{m\pi}{a}\right) W_m^M \right] \cos \frac{m\pi x}{a}$$

$$\equiv Q_{mx}^M \cos \frac{m\pi x}{a} \qquad (8.2.7d)$$

$$Q_y^M = K_s Gh \sum_{m=1}^{\infty} \left[\Phi_{ym} + \frac{dW_m^M}{dy} \right] \sin \frac{m\pi x}{a}$$

$$\equiv Q_{my}^M \sin \frac{m\pi x}{a} \qquad (8.2.7e)$$

8.2.3 Equilibrium Equations

For the classical (Kirchhoff) as well as the first-order (Mindlin) plate theories, the equilibrium equations have the same form, as given below:

$$-\left(\frac{\partial Q_x}{\partial x} + \frac{\partial Q_y}{\partial y}\right) = q \qquad (8.2.8)$$

$$\frac{\partial M_{xx}}{\partial x} + \frac{\partial M_{xy}}{\partial y} - Q_x = 0 \qquad (8.2.9)$$

$$\frac{\partial M_{yy}}{\partial y} + \frac{\partial M_{xy}}{\partial x} - Q_y = 0 \qquad (8.2.10)$$

However, in the Kirchhoff plate theory, Eqs. (8.2.9) and (8.2.10) define the shear forces Q_x and Q_y of Eq. (8.2.8) in terms of the bending moments, while in the Mindlin plate theory, Eqs. (8.2.9) and (8.2.10) provide additional moment equilibrium equations, and Q_x^M and Q_y^M are dependent variables much like M_{xx}^M and M_{yy}^M.

8.3 Bending Relationships
8.3.1 General Relationships

On the basis of load equivalence using Eqs. (8.2.8) to (8.2.10), one can write the following relationships

$$\frac{\partial Q_x^M}{\partial x} + \frac{\partial Q_y^M}{\partial y} = \frac{\partial Q_x^K}{\partial x} + \frac{\partial Q_y^K}{\partial y} \qquad (8.3.1)$$

$$\frac{\partial^2 M_{xx}^M}{\partial x^2} + 2\frac{\partial^2 M_{xy}^M}{\partial x \partial y} + \frac{\partial^2 M_{yy}^M}{\partial y^2} = \frac{\partial^2 M_{xx}^K}{\partial y^2} + 2\frac{\partial^2 M_{xy}^K}{\partial x \partial y} + \frac{\partial^2 M_{yy}^K}{\partial y^2} \qquad (8.3.2)$$

By substituting the stress resultants given in Eqs. (8.2.3a-e) and (8.2.7a-e) into Eqs. (8.3.1) and (8.3.2), the two load-equivalence relationships may be expressed as

$$K_s G h \left(\frac{M_m^M}{D} + \Lambda W_m^M\right) = \Lambda M_m^K \qquad (8.3.3)$$

$$\Lambda \left(\frac{M_m^M}{D}\right) = \Lambda \left(\frac{M_m^K}{D}\right) \qquad (8.3.4)$$

where the components of the moment sums (or Marcus moments) $\mathcal{M}$ in the two theories are defined by

$$\mathcal{M}_m^M \equiv \frac{M_{mxx}^M + M_{myy}^M}{1+\nu} = D\left[\frac{d\Phi_{ym}}{dy} - \left(\frac{m\pi}{a}\right)\Phi_{xm}\right] \quad (8.3.5)$$

$$\mathcal{M}_m^K \equiv \frac{M_{mxx}^K + M_{myy}^K}{1+\nu} = -D\left[\frac{d^2 W_m^K}{dy^2} - \left(\frac{m\pi}{a}\right)^2 W_m^K\right] = -D\Lambda W_m^K \quad (8.3.6)$$

and the operator Λ is defined as

$$\Lambda(\cdot) = \frac{d^2(\cdot)}{dy^2} - \left(\frac{m\pi}{a}\right)^2 (\cdot) \quad (8.3.7)$$

By solving Eq. (8.3.4), the relationship between the moment sums of Kirchhoff and Mindlin plates is given by

$$\frac{\mathcal{M}_m^M}{D} = \frac{\mathcal{M}_m^K}{D} + C_{1m}\sinh\frac{m\pi y}{a} + C_{2m}\cosh\frac{m\pi y}{a} \quad (8.3.8)$$

To solve for w_0^M in terms of w_0^K, we use Eq. (8.3.3) to obtain

$$\mathcal{M}_m^M = -D\Lambda\left(W_m^M - \frac{\mathcal{M}_m^K}{K_s Gh}\right) \quad (8.3.9)$$

The substitution of Eqs. (8.3.6) and (8.3.9) into Eq. (8.3.8) yields a differential equation, the solution of which leads to the following deflection relationship

$$W_m^M = W_m^K + \frac{\mathcal{M}_m^K}{K_s Gh} + \left(C_{3m} - C_{1m}\frac{ay}{2m\pi}\right)\cosh\frac{m\pi y}{a}$$
$$+ \left(C_{4m} - C_{2m}\frac{ay}{2m\pi}\right)\sinh\frac{m\pi y}{a} \quad (8.3.10)$$

where C_{im}, $i = 1, 2, 3, 4$ are constants to be evaluated from the boundary conditions.

It now remains to determine the Mindlin rotation relationships in terms of the Kirchhoff solution. The equilibrium equations (8.2.9) and

(8.2.10) and the stress resultant-displacement relations (8.2.3a-e) and (8.2.7a-e) are used to give

$$\Phi_{xm} + \left(\frac{m\pi}{a}\right)W_m^M = \frac{D}{K_sGh}\left[\frac{(1-\nu)}{2}\frac{d^2\Phi_{xm}}{dy^2} + \frac{(1+\nu)}{2}\left(\frac{m\pi}{a}\right)\frac{d\Phi_{ym}}{dy}\right.$$
$$\left. - \left(\frac{m\pi}{a}\right)^2 \Phi_{xm}\right] \quad (8.3.11)$$

and

$$\Phi_{ym} + \frac{dW_m^M}{dy} = \frac{D}{K_sGh}\left[\frac{d^2\Phi_{ym}}{dy^2} - \frac{(1-\nu)}{2}\left(\frac{m\pi}{a}\right)^2 \Phi_{ym}\right.$$
$$\left. - \frac{(1+\nu)}{2}\left(\frac{m\pi}{a}\right)\frac{d\Phi_{xm}}{dy}\right] \quad (8.3.12)$$

The transverse deflection may be eliminated from Eqs. (8.3.11) and (8.3.12) by first differentiating Eq. (8.3.11) with respect to y, and then substituting the expression for the derivative of the deflection into Eq. (8.3.12). By doing so, one obtains

$$\frac{d\Phi_{xm}}{dy} - \left(\frac{m\pi}{a}\right)\Phi_{ym} = \frac{D(1-\nu)}{2K_sGh}\left[-\left(\frac{m\pi}{a}\right)\frac{d^2\Phi_{ym}}{dy^2} + \left(\frac{m\pi}{a}\right)^3 \Phi_{ym}\right.$$
$$\left. + \frac{d^3\Phi_{xm}}{dy^3} - \left(\frac{m\pi}{a}\right)^2 \frac{d\Phi_{xm}}{dy}\right] \quad (8.3.13)$$

By letting

$$\lambda_m^2 = \left(\frac{m\pi}{a}\right)^2 + \frac{2K_sGh}{D(1-\nu)} \quad (8.3.14)$$

Eq. (8.3.13) may be written as

$$\frac{d^3\Phi_{xm}}{dy^3} - \lambda_m^2 \frac{d\Phi_{xm}}{dy} = \frac{m\pi}{a}\left(\frac{d^2\Phi_{ym}}{dy^2} - \lambda_m^2 \Phi_{ym}\right) \quad (8.3.15)$$

Solving Eq. (8.3.15), gives

$$\frac{d\Phi_{xm}}{dy} = \frac{m\pi}{a}\Phi_{ym} + C_{5m}\sinh\lambda_m y + C_{6m}\cosh\lambda_m y \quad (8.3.16)$$

Before proceeding further, it is noted that the substitution of Eqs. (8.3.5) and (8.3.6) into Eq. (8.3.8) yields

$$\left(\frac{d\Phi_{ym}}{dy} - \frac{m\pi}{a}\Phi_{xm}\right) = -\left(\frac{d^2 W_m^K}{dy^2} - \frac{m^2\pi^2}{a^2}W_m^K\right)$$
$$+ C_{1m}\sinh\frac{m\pi y}{a} + C_{2m}\cosh\frac{m\pi y}{a} \quad (8.3.17)$$

By differentiating Eq. (8.3.16) with respect to y and combining it with Eq. (8.3.17), it is found that

$$\frac{d^2\Phi_{xm}}{dy^2} - \frac{m^2\pi^2}{a^2}\Phi_{xm} = -\frac{m\pi}{a}\left(\frac{d^2 W_m^K}{dy^2} - \frac{m^2\pi^2}{a^2}W_M^K\right)$$
$$+ \lambda_m(C_{5m}\cosh\lambda_m y + C_{6m}\sinh\lambda_m y)$$
$$+ \frac{m\pi}{a}\left(C_{1m}\sinh\frac{m\pi y}{a} + C_{2m}\cosh\frac{m\pi y}{a}\right)$$
$$(8.3.18)$$

the solution of which is

$$\Phi_{xm} = -\frac{m\pi}{a}W_m^K + \lambda_m\frac{D(1-\nu)}{2K_s Gh}(C_{5m}\cosh\lambda_m y + C_{6m}\sinh\lambda_m y)$$
$$+ \left(C_{7m} + C_{2m}\frac{y}{2}\right)\sinh\frac{m\pi y}{a} + \left(C_{8m} + C_{1m}\frac{y}{2}\right)\cosh\frac{m\pi y}{a}$$
$$(8.3.19)$$

Also, by substituting Eq. (8.3.19) into Eq. (8.3.16), the corresponding expresion for Φ_{ym} is

$$\Phi_{ym} = -\frac{dW_m^K}{dy} + \frac{m\pi}{a}\frac{D(1-\nu)}{2K_s Gh}(C_{5m}\sinh\lambda + C_{6m}\cosh\lambda_m y)$$
$$+ \left(C_{7m} + C_{2m}\frac{y}{2} + C_{1m}\frac{a}{2m\pi}\right)\cosh\frac{m\pi y}{a}$$
$$+ \left(C_{8m} + C_{1m}\frac{y}{2} + C_{2m}\frac{a}{2m\pi}\right)\sinh\frac{m\pi y}{a} \quad (8.3.20)$$

By substituting Eqs. (8.3.10), (8.3.19), and (8.3.20) into Eq. (8.3.11), it is deduced that

$$C_{7m} = \frac{m\pi}{a}\left(\frac{D}{K_s Gh}C_{1m} - C_{4m}\right) \quad (8.3.21)$$

$$C_{8m} = \frac{m\pi}{a}\left(\frac{D}{K_s Gh}C_{2m} - C_{3m}\right) \quad (8.3.22)$$

In view of the foregoing deflection and rotation expressions, the relationships between solutions of Mindlin and Kirchhoff plates may be summarized as follows:

- *Deflection relationship*

$$w_0^M(x,y) = w_0^K(x,y) + \frac{M^K}{K_s Gh} + \sum_{m=1}^{\infty}\left[\left(C_{3m} - C_{1m}\frac{ay}{2m\pi}\right)\cosh\frac{m\pi y}{a}\right.$$
$$\left. + \left(C_{4m} - C_{2m}\frac{ay}{2m\pi}\right)\sinh\frac{m\pi y}{a}\right]\sin\frac{m\pi x}{a} \quad (8.3.23)$$

- *Rotation-slope relationships*

$$\phi_x(x,y) = -\frac{\partial w_0^K}{\partial x} + \sum_{m=1}^{\infty}\left[A_{1m}\left(C_{5m}\cosh\lambda_m y + C_{6m}\sinh\lambda_m y\right)\right.$$
$$+ \left(A_{2m} + C_{2m}\frac{y}{2}\right)\sinh\frac{m\pi y}{a}$$
$$\left. + \left(A_{3m} + C_{1m}\frac{y}{2}\right)\cosh\frac{m\pi y}{a}\right]\cos\frac{m\pi x}{a} \quad (8.3.24a)$$

$$\phi_y(x,y) = -\frac{\partial w_0^K}{\partial y} + \sum_{m=1}^{\infty}\left[B_{1m}\left(C_{5m}\sinh\lambda_m y + C_{6m}\cosh\lambda_m y\right)\right.$$
$$+ \left(B_{2m} + C_{2m}\frac{y}{2}\right)\cosh\frac{m\pi y}{a}$$
$$\left. + \left(B_{3m} + C_{1m}\frac{y}{2}\right)\sinh\frac{m\pi y}{a}\right]\sin\frac{m\pi x}{a} \quad (8.3.24b)$$

$$A_{1m} = \lambda_m\frac{D(1-\nu)}{2K_s Gh}, \quad A_{2m} = \frac{D}{K_s Gh}\frac{m\pi}{a}C_{1m} - \frac{m\pi}{a}C_{4m}$$
$$A_{3m} = \frac{D}{K_s Gh}\frac{m\pi}{a}C_{2m} - \frac{m\pi}{a}C_{3m}, \quad B_{1m} = \frac{m\pi}{a}\frac{D(1-\nu)}{2K_s Gh}$$
$$B_{2m} = \frac{D}{K_s Gh}\frac{m\pi}{a}C_{1m} - \frac{m\pi}{a}C_{4m} + \frac{a}{2m\pi}C_{1m}$$
$$B_{3m} = \frac{D}{K_s Gh}\frac{m\pi}{a}C_{2m} - \frac{m\pi}{a}C_{3m} + \frac{a}{2m\pi}C_{2m} \quad (8.3.25)$$

- *Moment relationships*

$$M_{xx}^M = M_{xx}^K + \nu D\sum_{m=1}^{\infty}\left[C_{1m}\sinh\frac{m\pi y}{a} + C_{2m}\cosh\frac{m\pi y}{a}\right]\sin\frac{m\pi x}{a}$$

$$- D(1-\nu) \sum_{m=1}^{\infty} \Big[M_{1m}^x \left(C_{5m} \cosh \lambda_m y + C_{6m} \sinh \lambda_m y \right)$$

$$+ \left(\frac{m\pi}{a}\right)^2 \left(M_{2m}^x + \frac{ay}{2m\pi} C_{2m} \right) \sinh \frac{m\pi y}{a}$$

$$+ \left(\frac{m\pi}{a}\right)^2 \left(M_{3m}^x + \frac{ay}{2m\pi} C_{1m} \right) \cosh \frac{m\pi y}{a} \Big] \sin \frac{m\pi x}{a}$$

$$\tag{8.3.26a}$$

$$M_{yy}^M = M_{yy}^K + D \sum_{m=1}^{\infty} \Big[C_{1m} \sinh \frac{m\pi y}{a} + C_{2m} \cosh \frac{m\pi y}{a} \Big] \sin \frac{m\pi x}{a}$$

$$+ D(1-\nu) \sum_{m=1}^{\infty} \Big[M_{1m}^y \left(C_{5m} \cosh \lambda_m y + C_{6m} \sinh \lambda_m y \right)$$

$$+ \left(\frac{m\pi}{a}\right)^2 \left(M_{2m}^y + \frac{ay}{2m\pi} C_{2m} \right) \sinh \frac{m\pi y}{a}$$

$$+ \left(\frac{m\pi}{a}\right)^2 \left(M_{3m}^y + \frac{ay}{2m\pi} C_{1m} \right) \cosh \frac{m\pi y}{a} \Big] \sin \frac{m\pi x}{a}$$

$$\tag{8.3.26b}$$

$$M_{xy}^M = M_{xy}^K + \frac{D(1-\nu)}{2} \sum_{m=1}^{\infty} \Big[M_{1m}^{xy} \left(C_{5m} \sinh \lambda_m y + C_{6m} \cosh \lambda_m y \right)$$

$$+ 2\left(\frac{m\pi}{a}\right)^2 \left(M_{2m}^{xy} + \frac{ay}{2m\pi} C_{2m} \right) \cosh \frac{m\pi y}{a}$$

$$+ 2\left(\frac{m\pi}{a}\right)^2 \left(M_{3m}^{xy} + \frac{ay}{2m\pi} C_{1m} \right) \sinh \frac{m\pi y}{a}$$

$$+ C_{1m} \cosh \frac{m\pi y}{a} + C_{2m} \sinh \frac{m\pi y}{a} \Big] \cos \frac{m\pi x}{a} \tag{8.3.26c}$$

$$M_{1m}^x = \lambda_m \frac{D(1-\nu)}{2K_s Gh} \left(\frac{m\pi}{a}\right), \quad M_{2m}^x = \frac{D}{K_s Gh} C_{1m} - C_{4m}$$

$$M_{3m}^x = \frac{D}{K_s Gh} C_{2m} - C_{3m}, \quad M_{1m}^y = \lambda_m \frac{D(1-\nu)}{2K_s Gh} \left(\frac{m\pi}{a}\right)$$

$$M_{2m}^y = \frac{D}{K_s Gh} C_{1m} - C_{4m}, \quad M_{3m}^y = \frac{D}{K_s Gh} C_{2m} - C_{3m}$$

$$M_{1m}^{xy} = \left[\lambda_m^2 + \left(\frac{m\pi}{a}\right)^2 \right] \frac{D(1-\nu)}{2K_s Gh}, \quad M_{2m}^{xy} = \frac{D}{K_s Gh} C_{1m} - C_{4m}$$

$$M_{3m}^{xy} = \frac{D}{K_s Gh} C_{2m} - C_{3m} \tag{8.3.27}$$

- *Shear force relationships*

$$Q_x^M = Q_x^K + \frac{D(1-\nu)}{2} \sum_{m=1}^{\infty} [\lambda_m (C_{5m} \cosh \lambda_m y + C_{6m} \sinh \lambda_m y)$$
$$+ \frac{2}{1-\nu} \left(\frac{m\pi}{a}\right) \left(C_{1m} \sinh \frac{m\pi y}{a} + C_{2m} \cosh \frac{m\pi y}{a}\right)] \cos \frac{m\pi x}{a}$$
(8.3.28)

$$Q_y^M = Q_y^K + \frac{D(1-\nu)}{2} \sum_{m=1}^{\infty} \left[\left(\frac{m\pi}{a}\right)(C_{5m} \sinh \lambda_m y + C_{6m} \cosh \lambda_m y)\right.$$
$$\left. + \frac{2}{1-\nu} \left(\frac{m\pi}{a}\right) \left(C_{1m} \cosh \frac{m\pi y}{a} + C_{2m} \sinh \frac{m\pi y}{a}\right)\right] \sin \frac{m\pi x}{a}$$
(8.3.29)

The foregoing relationships contain a total of six unknown constants C_{im}, $i = 1, 2, ...6$ which are dependent on the six boundary conditions at the two edges $y = -b/2$ and $y = +b/2$, that is, three boundary conditions for each edge. Below, these constants are evaluated for rectangular plates with various combinations of edge conditions for these two edges.

8.3.2 SSSS Plates

When all four edges $x = 0$, $y = b/2$, $x = a$, and $y = -b/2$ are simply supported, the Lévy solution reduces to the Navier solution with the following conditions at the edges $y = -b/2$ and $y = b/2$:

$$M_{yy}^M = M_{yy}^K = 0, \quad w_0^M = w_0^K = 0, \quad \phi_x = 0 \qquad (8.3.30)$$

In view of Eqs. (8.3.8), (8.3.23), (8.3.24) and (8.3.31), it can be shown that

$$C_{1m} = C_{2m} = C_{3m} = C_{4m} = C_{5m} = C_{6m} = 0 \qquad (8.3.31)$$

Thus, for simply supported plates, the relationships are given by

$$w_0^M = w_0^K + \frac{M^K}{K_s G h} \qquad (8.3.32)$$

$$\phi_x = -\frac{\partial w_0^K}{\partial x}, \quad \phi_y = -\frac{\partial w_0^K}{\partial y} \qquad (8.3.33)$$

$$M_{xx}^M = M_{xx}^K, \quad M_{yy}^M = M_{yy}^K, \quad M_{xy}^M = M_{xy}^K \qquad (8.3.34)$$

$$Q_x^M = Q_x^K, \quad Q_y^M = Q_y^K \qquad (8.3.35)$$

It can be shown that the above deflection relationship (8.3.32) is also valid for even other than Lévy-type loading conditions (see Wang and Alwis 1995).

8.3.3 SCSC Plates

For the clamped edges at $y = -b/2$ and $y = +b/2$, the boundary conditions are (see Figure 8.3.1)

$$w_0^M = w_0^K = 0, \quad \phi_x = \phi_y = \frac{\partial w_0^K}{\partial y} = 0 \qquad (8.3.36)$$

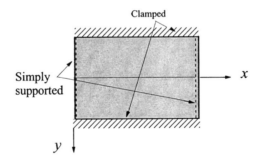

Figure 8.3.1. The SCSC plate and the coordinate system.

The substitution of the boundary conditions given in Eq. (8.3.36) into Eqs. (8.3.23) to (8.3.25) gives

$$C_{1m} = \frac{\Omega_m^-}{B_{1m}}\left(\coth\frac{m\pi b}{2a}\sinh\frac{\lambda_m b}{2} - \frac{m\pi}{a\lambda_m}\cosh\frac{\lambda_m b}{2}\right) \qquad (8.3.37a)$$

$$C_{2m} = \frac{\Omega_m^+}{B_{2m}}\left(\tanh\frac{m\pi b}{2a}\cosh\frac{\lambda_m b}{2} - \frac{m\pi}{a\lambda_m}\sinh\frac{\lambda_m b}{2}\right) \qquad (8.3.37b)$$

$$C_{3m} = C_{2m}\frac{ab}{4m\pi}\tanh\frac{m\pi b}{2a} - \Omega_m^+\operatorname{sech}\frac{m\pi b}{2a} \qquad (8.3.37c)$$

$$C_{4m} = C_{1m}\frac{ab}{4m\pi}\coth\frac{m\pi b}{2a} - \Omega_m^-\operatorname{csch}\frac{m\pi b}{2a} \qquad (8.3.37d)$$

$$C_{5m} = -\frac{2}{1-\nu}\left(\frac{m\pi}{a\lambda_m}\right)\operatorname{sech}\frac{\lambda_m b}{2}\left[C_{2m}\cosh\frac{m\pi b}{2a} + \frac{K_s Gh}{D}\Omega_m^+\right] \qquad (8.3.37e)$$

$$C_{6m} = -\frac{2}{1-\nu}\left(\frac{m\pi}{a\lambda_m}\right)\operatorname{csch}\frac{\lambda_m b}{2}\left[C_{1m}\sinh\frac{m\pi b}{2a} + \frac{K_s G h}{D}\Omega_m^-\right]$$
(8.3.37f)

where

$$\Omega_m^+ = \frac{\mathcal{M}_m^K(b/2) + \mathcal{M}_m^K(-b/2)}{2K_s G h}, \quad \Omega_m^- = \frac{\mathcal{M}_m^K(b/2) - \mathcal{M}_m^K(-b/2)}{2K_s G h}$$
(8.3.38a)

$$B_{1m} = A_{1m}\sinh\frac{m\pi b}{2a}\cosh\frac{\lambda_m b}{2} - A_{2m}\cosh\frac{m\pi b}{2a}\sinh\frac{\lambda_m b}{2}$$
$$+ A_{3m}\operatorname{csch}\frac{m\pi b}{2a}\sinh\frac{\lambda_m b}{2}$$
(8.3.38b)

$$B_{2m} = A_{1m}\cosh\frac{m\pi b}{2a}\sinh\frac{\lambda_m b}{2} - A_{2m}\sinh\frac{m\pi b}{2a}\cosh\frac{\lambda_m b}{2}$$
$$- A_{3m}\operatorname{sech}\frac{m\pi b}{2a}\cosh\frac{\lambda_m b}{2}$$
(8.3.38c)

$$A_{1m} = \frac{m\pi}{a\lambda_m}\frac{D}{K_s G h}, \quad A_{2m} = \frac{D}{K_s G h} + \frac{1}{2}\left(\frac{a}{m\pi}\right)^2, \quad A_{3m} = \frac{ab}{4m\pi}$$
(8.3.38d)

8.3.4 SFSF Plates

Consider a rectangular plate where the edges at $y = -b/2$ and $y = b/2$ are free. The boundary conditions on these edges are

$$M_{yy}^M = M_{yy}^K = 0, \quad Q_y^M = V_y^K = 0, \quad M_{xy}^M = 0 \qquad (8.3.39)$$

where $V_y^K = Q_y^K + \partial M_{xy}^K/\partial x$ is the effective Kirchhoff shear force.

In view of Eqs. (8.3.27b), (8.3.27c), (8.3.29) and (8.3.39), the constants are found to be

$$C_{1m} = \frac{1}{B_{1m}}\left\{\Phi_m^+\tanh\frac{m\pi b}{2a} + \Psi_m^+\left[\lambda_m\tanh\frac{\lambda_m b}{2} - A_{4m}\tanh\frac{m\pi b}{2a}\right]\right\}$$
(8.3.40a)

$$C_{2m} = \frac{1}{B_{2m}}\left\{\Phi_m^-\coth\frac{m\pi b}{2a} + \Psi_m^-\left[\lambda_m\coth\frac{\lambda_m b}{2} - A_{4m}\coth\frac{m\pi b}{2a}\right]\right\}$$
(8.3.40b)

$$C_{3m} = [\Phi_m^- - \Psi_m^- A_{4m}] \left(\frac{a}{m\pi}\right)^2 \operatorname{csch}\frac{m\pi b}{2a}$$

$$+ C_{2m} \left[\frac{ab}{4m\pi} \coth\frac{m\pi b}{2a} - \left(\frac{a}{m\pi}\right)^2 \frac{(1+\nu)}{2(1-\nu)}\right]$$

$$= -\Psi_m^- \lambda_m \left(\frac{a}{m\pi}\right)^2 \coth\frac{\lambda_m b}{2} \operatorname{sech}\frac{m\pi b}{2a}$$

$$+ C_{2m} \left[\frac{1}{1-\nu}\left(\frac{a}{m\pi}\right)^2 + \frac{D}{K_s Gh} + \frac{ab}{4m\pi} \tanh\frac{m\pi b}{2a}\right.$$

$$\left.- \left(\frac{a\lambda_m}{m\pi}\right)\frac{D}{K_s Gh} \coth\frac{\lambda_m b}{2} \tanh\frac{m\pi b}{2a}\right] \quad (8.3.40c)$$

$$C_{4m} = [\Phi_m^+ - \Psi_m^+ A_{4m}] \left(\frac{a}{m\pi}\right)^2 \operatorname{sech}\frac{m\pi b}{2a}$$

$$+ C_{1m} \left[\frac{ab}{4m\pi} \tanh\frac{m\pi b}{2a} - \left(\frac{a}{m\pi}\right)^2 \frac{(1+\nu)}{2(1-\nu)}\right]$$

$$= -\Psi_m^+ \lambda_m \left(\frac{a}{m\pi}\right)^2 \tanh\frac{\lambda_m b}{2} \operatorname{csch}\frac{m\pi b}{2a}$$

$$+ C_{1m} \left[\frac{1}{1-\nu}\left(\frac{a}{m\pi}\right)^2 + \frac{D}{K_s Gh} + \frac{ab}{4m\pi} \coth\frac{m\pi b}{2a}\right.$$

$$\left.- \left(\frac{a\lambda_m}{m\pi}\right)\frac{D}{K_s Gh} \tanh\frac{\lambda_m b}{2} \coth\frac{m\pi b}{2a}\right] \quad (8.3.40d)$$

$$C_{5m} = -\frac{2}{(1-\nu)} \operatorname{csch}\frac{\lambda_m b}{2} \left[C_{2m} \sinh\frac{m\pi b}{2a} + \Psi_m^- \frac{a}{m\pi} \frac{K_s Gh}{D}\right] \quad (8.3.40e)$$

$$C_{6m} = -\frac{2}{(1-\nu)} \operatorname{sech}\frac{\lambda_m b}{2} \left[C_{1m} \cosh\frac{m\pi b}{2a} + \Psi_m^+ \frac{a}{m\pi} \frac{K_s Gh}{D}\right] \quad (8.3.40f)$$

where

$$\Phi_m^+ = \frac{M_{mxy}^K(b/2) + M_{mxy}^K(-b/2)}{2D(1-\nu)}, \quad \Phi_m^- = \frac{M_{mxy}^K(b/2) - M_{mxy}^K(-b/2)}{2D(1-\nu)} \quad (8.3.41a)$$

$$\Psi_m^+ = \frac{Q_{my}^K(b/2) + Q_{my}^K(-b/2)}{2K_s Gh}, \quad \Psi_m^- = \frac{Q_{my}^K(b/2) - Q_{my}^K(-b/2)}{2K_s Gh} \quad (8.3.41b)$$

$$B_{1m} = A_{1m} \sinh\frac{m\pi b}{2a} - A_{2m} \tanh\frac{\lambda_m b}{2} \cosh\frac{m\pi b}{2a} + A_{3m} \operatorname{sech}\frac{m\pi b}{2a} \quad (8.3.41c)$$

$$B_{2m} = A_{1m} \cosh \frac{m\pi b}{2a} - A_{2m} \coth \frac{\lambda_m b}{2} \sinh \frac{m\pi b}{2a} - A_{3m} \operatorname{csch} \frac{m\pi b}{2a}$$
(8.3.41d)

$$A_{1m} = \frac{3+\nu}{2(1-\nu)} + \left(\frac{m\pi}{a}\right)^2 \frac{D}{K_s G h}, \quad A_{2m} = \lambda_m \left(\frac{m\pi}{a}\right) \frac{D}{K_s G h}$$

$$A_{3m} = \frac{m\pi b}{4a}, \quad A_{4m} = \frac{a}{2m\pi} \left[\lambda_m^2 + \left(\frac{m\pi}{a}\right)^2\right]$$
(8.3.41e)

8.3.5 SCSS Plates

Next, consider a rectangular plate with edges $x = 0$, $x = a$, and $y = -b/2$ are simply supported, while the edge $y = b/2$ is clamped. The boundary conditions are given by

$$M_{yy}^M = M_{yy}^K = 0, \quad w_0^M = w_0^K = 0, \quad \Psi_x = 0 \qquad (8.3.42a)$$

for the simply supported edge $y = -b/2$ and

$$w_0^M = w_0^K = 0, \quad \Psi_x = \Psi_y = \frac{\partial w_0^K}{\partial y} = 0 \qquad (8.3.42b)$$

for the clamped edge $y = b/2$. In view of Eqs. (8.3.23), (8.3.24), (8.3.25), (8.3.27a) and (8.3.42a,b), one obtains

$$C_{1m} = \frac{1}{B_{1m}} \left[\frac{m\pi}{a\lambda_m} \left(\Omega_m^+ \tanh \frac{\lambda_m b}{2} + \Omega_m^- \coth \frac{\lambda_m b}{2} \right) \right.$$
$$\left. - \left(\Omega_m^+ \tanh \frac{m\pi b}{2a} + \Omega_m^- \coth \frac{m\pi b}{2a} \right) \right]$$
(8.3.43a)

$$B_{1m} = 2A_{1m} \sinh \frac{m\pi b}{2a} + A_{2m} \cosh \frac{m\pi b}{a} \operatorname{sech} \frac{m\pi b}{2a}$$
$$- A_{1m} \left(\cosh^2 \frac{m\pi b}{2a} \operatorname{csch} \frac{m\pi b}{2a} + \sinh^3 \frac{m\pi b}{2a} \operatorname{sech}^2 \frac{m\pi b}{2a} \right)$$
$$- A_{3m} \sinh \frac{m\pi b}{2a} \coth \lambda_m b$$
(8.3.43b)

$$A_{1m} = \frac{ab}{4m\pi}, \quad A_{2m} = \frac{D}{K_s G h} + \frac{1}{2} \left(\frac{a}{m\pi}\right)^2, \quad A_{3m} = \frac{2m\pi}{a\lambda_m} \frac{D}{K_s G h}$$

$$C_{2m} = C_{1m} \tanh \frac{m\pi b}{2a}$$
(8.3.43c)

C_{3m} = right hand side of Eq. (8.3.37c)   (8.3.43d)
C_{4m} = right hand side of Eq. (8.3.37d)   (8.3.43e)
C_{5m} = right hand side of Eq. (8.3.37e)   (8.3.43f)
C_{6m} = right hand side of Eq. (8.3.37f)   (8.3.43g)

where Ω_m^+ and Ω_m^- have the same meaning as given in Eq. (8.3.38a)

8.3.6 SFSS Plates

Finally, consider a rectangular plate with edges $x = 0$, $a = 0$, and $y = -b/2$ simply supported and the edge $y = b/2$ free. The boundary conditions are given by

$$M_{yy}^M = M_{yy}^K, \quad w_0^M = w_0^K = 0, \quad \Psi_x = 0 \qquad (8.3.44a)$$

for the simply supported edge $y = -b/2$ and

$$M_{yy}^M = M_{yy}^K = 0 \quad Q_y^M = V_y^K = 0, \quad M_{xy}^M = 0 \qquad (8.3.44b)$$

for the free edge $y = b/2$.

In view of Eqs. (8.3.23), (8.3.24), (8.3.27a), (8.3.27b), (8.3.32) and (8.3.44a,b), the constants are found to be

$$C_{1m} = 0, \quad C_{2m} = 0 \qquad (8.3.45a)$$

$$C_{3m} = -\lambda_m \left(\frac{a}{m\pi}\right)^2 \frac{Q_{mh}^K(b/2)}{2K_s Gh} \tanh \lambda_m b \ \text{sech} \frac{m\pi b}{2a} \qquad (8.3.45b)$$

$$C_{4m} = -\lambda_m \left(\frac{a}{m\pi}\right)^2 \frac{Q_{mn}^K(b/2)}{2K_s Gh} \tanh \lambda_m b \ \text{csch} \frac{m\pi b}{2a} \qquad (8.3.45c)$$

$$C_{5m} = -\frac{2}{1-\nu} \left(\frac{a}{m\pi}\right) \frac{Q_{my}^K(b/2)}{D} \sinh \frac{\lambda_m b}{2} \ \text{sech} \lambda_m b \qquad (8.3.45d)$$

$$C_{6m} = -\frac{2}{1-\nu} \left(\frac{a}{m\pi}\right) \frac{Q_{my}^K(b/2)}{D} \cosh \frac{\lambda_m b}{2} \ \text{sech} \lambda_m b \qquad (8.3.45e)$$

8.4 Numerical Results

The relationships developed herein can be used to furnish the deflection, rotations and stress-resultants of the Mindlin plate theory upon supplying the corresponding Kirchhoff plate solutions. This is illustrated below for the Mindlin plate deflection using the example of a uniformly loaded SCSC plate. Numerical results for SFSF plates are also presented.

8.4.1 SCSC Plates

Based on the Kirchhoff plate theory, the transverse deflection for an SCSC plate (see Figure 8.3.1) under a uniform load q_0, is given by (Mansfield 1989)

$$w_0^K = \frac{a^4}{D\pi^4} \sum_{m=1}^{\infty} \frac{q_m}{m^4} \left(1 + A_m \cosh \frac{m\pi y}{a} + B_m \frac{m\pi y}{a} \sinh \frac{m\pi y}{a}\right) \sin \frac{m\pi x}{a} \quad (8.4.1)$$

where

$$q_m = \frac{2q_0}{m\pi}[1 - (-1)^m] \quad (8.4.2a)$$

$$A_m = -\frac{1 + \frac{m\pi b}{2a}\coth\frac{m\pi b}{2a}}{\cosh\frac{m\pi b}{2a} + \frac{m\pi b}{2a}\operatorname{csch}\frac{m\pi b}{2a}} \quad (8.4.2b)$$

$$B_m = \frac{1}{\cosh\frac{m\pi b}{2a} + \frac{m\pi b}{2a}\operatorname{csch}\frac{m\pi b}{2a}} \quad (8.4.2c)$$

The substitution of Eq. (8.4.1) into Eq. (8.3.6) yields the moment sum for the Kirchhoff plate, which is

$$\mathcal{M}^K = \sum_{m=1}^{\infty} q_m \left(\frac{a}{m\pi}\right)^2 \left[1 - 2B_m \cosh \frac{m\pi y}{a}\right] \sin \frac{m\pi x}{a} \quad (8.4.3)$$

Under symmetric loading, the Mindlin deflection is symmetrical about the x–axis while the Mindlin rotation Ψ_y must take on the form of an odd function. Correspondingly, the terms in Eq. (8.3.36) become

$$\Omega_m^- = 0, \quad \Omega_m^+ = \frac{\mathcal{M}_m^K(b/2)}{K_s G h} \quad (8.4.4a,b)$$

In view of Eqs. (8.3.23), (8.3.35), (8.4.1) to (8.4.4), the deflection of the Mindlin plate is thus given by

$$w_0^M = \sum_{m=1}^{\infty} \frac{q_m}{D} \left(\frac{a}{m\pi}\right)^4 \left(1 + A_m \cosh \frac{m\pi y}{a} + B_m \frac{m\pi y}{a} \sinh \frac{m\pi y}{a}\right) \sin \frac{m\pi x}{a}$$

$$+ \sum_{m=1}^{\infty} \frac{q_m}{K_s G h} \left(\frac{a}{m\pi}\right)^2 \left[1 - 2B_m \cosh \frac{m\pi y}{a} + \left(1 - 2B_m \cosh \frac{m\pi b}{2a}\right) \times \right.$$

$$\left.\left(\bar{\xi}_m \cosh \frac{m\pi y}{a} - \xi_m \frac{ay}{2m\pi} \sinh \frac{m\pi y}{a}\right)\right] \sin \frac{m\pi x}{a} \quad (8.4.5)$$

where

$$\xi_m = \frac{1}{B_{1m}}\left(\tanh\frac{m\pi b}{2a}\cosh\frac{\lambda_m b}{2} - \frac{m\pi}{a\lambda_m}\sinh\frac{\lambda_m b}{2}\right) \quad (8.4.6a)$$

$$\bar{\xi}_m = \xi_m\frac{ab}{4m\pi}\tanh\frac{m\pi b}{2a} - \text{sech}\frac{m\pi b}{2a} \quad (8.4.6b)$$

$$B_{1m} = A_{1m}\cosh\frac{m\pi b}{2a}\sinh\frac{\lambda_m b}{2} - A_{2m}\sinh\frac{m\pi b}{2a}\cosh\frac{\lambda_m b}{2}$$
$$- A_{3m}\,\text{sech}\frac{m\pi b}{2a}\cosh\frac{\lambda_m b}{2} \quad (8.4.6c)$$

$$A_{1m} = \frac{m\pi}{a\lambda_m}\frac{D}{K_sGh},\quad A_{2m} = \frac{D}{K_sGh} + \frac{1}{2}\left(\frac{a}{m\pi}\right)^2,\quad A_{3m} = \frac{ab}{4m\pi}$$
$$(8.4.6d)$$

Table 8.4.1 contains the non-dimensionalized maximum deflection, $\bar{w} = w_0^M(a/2,0)D/(q_0a^4)$ of square plates with a clamped boundary on two sides (SCSC) for two different thickness-to-side ratios. It can be observed that the deflection values are in agreement with those obtained using *ABAQUS* (1997), thus confirming the correctness of the derived relationship. The results, however, differ from those determined by Cooke and Levinson (1983), which are in error.

The non-dimensionalized stress resultants of square SCSC plates are presented in Table 8.4.2 for different thicknesses. The bending moments and shear forces are nondimensionalized as follows:

$$\bar{M} = 10\times\frac{M}{q_0a^2},\quad \bar{Q} = \frac{Q}{q_0a} \quad (8.4.7)$$

Table 8.4.1. Maximum deflection parameter $\bar{w}$ of uniformly loaded square SCSC Mindlin plates ($\nu = 0.3$, $K_s = 5/6$, and $m = 40$).

$\frac{h}{a}$	Cooke and Levinson (1983)	ABAQUS [1]	Eq.(8.4.5)
0.1	0.00213	0.00221	0.00221
0.2	0.00276	0.00302	0.00302

[1] Solution obtained with 40 × 40 mesh of S8R shell elements.

Table 8.4.2. Non-dimensionalized stress resultants of uniformly loaded square SCSC Mindlin plates ($\nu = 0.3$ and $K_s = 5/6$).

$(\frac{x}{a}, \frac{y}{a})$	Resultant	CPT	$\frac{h}{a} = 0.02$	$\frac{h}{a} = 0.1$	$\frac{h}{a} = 0.2$
(0.5, 0.0)	$\bar{M}_{xx}$	0.244	0.244	0.258	0.292
(0.5, 0.0)	$\bar{M}_{yy}$	0.332	0.332	0.333	0.331
(0.5, 0.5)	$\bar{M}_{yy}$	0.698	0.698	0.680	0.627
(1.0, 0.0)	$\bar{Q}_x$	0.239	0.240	0.243	0.251
(0.5, 0.5)	$\bar{Q}_y$	0.516	0.513	0.500	0.475

8.4.2 SFSF Plates

Numerical results of the deflections and stress resultants for SFSF plates are included in Tables 8.4.3 and 8.4.4. The same nondimensionalizations used for the SCSC plates are also used here.

Table 8.4.3. Maximum deflection parameter $\bar{w} \times 10$ of uniformly loaded square SFSF Mindlin plates ($\nu = 0.3$, $K_s = 5/6$, and $m = 40$).

$\frac{h}{a}$	At the center of the plate			At mid-span of free edge		
	Dong (1993)	Dong (1994)	Present Results	Dong (1993)	Dong (1994)	Present Results
0.10	0.1346	0.1340	0.1346	0.1562	0.1549	0.1560
0.15	0.1391	0.1385	0.1391	0.1617	0.1607	0.1616
0.20	0.1454	0.1448	0.1454	0.1690	0.1679	0.1690
0.25	0.1535	0.1528	0.1535	0.1781	0.1771	0.1781
0.30	0.1633	0.1627	0.1633	0.1890	0.1879	0.1889

Table 8.4.4. Non-dimensionalized stress resultants of uniformly loaded square SCSC Mindlin plates ($\nu = 0.3$ and $K_s = 5/6$).

$(\frac{x}{a}, \frac{y}{a})$	Resultant	CPT	$\frac{h}{a} = 0.02$	$\frac{h}{a} = 0.1$	$\frac{h}{a} = 0.2$
(0.5, 0.0)	$\bar{M}_{xx}$	0.123	0.123	0.122	0.123
(0.5, 0.0)	$\bar{M}_{yy}$	0.271	0.268	0.256	0.237
(1.0, 0.0)	$\bar{Q}_x$	0.464	0.463	0.460	0.457

Problems

8.1 Establish the relations in Eqs. (8.3.3) and (8.3.4).

8.2 Verify the relation in Eqs. (8.3.10).

8.3 Verify the relation in Eqs. (8.3.11).

8.4 From Eq. (7.2.6a) we have

$$w_0^M = w_0^K + \frac{\mathcal{M}^K}{K_s Gh} + \Psi - \Phi \qquad (i)$$

where $\Psi(x,y)$ and $\Phi(x,y)$ are harmonic and biharmonic functions, respectively (i.e., Ψ satisfies the Laplace equation $\nabla^2 \Psi = 0$ and Φ satisfies the biharmonic equation $\nabla^4 \Phi = 0$). Suppose that $\Omega(x,y) = \partial \phi_x / \partial y - \partial \phi_y / \partial x$ is the solution of the equation [see Eq. (i) of Problem 6.6]

$$\nabla^2 \Omega = c^2 \Omega, \quad c^2 = \frac{2K_s Gh}{D(1-\nu)} \qquad (ii)$$

Then use Eqs. (6.3.13d,e), (7.2.4), and (i) in Eqs. (iii) and (iv) of Problem 6.5 to show that

$$\phi_x^M = -\frac{\partial w_0^K}{\partial x} + \frac{\partial}{\partial x}\left[\frac{D}{K_s Gh}\left(\nabla^2 \Phi\right) + \Phi - \Psi\right] + \frac{1}{c^2}\frac{\partial \Omega}{\partial y} \qquad (iii)$$

$$\phi_y^M = -\frac{\partial w_0^K}{\partial y} + \frac{\partial}{\partial y}\left[\frac{D}{K_s Gh}\left(\nabla^2 \Phi\right) + \Phi - \Psi\right] - \frac{1}{c^2}\frac{\partial \Omega}{\partial x} \qquad (iv)$$

8.5 Use Eqs. (i), (iii), and (iv) of Problem 8.4 in Eq. (6.3.13a-d) to establish the following relationships:

$$M_{xx}^M = M_{xx}^K - D(1-\nu)\frac{\partial}{\partial y}\left(\frac{\partial \Lambda}{\partial y} - \frac{1}{c^2}\frac{\partial \Omega}{\partial x}\right) + D\nabla^2 \Phi \qquad (i)$$

$$M_{yy}^M = M_{yy}^K - D(1-\nu)\frac{\partial}{\partial x}\left(\frac{\partial \Lambda}{\partial x} + \frac{1}{c^2}\frac{\partial \Omega}{\partial y}\right) + D\nabla^2 \Phi \qquad (ii)$$

$$M_{xy}^M = M_{xy}^K + D(1-\nu)\left[\frac{\partial^2 \Lambda}{\partial x \partial y} + \frac{1}{2c^2}\left(\frac{\partial^2 \Omega}{\partial y^2} - \frac{\partial^2 \Omega}{\partial x^2}\right)\right] \qquad (iii)$$

$$Q_x^M = Q_x^K + D\frac{\partial}{\partial x}\left(\nabla^2 \Phi\right) + \frac{D(1-\nu)}{2}\frac{\partial \Omega}{\partial y} \qquad (iv)$$

$$Q_y^M = Q_y^K + D\frac{\partial}{\partial y}\left(\nabla^2 \Phi\right) - \frac{D(1-\nu)}{2}\frac{\partial \Omega}{\partial x} \qquad (v)$$

where Ω is the function defined in Problem 8.4 and

$$\Lambda = \frac{D}{K_s Gh}\left(\nabla^2 \Phi\right) + \Phi - \Psi \qquad (vi)$$

Chapter 9

Bending Relationships for Circular and Annular Plates

In this chapter exact relationships between the bending solutions of the classical plate theory (CPT) and the Mindlin (FSDT) and Reddy (TSDT) plate theories for circular and annular plates are developed. Since both the CPT and FSDT are fourth-order theories, the relationships are algebraic. However, since the TSDT is a sixth-order theory and the CPT is a fourth-order theory, the exact relationships between deflections, slopes, moments, and shear forces of the two theories can only be developed by solving an additional second-order differential equation. Here, a second-order differential equation in terms of the transverse shear force Q_r is developed. Upon having the solution of this equation, the exact relationships between the deflections, slopes, bending moments, and shear forces of the two theories (CPT and TSDT) are established.

9.1 Governing Equations

For axisymmetric bending of circular and annular plates, it is expedient to formulate the problem in the polar coordinate system. The r coordinate is taken radially outward from the center of the plate, the z coordinate is taken along the thickness (or height) of the plate and the θ coordinate is taken along a circumference of the plate (see Figure 9.1.1). In a general case where the applied loads and geometric boundary conditions are not axisymmetric, the displacements (u_r, u_θ, w) along the coordinates (r, θ, z) are functions of r, θ, and z. Here, we assume that the applied loads and boundary conditions are independent of the θ coordinate, i.e. axisymmetric, so that the displacement u_θ is identically zero and (u_r, w) are only functions of r and z. The displacement fields of the three theories (CPT, FSDT, and TSDT) are similar to those in Eqs. (6.1.1a,b), (6.1.2a-c), and (6.1.4a-c).

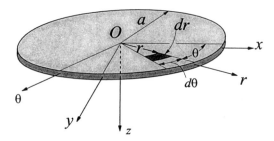

Figure 9.1.1. A circular plate with rectangular (x, y, z) and cylindrical (r, θ, z) coordinate systems.

Based on the polar coordinate system (r, θ), the equations of equilibrium and stress resultant-displacement relations of the CPT, FSDT, and TSDT are summarized below (see Reddy 1999a, Reddy and Wang 1997) for axisymmetric bending and constant material and geometric properties (see Figure 9.1.2 for the meaning of the stress resultants per unit length of a general circular plate in polar coordinates).

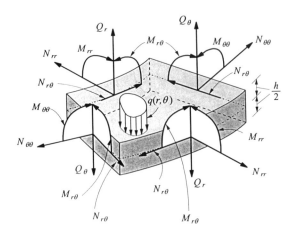

Figure 9.1.2. Forces and moments of a circular plate.

BENDING RELATIONSHIPS FOR CIRCULAR AND ANNULAR PLATES

CPT:

$$\frac{d}{dr}\left(rQ_r^K\right) = -rq, \quad rQ_r^K \equiv \frac{d}{dr}\left(rM_{rr}^K\right) - M_{\theta\theta}^K \tag{9.1.1}$$

$$M_{rr}^K = -D\left(\frac{d^2w_0^K}{dr^2} + \frac{\nu}{r}\frac{dw_0^K}{dr}\right), \quad M_{\theta\theta}^K = -D\left(\nu\frac{d^2w_0^K}{dr^2} + \frac{1}{r}\frac{dw_0^K}{dr}\right) \tag{9.1.2}$$

FSDT:

$$\frac{d}{dr}\left(rQ_r^M\right) = -rq, \quad rQ_r^M = \frac{d}{dr}\left(rM_{rr}^M\right) - M_{\theta\theta}^M \tag{9.1.3}$$

$$M_{rr}^M = D\left(\frac{d\phi_r^M}{dr} + \frac{\nu}{r}\phi_r^M\right), \quad M_{\theta\theta}^M = D\left(\nu\frac{d\phi_r^M}{dr} + \frac{1}{r}\phi_r^M\right) \tag{9.1.4a}$$

$$Q_r^M = K_s Gh\left(\phi_r^M + \frac{dw_0^M}{dr}\right) \tag{9.1.4b}$$

TSDT:

$$r\left(Q_r^R - \frac{4}{h^2}R_r^R\right) = \frac{d}{dr}\left(rM_{rr}^R - \frac{4}{3h^2}rP_{rr}^R\right) - \left(M_{\theta\theta}^R - \frac{4}{3h^2}P_{\theta\theta}^R\right) \tag{9.1.5a}$$

$$\frac{d}{dr}\left(rQ_r^R - \frac{4}{h^2}rR_r^R\right) + \frac{4}{3h^2}\left[\frac{d^2}{dr^2}\left(rP_{rr}^R\right) - \frac{dP_{\theta\theta}^R}{dr}\right] = -rq \tag{9.1.5b}$$

$$M_{rr}^R = \frac{4D}{5}\left(\frac{d\phi_r^R}{dr} + \frac{\nu}{r}\phi_r^R\right) - \frac{D}{5}\left(\frac{d^2w_0^R}{dr^2} + \frac{\nu}{r}\frac{dw_0^R}{dr}\right) \tag{9.1.6a}$$

$$M_{\theta\theta}^R = \frac{4D}{5}\left(\nu\frac{d\phi_r^R}{dr} + \frac{1}{r}\phi_r^R\right) - \frac{D}{5}\left(\nu\frac{d^2w_0^R}{dr^2} + \frac{1}{r}\frac{dw_0^R}{dr}\right) \tag{9.1.6b}$$

$$P_{rr}^R = \frac{4h^2 D}{35}\left(\frac{d\phi_r^R}{dr} + \frac{\nu}{r}\phi_r^R\right) - \frac{h^2 D}{28}\left(\frac{d^2w_0^R}{dr^2} + \frac{\nu}{r}\frac{dw_0^R}{dr}\right) \tag{9.1.6c}$$

$$P_{\theta\theta}^R = \frac{4h^2 D}{35}\left(\nu\frac{d\phi_r^R}{dr} + \frac{1}{r}\phi_r^R\right) - \frac{h^2 D}{28}\left(\nu\frac{d^2w_0^R}{dr^2} + \frac{1}{r}\frac{dw_0^R}{dr}\right) \tag{9.1.6d}$$

$$Q_r^R = \frac{2Gh}{3}\left(\phi_r^R + \frac{dw_0^R}{dr}\right), \quad R_r^R = \frac{Gh^3}{30}\left(\phi_r^R + \frac{dw_0^R}{dr}\right) \tag{9.1.6e}$$

9.2 Relationships Between CPT and FSDT
9.2.1 General Relationships

The deflection, bending moment and shear force of the FSDT can be expressed in terms of the corresponding quantities of CPT for axisymmetric bending of isotropic circular plates. The relationships are established using load equivalence (Reddy and Wang 1997).

We introduce the moment sum

$$\mathcal{M} = \frac{M_{rr} + M_{\theta\theta}}{1+\nu} \tag{9.2.1}$$

Using Eqs. (9.1.2a) and (9.1.2b) in Eq. (9.2.1), we can show that

$$\mathcal{M}^K = -D\left(\frac{d^2 w_0^K}{dr^2} + \frac{1}{r}\frac{dw_0^K}{dr}\right) = -D\frac{1}{r}\frac{d}{dr}\left(r\frac{dw_0^K}{dr}\right) \tag{9.2.2}$$

and

$$\frac{1}{r}\frac{d}{dr}\left(r\frac{d\mathcal{M}^K}{dr}\right) = -q \tag{9.2.3}$$

We can establish the following equality using the definition (9.2.1) and Eqs. (9.1.2a) and (9.1.2b):

$$r\frac{d\mathcal{M}^K}{dr} = \frac{d}{dr}\left(rM_{rr}^K\right) - M_{\theta\theta}^K = rQ_r^K \tag{9.2.4}$$

Similarly, we have

$$\mathcal{M}^M = D\left(\frac{d\phi_r^M}{dr} + \frac{1}{r}\phi_r^M\right) = D\frac{1}{r}\frac{d}{dr}\left(r\phi_r^M\right) \tag{9.2.5}$$

and

$$\frac{1}{r}\frac{d}{dr}\left(r\frac{d\mathcal{M}^M}{dr}\right) = -q \tag{9.2.6}$$

$$r\frac{d\mathcal{M}^M}{dr} = \frac{d}{dr}\left(rM_{rr}^M\right) - M_{\theta\theta}^M = rQ_r^M \tag{9.2.7}$$

From Eqs. (9.1.1a), (9.1.1b) and (9.1.3b), it follows that

$$rQ_r^M = rQ_r^K + C_1 \tag{9.2.8}$$

and from Eqs. (9.2.4), (9.2.7) and (9.2.8), we have

$$r\frac{d\mathcal{M}^M}{dr} = r\frac{d\mathcal{M}^K}{dr} + C_1 \tag{9.2.9}$$

or

$$\mathcal{M}^M = \mathcal{M}^K + C_1 \log r + C_2 \tag{9.2.10}$$

where C_1 and C_2 are constants of integration.

Next, from Eqs. (9.2.4), (9.2.7) and (9.2.10), we have

$$\phi_r^M = -\frac{dw_0^K}{dr} + \frac{C_1 r}{4D}(2\log r - 1) + \frac{C_2 r}{2D} + \frac{C_3}{rD} \tag{9.2.11}$$

In view of Eqs. (9.1.2a), (9.1.2b), (9.1.4a), (9.1.4b) and (9.2.11), one can readily obtain the following bending moment relationships

$$M_{rr}^M = M_{rr}^K + C_1\left(\frac{1+\nu}{2}\log r + \frac{1-\nu}{4}\right) + C_2\frac{1+\nu}{2} - C_3\frac{1-\nu}{r^2} \tag{9.2.12a}$$

$$M_{\theta\theta}^M = M_{\theta\theta}^K + C_1\left(\frac{1+\nu}{2}\log r - \frac{1-\nu}{4}\right) + C_2\frac{1+\nu}{2} + C_3\frac{1-\nu}{r^2} \tag{9.2.12b}$$

Finally, from Eqs. (9.1.4c), (9.2.10) and (9.2.11), we obtain

$$\frac{dw_0^M}{dr} = -\phi_r^M + \frac{1}{K_sGh}\left(Q_r^K + \frac{C_1}{r}\right) \tag{9.2.13}$$

and noting that $Q_r^K = d\mathcal{M}^K/dr$, we have

$$w_0^M = w_0^K + \frac{\mathcal{M}^K}{K_sGh} + \frac{C_1 r^2}{4D}(1-\log r) + \frac{C_1}{K_sGh}\log r - \frac{C_2 r^2}{4D} - \frac{C_3 \log r}{D} + \frac{C_4}{D} \tag{9.2.14}$$

The four constants of integration are determined using the boundary conditions. The boundary conditions for various cases are given below.

Free edge

$$rQ_r^M = rQ_r^K = 0, \quad rM_{rr}^M = rM_{rr}^K = 0 \tag{9.2.15}$$

Simply supported edge

$$w_0^M = w_0^K = 0, \quad rM_{rr}^M = rM_{rr}^K = 0 \tag{9.2.16}$$

Clamped edge

$$w_0^M = w_0^K = 0, \quad \phi_r^M = \frac{dw_0^K}{dr} = 0 \tag{9.2.17}$$

Solid circular plate at r = 0 (i.e. at the plate center)

$$\phi_r^M = \frac{dw_0^K}{dr} = 0, \quad C_1 = 0 \tag{9.2.18}$$

Note that the first condition in Eq. (9.2.18) is due to symmetry, and the second condition follows directly from Eq. (9.2.8) with $r = 0$.

In the sequel, the constants C_1 to C_4 for axisymmetric plates with different boundary conditions are determined. First consider statically determinate plate problems, i.e. (1) circular plates with (a) simply supported edge (S plate) and (b) clamped edge (C plate), (2) annular plates with (a) one edge free and the other simply supported (F-S or S-F plate) and (b) one edge free and the other edge clamped (F-C or C-F) plate. By using the appropriate boundary conditions and substituting into Eqs. (9.2.8), (9.2.11), (9.2.12a) and (9.2.14), it is a straightforward matter to show that the constants have the same form for this group of statically determinate problems. Thus

$$C_1 = C_2 = C_3 = 0 \text{ and } C_4 = -\frac{\bar{\mathcal{M}}^K}{K_s G h} \tag{9.2.19}$$

where $\bar{\mathcal{M}}^K$ is the Marcus moment at the simply supported or clamped edge of the Kirchhoff plate and is given as follows:

Simply supported edge of S and F-S plates

$$\bar{\mathcal{M}}^K = \mathcal{M}_0^K = \frac{M_{\theta\theta}^K(R_0)}{1+\nu} = -\frac{D(1-\nu)}{R_0}\frac{dw_0^K}{dr}\bigg|_{R_0} = \frac{D(1-\nu)}{\nu}\frac{d^2 w_0^K}{dr^2}\bigg|_{R_0} \tag{9.2.20a}$$

Simply supported edge of S-F plate

$$\bar{\mathcal{M}}^K = \mathcal{M}_i^K = \frac{M_{\theta\theta}^K(R_i)}{1+\nu} = -\frac{D(1-\nu)}{R_0}\frac{dw_0^K}{dr}\bigg|_{R_i} = \frac{D(1-\nu)}{\nu}\frac{d^2 w_0^K}{dr^2}\bigg|_{R_i} \tag{9.2.20b}$$

Clamped edge of C and F-C plates

$$\bar{\mathcal{M}}^K = \mathcal{M}_0^K = -D\frac{d^2 w_0^K}{dr^2}\bigg|_{R_0} \tag{9.2.20c}$$

Clamped edge of C-F plate

$$\bar{\mathcal{M}}^K = \mathcal{M}_i^K = -D\frac{d^2 w_0^K}{dr^2}\bigg|_{R_i} \tag{9.2.20d}$$

In the expressions for $\mathcal{M}^K$ given above, R_0 is the radius of the circular plate or the outer radius of the annular plate whereas R_i is the inner radius of the annular plate. In order to distinguish between the F-S and S-F plates, as well as the F-C and C-F plates (where the two letters denote the boundary conditions at the inner and the outer edges, respectively), the subscripts 'i' and '0' are used to represent values at the inner and outer edges, respectively.

Next we consider statically indeterminate problems.

For an annular plate with simply supported inner and outer edges (S-S plate), the boundary conditions are given in Eqs. (9.2.16a) and (9.2.16b) for $r = R_0$ and $r = R_i$. Substitution of Eqs. (9.2.12a) and (9.2.14) into these boundary conditions yields

$$C_0 = \frac{(3+\nu)(R_i^2 - R_0^2)}{8(1+\nu)} - \frac{D}{K_s G h} \log \frac{R_0}{R_i}$$
$$+ \frac{(1+\nu)R_0^2 R_i^2}{2(1-\nu)(R_i^2 - R_0^2)} \left(\log \frac{R_0}{R_i}\right)^2 \tag{9.2.21a}$$

$$C_1 = D\left[\frac{\mathcal{M}_0^K - \mathcal{M}_i^K}{K_s G h}\right](C_0)^{-1} \tag{9.2.21b}$$

$$C_2 = C_1\left[\frac{R_0^2 \log R_0 - R_i^2 \log R_i}{R_i^2 - R_0^2} - \frac{(1-\nu)}{2(1+\nu)}\right] \tag{9.2.21c}$$

$$C_3 = C_1\left[\frac{(1+\nu)R_0^2 R_i^2}{2(1-\nu)(R_i^2 - R_0^2)} \log \frac{R_0}{R_i}\right] \tag{9.2.21d}$$

$$C_4 = C_1\left[-\frac{(3+\nu)(R_i^2 + R_0^2)}{16(1+\nu)} - \frac{D}{2K_s G h} \log(R_0 R_i)\right.$$
$$\left.+ \frac{R_0^2 R_i^2}{4(R_i^2 - R_0^2)} \left(\frac{1+\nu}{1-\nu} \log(R_0 R_i) + 1\right) \log \frac{R_0}{R_i}\right] - \frac{\mathcal{M}_0^K + \mathcal{M}_i^K}{2K_s G h}$$
$$\tag{9.2.21e}$$

For an annular plate with clamped inner and outer edges (C-C plate), the boundary conditions are given in Eqs. (9.2.17a) and (9.2.27b) for $r = R_0$ and $r = R_i$. Substitution of Eqs. (9.2.11) and (9.2.14) into these boundary conditions yields

$$C_0 = \frac{(R_i^2 - R_0^2)}{8} - \frac{D}{K_s Gh} \log \frac{R_0}{R_i} - \frac{R_0^2 R_i^2}{2(R_i^2 - R_0^2)} \left(\log \frac{R_0}{R_i} \right)^2 \quad (9.2.22\text{a})$$

$$C_1 = D \left[\frac{\mathcal{M}_0^K - \mathcal{M}_i^K}{K_s Gh} \right] (C_0)^{-1} \quad (9.2.22\text{b})$$

$$C_2 = C_1 \left[\frac{R_0^2 \log R_0 - R_i^2 \log R_i}{R_i^2 - R_0^2} + \frac{1}{2} \right] \quad (9.2.22\text{c})$$

$$C_3 = C_1 \left[-\frac{R_0^2 R_i^2}{2(R_i^2 - R_0^2)} \log \frac{R_0}{R_i} \right] \quad (9.2.22\text{d})$$

$$C_4 = C_1 \left\{ -\frac{(R_i^2 + R_0^2)}{16} - \frac{D}{2K_s Gh} \log(R_0 R_i) \right.$$
$$\left. + \frac{R_0^2 R_i^2 [1 - \log(R_0 R_i)]}{4(R_i^2 - R_0^2)} \log \frac{R_0}{R_i} \right\} - \frac{\mathcal{M}_0^K + \mathcal{M}_i^K}{2K_s Gh} \quad (9.2.22\text{e})$$

For an annular plate with simply supported inner edge and clamped outer edge (S-C plate), the boundary conditions are given in Eq. (9.2.17) for $r = R_0$ and Eq. (9.2.16) for $r = R_i$. The substitution of Eqs. (9.2.11), (9.2.12a), (9.2.14) into these boundary conditions yields

$$C_0 = \left(\frac{R_0^2 - R_i^2}{8} \right) \left[(1 - \nu) R_0^2 + (3 + \nu) R_i^2 \right]$$
$$- \frac{R_i^2 R_0^2}{8} \left[2 - (1 + \nu) \log \frac{R_0}{R_i} \right] \log \frac{R_0}{R_i}$$
$$+ \frac{D}{K_s Gh} \log \frac{R_0}{R_i} \left[(1 + \nu) R_i^2 + (1 - \nu) R_0^2 \right] \quad (9.2.23\text{a})$$

$$C_1 = D \left[\frac{\mathcal{M}_i^K - \mathcal{M}_0^K}{K_s Gh} \right] \left[(1 + \nu) R_i^2 + (1 - \nu) R_0^2 \right] (C_0)^{-1} \quad (9.2.23\text{b})$$

$$C_2 = C_1 \left[\frac{(1 - \nu)(R_0^2 - R_i^2) - 2(1 - \nu) R_0^2 \log R_0 - 2(1 + \nu) R_i^2 \log R_i}{2(1 + \nu) R_i^2 + 2(1 - \nu) R_0^2} \right] \quad (9.2.23\text{c})$$

$$C_3 = C_1 \frac{R_0^2 R_i^2}{2} \left[\frac{1 + (1 + \nu) \log \frac{R_i}{R_0}}{(1 + \nu) R_i^2 + (1 - \nu) R_0^2} \right] \quad (9.2.23\text{d})$$

$$C_4 = -C_1 \left\{ \frac{(R_i^2 + R_0^2)\left[(3+\nu)R_i^2 + (1-\nu)R_0^2\right] + 4\nu R_0^2 R_i^2 \log\frac{R_i}{R_0}}{16\left[(1+\nu)R_i^2 + (1-\nu)R_0^2\right]} \right.$$

$$+ \frac{D}{2K_s Gh} \log(R_0 R_i) - \frac{R_0^2 R_i^2}{4} \left[\frac{1 + (1+\nu)\log\frac{R_i}{R_0}}{(1+\nu)R_i^2 + (1-\nu)R_0^2}\right]$$

$$\left. \times \log(R_0 R_i) \right\} - \frac{\mathcal{M}_0^K + \mathcal{M}_i^K}{2K_s Gh} \qquad (9.2.23e)$$

For an annular plate with clamped inner edge and simply supported outer edge (C-S plate), the constants C_1 to C_4 are the same as those above except for the interchange in the subscripts 'i' and '0'.

In view of the foregoing expressions for the constants C_1 to C_4, the relationships for the shear forces, bending moments, deflection gradient and deflection may be obtained, respectively from Eqs. (9.2.8), (9.2.12a), (9.2.12b), (9.2.11) and (9.2.14). These relationships are summarized below:

S, C, F-S, F-C, S-F and C-F plates:

$$Q_r^M = Q_r^K \qquad (9.2.24)$$
$$M_{rr}^M = M_{rr}^K \qquad (9.2.25)$$
$$M_{\theta\theta}^M = M_{\theta\theta}^K \qquad (9.2.26)$$
$$\phi_r^M = -\frac{dw_0^K}{dr} \qquad (9.2.27)$$
$$w_0^M = w_0^K + \frac{\mathcal{M}^K - \bar{\mathcal{M}}^K}{\mathcal{M}^2 Gh} \qquad (9.2.28)$$

where

$$\mathcal{M}^K = \begin{cases} \mathcal{M}_0^K & \text{for S, C, F-S, F-C plates} \\ \mathcal{M}_i^K & \text{for S-F, C-F plates} \end{cases}$$

S-S, C-C, S-C, C-S plates:

$$Q_r^M = Q_r^K + \frac{C_1}{r} \qquad (9.2.29)$$

$$M_{rr}^M = M_{rr}^K + C_1\left(\frac{1+\nu}{2}\log r + \frac{1-\nu}{4}\right) + C_2\frac{1+\nu}{2} - C_3\frac{1-\nu}{r^2} \qquad (9.2.30)$$

$$M_{\theta\theta}^M = M_{\theta\theta}^K + C_1\left(\frac{1+\nu}{2}\log r - \frac{1-\nu}{4}\right) + C_2\frac{1+\nu}{2} + C_3\frac{1-\nu}{r^2} \qquad (9.2.31)$$

$$\phi_r^M = -\frac{dw_0^K}{dr} + \frac{C_1 r}{4D}(2\log r - 1) + \frac{C_2 r}{2D} + \frac{C_3}{rD} \qquad (9.2.32)$$

$$w_0^M = w_0^K + \frac{\mathcal{M}^K}{K_s Gh} + \frac{C_1 r^2}{4D}(1 - \log r) + \frac{C_1}{K_s Gh}\log r$$
$$- \frac{C_2 r^2}{4D} - \frac{C_3 \log r}{D} + \frac{C_4}{D} \qquad (9.2.33)$$

where the constants C_1 to C_4 are given in Eqs. (9.2.21a-e), (9.2.22a-e) and (9.2.23a-e) for S-S, C-C and S-C plates, respectively. Those for C-S plates are obtained by interchanging the subscripts 'i' and '0' in Eqs. (9.2.23a-e).

It may be observed that the stress-resultants of Mindlin plates and the corresponding Kirchhoff plates are equal to each other for statically determinate plates. In such plates, the deflection component w_s due to transverse shear deformation is given by

$$w_s = \frac{\mathcal{M}^K}{K_s Gh} - C \qquad (9.2.34)$$

where C is a constant given by the Marcus moment at the simply supported edge divided by $K_s Gh$. However, for the statically indeterminate plates, the stress-resultants of these Mindlin plates are obviously not equal to their Kirchhoff counterparts. Thus Eq. (9.2.34) no longer applies and the more complicated form given by Eq. (9.2.33) is necessary. It is to be noted that Panc (1975) and Barrett and Ellis (1988) presented the more restrictive expression for w_s given in Eq. (9.2.34) for general rotationally symmetric bending of axisymmetric plates. Although their restrictive form is correct for statically determinate plates, it is not correct for statically indeterminate plates as shown by the foregoing derivations.

9.2.2 Examples

The use of the foregoing relationships is illustrated with the following circular plate examples.

Circular plates under axisymmetric partial uniform load over inner portion

Consider a circular plate under a uniformly distributed load q_0 over the inner portion of the plate $r \leq \alpha R$ ($0 < \alpha \leq 1$) as shown in Figure

9.2.1. The Kirchhoff solution for such loaded plates is given by (see Szilard 1974 and Reddy 1999). The case of uniformly distributed load on the entire plate is obtained by setting $\alpha = 1$ in the first part of the solution and omitting the second.

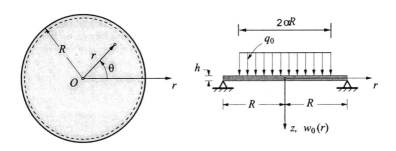

Figure 9.2.1. Circular plate under partial uniformly distributed load.

For $0 \leq r \leq \alpha R$:

$$w_0^K = \frac{q_0 R^4}{64 D} \left\{ \left(\frac{r}{R}\right)^4 + \alpha^2 \left[4 - 5\alpha^2 + 4\left(2 + \alpha^2\right) \log \alpha\right] \right.$$
$$\left. + 2\frac{\alpha^2}{1+\nu}\left[1 - \left(\frac{r}{R}\right)^2\right]\left[4 - (1-\nu)\alpha^2 - 4(1+\nu)\log\alpha\right] \right\}$$

(9.2.35a)

for simply supported plate, and

$$w_0^K = \frac{q_0 R^4}{64 D}\left[\left(\frac{r}{R}\right)^4 + \alpha^2\left(4 - 3\alpha^2 + 4\alpha^2 \log \alpha\right)\right.$$
$$\left. - 2\alpha^2 \left(\frac{r}{R}\right)^2 \left(\alpha^2 - 4 \log \alpha\right)\right]$$

(9.2.35b)

for clamped plate.

For $\alpha R \leq r \leq R$:

$$w_0^K = \frac{q_0 \alpha^2 R^4}{32 D}\left\{2\left[\alpha^2 + 2\left(\frac{r}{R}\right)^2\right]\log\frac{r}{R}\right.$$

$$+\frac{2(3+\nu)-\alpha^2(1-\nu)}{1+\nu}\left[1-\left(\frac{r}{R}\right)^2\right]\right\} \quad (9.2.36a)$$

for simply supported plate, and

$$w_0^K = \frac{q_0\alpha^2 R^4}{32D}\left\{2\left[\alpha^2+2\left(\frac{r}{R}\right)^2\right]\log\left(\frac{r}{R}\right)+(2+\alpha^2)\left[1-\left(\frac{r}{R}\right)^2\right]\right\} \quad (9.2.36b)$$

for clamped plate.

Substituting Eqs. (9.2.35a,b) and (9.2.36a,b) into Eqs. (9.2.2) and (9.2.28) yields the corresponding Mindlin plate deflection

$$w_0^M = w_0^K + \begin{cases} \frac{q_0 R^2}{4K_s Gh}\left[\alpha^2(1-2\log\alpha)-\left(\frac{r}{R}\right)^2\right], & 0 \le r \le \alpha R \\ \frac{q_0 R^2 \alpha^2}{2K_s Gh}\log\frac{R}{r}, & \alpha R \le r \le R \end{cases} \quad (9.2.37)$$

Note that the deflection component due to the transverse shear deformation given in Eq. (9.2.37) applies to both simply supported and clamped plates. In fact, the deflection component due to shear deformation is the same regardless of the supported edge being simply supported or clamped.

Circular plates under axisymmetric linearly varying load

Consider a circular plate under an axisymmetric linearly varying load $q = q_0(1-r/R)$ (set $q_1 = 0$ in Figure 9.2.2). The Kirchhoff solution for such loaded plates is given by (see Szilard 1974 and Reddy 1999).

$$w_0^K = \frac{q_0 R^4}{14400D}\left[\frac{3(183+43\nu)}{1+\nu} - \frac{10(71+29\nu)}{1+\nu}\left(\frac{r}{R}\right)^2 \right.$$
$$\left. +255\left(\frac{r}{R}\right)^4 - 64\left(\frac{r}{R}\right)^5\right] \quad (9.2.38a)$$

for simply supported plate, and

$$w_0^K = \frac{q_0 R^4}{14400D}\left[129 - 290\left(\frac{r}{R}\right)^2 + 225\left(\frac{r}{R}\right)^4 - 64\left(\frac{r}{R}\right)^5\right] \quad (9.2.38b)$$

for clamped plate.

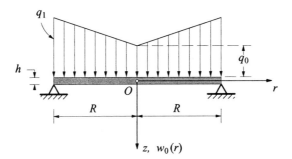

Figure 9.2.2. Circular plate under axisymmetric linearly varying load.

Substituting Eqs. (9.2.38) into Eqs. (9.2.6) and (9.2.28) yields the corresponding Mindlin plate deflection

$$w_0^M = w_0^K + \frac{q_0 R^4}{36 K_s Gh}\left[5 - 9\left(\frac{r}{R}\right)^2 + 4\left(\frac{r}{R}\right)^3\right] \qquad (9.2.39)$$

and the maximum deflection occurs at $r = 0$

$$w_{max}^M = w_{max}^K + \frac{5 q_0 R^4}{36 K_s Gh} \qquad (9.2.40)$$

9.3 Relationships between CPT and TSDT

9.3.1 General Relationships

Here we develop the relationships between the bending solutions of CPT and TSDT. At the outset, we note that both the classical and the first-order shear deformation plate theories are fourth order theories, whereas Reddy's third-order shear deformation plate theory is a sixth-order theory. The order referred to here is the total order of all equations of equilibrium expressed in terms of the generalized displacements. The third-order plate theory is governed by a fourth order differential equation in w_0^R and a second order equation in ϕ_r^R. Therefore, the relationships between the solutions of two different order theories can only be established by solving an additional second-order equation.

First we note that Eqs. (9.1.5a) and (9.1.5b) together yield

$$-\frac{d^2}{dr^2}\left(rM_{rr}^R\right) + \frac{dM_{\theta\theta}^R}{dr} = rq \qquad (9.3.1)$$

Defining the effective shear force V_r^R as

$$rV_r^R = r\left(Q_r^R - \frac{4}{h^2}R_r^R\right) + \frac{4}{3h^2}\left[\frac{d}{dr}\left(rP_{rr}^R\right) - P_{\theta\theta}^R\right] \qquad (9.3.2)$$

Eq. (9.1.5a) may be written as

$$-\frac{d^2}{dr^2}\left(rM_{rr}^R\right) + \frac{dM_{\theta\theta}^R}{dr} + rV_r^R = 0 \qquad (9.3.3)$$

From Eqs. (9.3.2) and (9.1.5b), we have

$$\frac{d}{dr}\left(rV_r^R\right) = rq \qquad (9.3.4)$$

Hence it follows, from Eqs. (9.2.1a), (9.2.1b) and (9.3.4) that

$$rV_r^R = rQ_r^K + C_1 \qquad (9.3.5)$$

Next, we introduce the moment and higher-order moment sums

$$\mathcal{M}^R = \frac{M_{rr}^R + M_{\theta\theta}^R}{1+\nu}, \quad \mathcal{P}^R = \frac{P_{rr}^R + P_{\theta\theta}^R}{1+\nu} \qquad (9.3.6)$$

Using the definitions (9.3.6) and Eqs. (9.1.6a), (9.1.6b), (9.1.6c) and (9.1.6d), one can show that

$$\mathcal{M}^R = \frac{4D}{5}\frac{1}{r}\frac{d}{dr}\left(r\phi_r^R\right) - \frac{D}{5}\frac{1}{r}\frac{d}{dr}\left(r\frac{dw_0^R}{dr}\right) \qquad (9.3.7)$$

$$\mathcal{P}^R = \frac{4Dh^2}{35}\frac{1}{r}\frac{d}{dr}\left(r\phi_r^R\right) - \frac{Dh^2}{28}\frac{1}{r}\frac{d}{dr}\left(r\frac{dw_0^R}{dr}\right) \qquad (9.3.8)$$

$$r\frac{d\mathcal{M}^R}{dr} = \frac{d}{dr}\left(rM_{rr}^R\right) - M_{\theta\theta}^R \qquad (9.3.9)$$

$$r\frac{d\mathcal{P}^R}{dr} = \frac{d}{dr}\left(rP_{rr}^R\right) - P_{\theta\theta}^R \qquad (9.3.10)$$

BENDING RELATIONSHIPS FOR CIRCULAR AND ANNULAR PLATES

Substituting for $r\phi_r^R$ from Eq. (9.1.6e) into Eqs. (9.3.7) and (9.3.8), we arrive at

$$r\mathcal{M}^R = D\frac{1}{r}\frac{d}{dr}\left(rQ_r^R\right) - D\frac{d}{dr}\left(r\frac{dw_0^R}{dr}\right) \qquad (9.3.11)$$

$$r\mathcal{P}^R = D\frac{1}{r}\frac{d}{dr}\left(rQ_r^R\right) - D\frac{d}{dr}\left(r\frac{dw_0^R}{dr}\right) \qquad (9.3.12)$$

Now solving Eq. (9.3.11) for $(d/dr)(rdw_0^R/dr)$, we obtain

$$\frac{d}{dr}\left(r\frac{dw_0^R}{dr}\right) = -\frac{1}{D}r\mathcal{M}^R + \frac{h^2}{5(1-\nu)D}\frac{d}{dr}\left(rQ_r^R\right) \qquad (9.3.13)$$

Substituting the result into Eq. (9.3.12), we obtain

$$r\mathcal{P}^R = -\frac{3h^4}{700(1-\nu)}\frac{d}{dr}\left(rQ_r^R\right) + \frac{3h^2}{20}\left(r\mathcal{M}^R\right) \qquad (9.3.14)$$

From Eqs. (9.2.4), (9.3.3), (9.3.6), and (9.3.9), we obtain the result

$$\mathcal{M}^R = \mathcal{M}^K + C_1 \log r + C_2 \qquad (9.3.15)$$

Next we use Eqs. (9.2.2), (9.3.7), and (9.3.10) to arrive at

$$\frac{4}{5}D\phi_r^R - \frac{D}{5}\frac{dw_0^R}{dr} = -D\frac{dw_0^K}{dr} + \frac{C_1 r}{4}(2\log r - 1) + \frac{C_2 r}{2} + \frac{C_3}{r} \qquad (9.3.16)$$

From Eqs. (9.3.2) and (9.3.3), we have

$$r\left(Q_r^R - \beta R_r\right) = \frac{d}{dr}\left(rM_r^R\right) - M_{\theta\theta}^R - \frac{4}{3h^2}\left[\frac{d}{dr}\left(rP_{rr}^R\right) - P_{\theta\theta}^R\right] \qquad (9.3.17)$$

Substituting Eqs. (9.3.11) and (9.3.12), and $R_r = \{2h^2/[15(1-\nu)]\}Q_r^R$ into Eq. (9.3.17), we obtain

$$\frac{1-5\nu}{5(1-\nu)}\left(rQ_r^R\right) = \frac{4}{5}r\frac{dM^R}{dr} + \frac{2h^2}{350(1-\nu)}r\frac{d}{dr}\left[\frac{1}{r}\frac{d}{dr}\left(rQ_r^R\right)\right] \qquad (9.3.18)$$

and using Eq. (9.3.9), we have

$$r\frac{d}{dr}\left[\frac{1}{r}\frac{d}{dr}\left(rQ_r^R\right)\right] - 105(1-5\nu)h^2\left(rQ_r^R\right)$$
$$+ \frac{420(1-\nu)}{h^2}\left(rQ_r^R + C_1\right) = 0 \qquad (9.3.19)$$

Thus a second-order equation must be solved to determine the shear force.

Next, we derive the relationships between deflections w_0^R and w_0^K and rotation ϕ_r^R and slope $-dw_0^K/dr$. Replacing $\mathcal{M}^R$ in terms of $\mathcal{M}^K$ by means of Eq. (9.3.15), and then using Eq. (9.2.2), Eq. (9.3.13) can be written as

$$\frac{d}{dr}\left(r\frac{dw_0^R}{dr}\right) = \frac{d}{dr}\left(r\frac{dw_0^K}{dr}\right) - \frac{r}{D}(C_1 \log r + C_2) + \frac{6}{5Gh}\frac{d}{dr}\left(rQ_r^R\right) \tag{9.3.20}$$

Integrating twice with respect to r, we obtain

$$\frac{dw_0^R}{dr} = \frac{dw_0^K}{dr} - \frac{1}{D}\left[\frac{C_1 r}{4}(2\log r - 1) + \frac{C_2 r}{2} + \frac{C_3}{r}\right] + \frac{6}{5Gh}Q_r^R \tag{9.3.21}$$

$$w_0^R = w_0^K - \frac{1}{D}\left[\frac{C_1 r^2}{4}(\log r - 1) + \frac{C_2 r^2}{4} + C_3 \log r + C_4\right]$$
$$+ \frac{6}{5Gh}\int Q_r^R \, dr \tag{9.3.22}$$

Finally, using Eq. (9.3.21) in Eq. (9.3.16) we obtain

$$\phi_r^R = -\frac{dw_0^K}{dr} + \frac{1}{D}\left[\frac{C_1 r}{4}(2\log r - 1) + \frac{C_2 r}{2} + \frac{C_3}{r}\right] + \frac{3}{10Gh}Q_r^R \tag{9.3.23}$$

It is informative to discuss various types of boundary conditions in terms of the dependent variables for the third-order theory. Since a second-order equation for Q_r^R must be solved to determine solutions of the third-order theory, it is also useful to have the boundary conditions on Q_r^R for various types of edge supports. These are listed below.

Clamped edge

$$\phi_r^R = 0, \quad \frac{dw_0^R}{dr} = 0 \text{ which imply } Q_r^R = 0 \tag{9.3.24a}$$
$$w_0^R = 0 \tag{9.3.24b}$$

Simply supported edge

$$M_{rr}^R = 0, \quad P_{rr}^R = 0 \text{ which imply } r\frac{dQ_r^R}{dr} + \nu Q_r^R = 0 \tag{9.3.25a}$$
$$w_0^R = 0 \tag{9.3.25b}$$

Free edge

$$M_{rr}^R = 0, \quad P_{rr}^R = 0 \quad \text{which imply} \quad r\frac{dQ_r^R}{dr} + \nu Q_r^R = 0 \quad (9.3.26a)$$

$$rV_r^R = r\bar{Q}_r^R + \left(\frac{4}{3h^2}\right) r\frac{d\mathcal{P}^R}{dr} = 0 \quad (9.3.26b)$$

For solid circular plates, we have the additional "boundary conditions" at the center of the plate (i.e., at $r = 0$):

$$Q_r^R = 0, \quad \phi_r^R = 0, \quad \frac{dw_0^R}{dr} = 0, \quad rV_r^R = r\bar{Q}_r^R + \left(\frac{4}{3h^2}\right) r\frac{d\mathcal{P}}{dr} = 0 \quad (9.3.27)$$

For annular plates, the boundary conditions at the inner edge are given by the type of edge support there.

9.3.2 An Example

Here we present an example to illustrate the derivation of the solutions of the third-order theory using the relationships developed between the CPT and TSDT. First note that Eq. (9.3.19) can be expressed in the alternative form

$$\frac{d^2 Q_r^R}{dr^2} + \frac{1}{r}\frac{dQ_r^R}{dr} - \left(\frac{1}{r^2} + \xi\right) Q_r^R = -\xi \left(Q_r^K + \frac{C_1}{r}\right) \quad (9.3.28)$$

where

$$\xi = \frac{420(1 - \nu)}{h^2} \quad (9.3.29)$$

The solution to the homogeneous differential equation

$$\frac{d^2 Q_r^R}{dr^2} + \frac{1}{r}\frac{dQ_r^R}{dr} - \left(\frac{1}{r^2} + \xi\right) Q_r^R = 0 \quad (9.3.30)$$

is given by

$$Q_r^R(r) = C_5 I_1(\sqrt{\xi} r) + C_6 K_1(\sqrt{\xi} r) \quad (9.3.31)$$

where I_1 and K_1 are the first-order modified Bessel functions of the first and second kind, respectively.

Consider a solid circular plate under uniformly distributed load of intensity q_0 and clamped at the edge. For this case, the boundary conditions at $r = R_0$ give $Q_r^R(R_0) = 0$ and $C_2 = 0$, and those at $r = 0$ give $Q_r^R(0) = 0$ and $C_1 = C_3 = 0$. Then the general solution to Eq. (9.3.28) is given by

$$Q_r^R(r) = C_5 I_1(\sqrt{\xi} r) + C_6 K_1(\sqrt{\xi} r) - \frac{q_0 r}{2} \qquad (9.3.32)$$

Using the boundary conditions on Q_r^R, we obtain

$$C_6 = 0, \quad C_5 = \frac{q_0 R_0}{2 I_1(\sqrt{\xi} R_0)} \qquad (9.3.33)$$

Hence the solution becomes

$$Q_r^R(r) = \frac{q_0 R_0}{2} \left[\frac{I_1(\sqrt{\xi} r)}{I_1(\sqrt{\xi} R_0)} - \frac{r}{R_0} \right] \qquad (9.3.34a)$$

and

$$\int Q_r^R \, dr = \left(\frac{q_0 R_0^2}{4} \right) \left[\frac{2 I_0(\sqrt{\xi} r)}{R_0 I_1(\sqrt{\xi} R_0) \sqrt{\xi}} - \left(\frac{r}{R_0} \right)^2 \right] \qquad (9.3.34b)$$

Then the exact deflection of the TSDT plate is given by

$$w_0^R(r) = w_0^K(r) + \frac{6}{5Gh} \left(\frac{q_0 R_0^2}{4} \right) \left[\frac{2 I_0(\sqrt{\xi} r)}{R_0 I_1(\sqrt{\xi} R_0) \sqrt{\xi}} - \left(\frac{r}{R_0} \right)^2 \right] - \frac{C_4}{D} \qquad (9.3.35)$$

where the constant C_4 is evaluated using the boundary conditions $w_0^R = w_0^K = 0$ at $r = R_0$

$$C_4 = \frac{h^2}{5(1-\nu)} \left(\frac{q_0 R_0^2}{4} \right) \left[\frac{2 I_0(\sqrt{\xi} R_0)}{R_0 I_1(\sqrt{\xi} R_0) \sqrt{\xi}} - 1 \right] \qquad (9.3.36)$$

Note that the deflection $w_0^K(r)$ of the classical plate theory for the problem is given by setting $\alpha = 1$ in Eq. (9.2.35b). The maximum deflection is

$$w_{max}^R = \frac{q_0 R_0^4}{64 D} + \frac{q_0 R_0^2 h^2}{20 D(1-\nu)} \qquad (9.3.37a)$$

$$= \frac{q_0 R_0^4}{64 D} + \frac{6 q_0 R_0^2}{20 Gh} \qquad (9.3.37b)$$

BENDING RELATIONSHIPS FOR CIRCULAR AND ANNULAR PLATES 171

Comparing w_{max}^R with w_{max}^M from Eq. (9.2.37) (set $\alpha = 1$ in the first line), we note that for $K_s = 5/6$ the maximum deflection predicted by the first-order shear deformation theory coincides with that predicted by the third-order plate theory. Of course, the third-order theory does not require a shear correction coefficient. Further, the comparison of the solutions of the first-order and third-order theories for different boundary conditions and loads may lead to different shear correction factors.

9.4 Closure

The relationships developed herein between the CPT and shear deformation theories (FSDT and TSDT) facilitate actual derivation of the exact solutions of the first-order and third-order theories whenever the corresponding CPT solutions are available. It is also possible to develop finite element models of circular and annular plates based on the FSDT and TSDT using the finite element model of the CPT, as was illustrated by Reddy and Wang (1998) for the FSDT. The stiffness matrix of the shear deformable elements are also 4×4 for the pure bending case, and the finite elements are free from shear locking phenomenon (see Reddy 1998 and 1999b) experienced by the conventional shear deformable finite elements. It is also possible to develop the shear correction factors required in the first-order shear deformation theory using the relationships between CPT, FSDT, and TSDT. Such factors may depend on the boundary conditions as well as the applied transverse loads.

Problems

9.1 Assume the following displacement field for the classical plate theory (CPT) in polar coordinates (for pure bending case):

$$u_r(r, \theta, z) = -z \frac{\partial w_0}{\partial r}$$
$$u_\theta(r, \theta, z) = -z \left(\frac{1}{r} \frac{\partial w_0}{\partial \theta} \right)$$
$$u_z(r, \theta, z) = w_0(r, \theta) \tag{i}$$

where (u_r, u_θ, u_z) are the displacements along the three coordinate directions (r, θ, z), respectively. Show that the linear strains of the theory are given by

$$\varepsilon_{rr} = z\varepsilon_{rr}^{(1)}, \quad \varepsilon_{\theta\theta} = z\varepsilon_{\theta\theta}^{(1)}, \quad 2\varepsilon_{r\theta} = z\gamma_{r\theta}^{(1)} \qquad (ii)$$

where

$$\varepsilon_{rr}^{(1)} = -\frac{\partial^2 w_0}{\partial r^2}$$

$$\varepsilon_{\theta\theta}^{(1)} = -\frac{1}{r}\left(\frac{\partial w_0}{\partial r} + \frac{1}{r}\frac{\partial^2 w_0}{\partial \theta^2}\right)$$

$$\gamma_{r\theta}^{(1)} = -\frac{2}{r}\left(\frac{\partial^2 w_0}{\partial r \partial \theta} - \frac{1}{r}\frac{\partial w_0}{\partial \theta}\right) \qquad (iii)$$

9.2 Show that the virtual work statement for the Kirchhoff plate in the polar coordinates is given by (see Problem 9.1)

$$0 = \int_{\Omega_0}\left[-M_{\theta\theta}\frac{1}{r}\left(\frac{\partial \delta w_0}{\partial r} + \frac{1}{r}\frac{\partial^2 \delta w_0}{\partial \theta^2}\right)\right.$$
$$\left. - 2M_{r\theta}\frac{1}{r}\left(\frac{\partial^2 \delta w_0}{\partial r \partial \theta} - \frac{1}{r}\frac{\partial \delta w_0}{\partial \theta}\right) - q\delta w_0\right] r\, dr\, d\theta \qquad (i)$$

where the moment resultants are defined by

$$M_{rr} = \int_{-\frac{h}{2}}^{\frac{h}{2}} \sigma_{rr} z\, dz, \quad M_{\theta\theta} = \int_{-\frac{h}{2}}^{\frac{h}{2}} \sigma_{rr} z\, dz, \quad M_{r\theta} = \int_{-\frac{h}{2}}^{\frac{h}{2}} \sigma_{r\theta} z\, dz \qquad (ii)$$

9.3 Show that the Euler–Lagrange equation associated with the virtual work statement of Problem 9.2 is

$$-\frac{1}{r}\left[\frac{\partial^2}{\partial r^2}(rM_{rr}) - \frac{\partial M_{\theta\theta}}{\partial r} + \frac{1}{r}\frac{\partial^2 M_{\theta\theta}}{\partial \theta^2} + 2\frac{\partial^2 M_{r\theta}}{\partial r \partial \theta} + \frac{2}{r}\frac{\partial M_{r\theta}}{\partial \theta}\right] = q \qquad (i)$$

Introduce the transverse shear forces acting on the rz–plane and θz–plane as

$$Q_r = \frac{1}{r}\left[\frac{\partial}{\partial r}(rM_{rr}) + \frac{\partial M_{r\theta}}{\partial \theta} - M_{\theta\theta}\right] \qquad (ii)$$

$$Q_\theta = \frac{1}{r}\left[\frac{\partial}{\partial r}(rM_{r\theta}) + \frac{\partial M_{\theta\theta}}{\partial \theta} + M_{r\theta}\right] \qquad (iii)$$

and express Eq. (i) in the form
$$-\frac{1}{r}\left[\frac{\partial}{\partial r}(rQ_r) + \frac{\partial Q_\theta}{\partial \theta}\right] = q \quad (iv)$$
Simplify Eqs. (i)–(iv) for the axisymmetric case.

9.4 Show that Eqs. (9.1.1) and (9.1.2) can be combined to express them solely in terms of the deflection $w_0^K(r)$ as
$$\frac{D}{r}\frac{d}{dr}\left\{r\frac{d}{dr}\left[\frac{1}{r}\frac{d}{dr}\left(r\frac{dw_0^K}{dr}\right)\right]\right\} = q(r) \quad (i)$$
Hint: Use the identity
$$\frac{d}{dr}\left(r\frac{d^2 w_0^K}{dr^2}\right) = r\frac{d}{dr}\left[\frac{1}{r}\frac{d}{dr}\left(r\frac{dw_0^K}{dr}\right)\right] + \frac{1}{r}\frac{dw_0^K}{dr} \quad (ii)$$

9.5 Show that the deflection, slope, and bending moments of a simply supported circular plate under linearly varying load (see Figure 9.2.2)
$$q(r) = q_0 + \frac{q_1 - q_0}{R}r \quad (i)$$
are given by
$$Dw_0^K(r) = F(r) + K_1 \frac{r^2}{4} + K_2 \quad (ii)$$
$$D\frac{dw_0^K}{dr} = F'(r) + K_1 \frac{r}{2} \quad (iii)$$
$$M_{rr}^K = -\left(F''(r) + \frac{\nu}{r}F'(r)\right) - \frac{1+\nu}{2}K_1 \quad (iv)$$
$$M_{\theta\theta}^K = -\left(\nu F''(r) + \frac{1}{r}F'(r)\right) - \frac{1+\nu}{2}K_1 \quad (v)$$
where K_1 and K_2 are constants to be determined using the boundary conditions at $r = R$. In particular, show that
$$F(r) = \frac{q_0 r^4}{64} + \left(\frac{q_1 - q_0}{a}\right)\frac{r^5}{225} \quad (vi)$$
$$F'(r) = \frac{q_0 r^3}{16} + \left(\frac{q_1 - q_0}{R}\right)\frac{r^4}{45} \quad (vii)$$
$$F''(r) = \frac{3q_0 r^2}{16} + \left(\frac{q_1 - q_0}{R}\right)\frac{4r^3}{45} \quad (viii)$$
$$K_1 = -\frac{2R^2}{(1+\nu)}\left[\frac{3+\nu}{16}q_0 + \frac{4+\nu}{45}(q_1 - q_0)\right] \quad (ix)$$
$$K_2 = \frac{R^4}{(1+\nu)}\left[\frac{5+\nu}{64}q_0 + \frac{6+\nu}{150}(q_1 - q_0)\right] \quad (x)$$

9.6 Repeat Problem 9.5 for the clamped boundary condition at $r = R$.

9.7 Consider a simply supported annular plate of inner radius R_i and outer radius R_0, loaded with uniformly distributed load of intensity q_0. The boundary conditions are

$$\text{At } r = R_i: \quad M_{rr}^K = 0, \quad (rQ_r^K) = 0 \qquad (i)$$
$$\text{At } r = R_0: \quad w_0^K = 0, \quad M_{rr}^K = 0 \qquad (ii)$$

Show that the deflections and bending moments are given by

$$w_0^K(r) = \frac{q_0 R_0^4}{64D}\left\{-\left[1-\left(\frac{r}{R_0}\right)^4\right] + \frac{2\alpha_1}{1+\nu}\left[1-\left(\frac{r}{R_0}\right)^2\right]\right.$$
$$\left. - \frac{4\alpha_2 \beta^2}{1-\nu}\log\left(\frac{r}{R_0}\right)\right\} \qquad (iii)$$

$$M_{rr}^K = \frac{q_0 R_0^2}{16}\left\{(3+\nu)\left[1-\left(\frac{r}{R_0}\right)^2\right] - \beta^2(3+\nu)\left[1+\left(\frac{r}{R_0}\right)^2\right]\right.$$
$$\left. + 4(1+\nu)\beta^2\kappa\left[1-\left(\frac{r}{R_0}\right)^2\right] + 4(1+\nu)\beta^2\log\left(\frac{r}{R_0}\right)\right\} \qquad (iv)$$

$$M_{\theta\theta}^K = \frac{q_0 R_0^2}{16}\left\{(3+\nu)\left[1-\left(\frac{r}{R_0}\right)^2\right] + 4(1+\nu)\beta^2\kappa\left[1+\left(\frac{r}{R_0}\right)^2\right]\right.$$
$$\left. + \beta^2\left[(5\nu-1) + (3+\nu)\left(\frac{r}{R_0}\right)^2\right]\right\} \qquad (v)$$

where

$$\alpha_1 = (3+\nu)(1-\beta^2) - 4(1+\nu)\beta^2\kappa$$
$$\alpha_2 = (3+\nu) + 4(1+\nu)\kappa \qquad (vi)$$
$$\kappa = \frac{\beta^2}{1-\beta^2}\log\beta, \quad \beta = \frac{R_i}{R_0} \qquad (vii)$$

9.8 Show that the deflection of a clamped (at the outer edge) circular plate under linearly varying load, $q = q_0(1 - r/R_0)$ is

$$w_0^K(r) = \frac{q_0 R_0^4}{14400D}\left(129 - 290\frac{r^2}{R_0^2} + 225\frac{r^4}{R_0^4} - 64\frac{r^5}{R_0^5}\right) \qquad (i)$$

BENDING RELATIONSHIPS FOR CIRCULAR AND ANNULAR PLATES   175

The relationships presented in this chapter can be used to develop finite element models for axisymmetric bending of isotropic circular plates that contains the classical plate element as a special case. Such elements were developed for beams and axisymmetric bending of circular plates by Reddy (1993) and Reddy, Wang, and Lam (1997). We illustrate the procedure for the FSDT using a set of problems.

9.9 Show that the general solution of Eqs. (9.1.3) and (9.1.4a,b) is

$$w_0^M(r) = \frac{1}{4D}\left\{\left[\frac{4D}{GAK_s}\ln r - r^2(\ln r - 1)\right]C_1 - C_2 r^2\right.$$
$$\left. - 4C_3 \ln r - 4C_4\right\}$$
$$= \hat{C}_1 + \hat{C}_2 r^2 + \hat{C}_3 \ln r + \hat{C}_4 r^2 \ln r \qquad (i)$$

$$\phi_r^M(r) = \frac{1}{4D}\left[C_1 r(2\ln r - 1) + 2C_2 r + 4\frac{C_3}{r}\right]$$
$$= -2\hat{C}_2 r - \frac{\hat{C}_3}{r} - \hat{C}_4\left[r(1 + 2\ln r) + \frac{1}{r}\Gamma\right] \qquad (ii)$$

where $D = Eh^3/12(1-\nu^2)$, $\Gamma = (4D/GAK_s)$, and C_i are constants of integration. The classical plate theory solution is obtained from (i) and (ii) by setting $\Gamma = 0$.

9.10 Consider a typical finite element located between $r_a \leq r \leq r_b$. Let the generalized displacements at nodes 1 and 2 of the element be defined as

$$w_0^M(r_a) = \Delta_1, \quad \phi_r(r_a) \equiv \Delta_2$$
$$w_0^M(r_b) = \Delta_3, \quad \phi_r(r_b) \equiv \Delta_4 \qquad (i)$$

where ϕ_r denotes the slope (positive clockwise), which has different meanings in different theories, as defined below:

$$\phi_r = \begin{cases} -\frac{dw_0}{dr} & \text{for CPT} \\ \phi & \text{for FSDT} \end{cases} \qquad (ii)$$

Next, let Q_1 and Q_3 denote the shear forces (i.e., values of rQ_r) at nodes 1 and 2, respectively, and Q_2 and Q_4 the bending moments (i.e., values of rM_{rr}) at nodes 1 and 2, respectively.

Using Eqs. (i) and (ii) of Problem 9.9, relate the nodal degrees of freedom Δ_i defined in Eqs. (i) and (ii) to the constants $\hat{C}_i$. In particular, show that

$$\begin{Bmatrix} \Delta_1 \\ \Delta_2 \\ \Delta_3 \\ \Delta_4 \end{Bmatrix} = \begin{bmatrix} 1 & r_a^2 & \ln r_a & r_a^2 \ln r_a \\ 0 & -2r_a & -\frac{1}{r_a} & -r_a(1+2\ln r_a) - \frac{1}{r_a}\Gamma \\ 1 & r_b^2 & \ln r_b & r_b^2 \log r_b \\ 0 & -2r_b & -\frac{1}{r_b} & -r_b(1+2\ln r_b) - \frac{1}{r_b}\Gamma \end{bmatrix} \begin{Bmatrix} \hat{C}_1 \\ \hat{C}_2 \\ \hat{C}_3 \\ \hat{C}_4 \end{Bmatrix} \quad (iii)$$

Similarly, relate the nodal forces Q_i to the constants $\hat{C}_i$:

$$Q_1 \equiv -2\pi \left(rQ_r^F\right)_{r=r_a} = 8\pi D \hat{C}_4$$

$$Q_2 \equiv 2\pi \left(-r M_{rr}^F\right)_{r=r_a}$$
$$= 2\pi D \left\{ 2(1+\nu)r_a \hat{C}_2 - \frac{(1-\nu)}{r_a}\hat{C}_3 + \left[\Lambda_a - \frac{(1-\nu)}{r_a}\Gamma\right]\hat{C}_4 \right\}$$

$$Q_3 \equiv 2\pi \left(rQ_r^F\right)_{r=r_b} = -8\pi D \hat{C}_4$$

$$Q_4 \equiv 2\pi \left(r M_{rr}^F\right)_{r=r_b}$$
$$= -2\pi D \left\{ 2(1+\nu)r_b \hat{C}_2 - \frac{(1-\nu)}{r_b}\hat{C}_3 + \left[\Lambda_b - \frac{(1-\nu)}{r_b}\Gamma\right]\hat{C}_4 \right\}$$

9.11 Using the relations developed in Problem 9.10

$$\{Q\} = [G]\{\hat{C}\}, \quad \{\Delta\} = [H]\{\hat{C}\} \quad (i)$$

derive the stiffness matrix $[K]$, defined by

$$\{Q\} = [G]\{\hat{C}\} = [G][H]^{-1}\{\Delta\} \equiv [K]\{\Delta\} \quad (ii)$$

where

$$\Lambda_a = [2(1+\nu)\ln r_a + (3+\nu)]\,r_a, \quad \Lambda_b = [2(1+\nu)\ln r_b + (3+\nu)]\,r_b \quad (iii)$$

$$[G] = 2\pi D \begin{bmatrix} 0 & 0 & 0 & 4 \\ 0 & 2(1+\nu)r_a & -\frac{(1-\nu)}{r_a} & \left[\Lambda_a - \frac{(1-\nu)}{r_a}\Gamma\right] \\ 0 & 0 & 0 & -4 \\ 0 & -2(1+\nu)r_b & \frac{(1-\nu)}{r_b} & -\left[\Lambda_b - \frac{(1-\nu)}{r_b}\Gamma\right] \end{bmatrix} \quad (iv)$$

The stiffness matrix of the classical plate theory is obtained from $[K]$ by setting $\Gamma = 0$.

Chapter 10

Bending Relationships For Sectorial Plates

This chapter presents exact relationships between the bending solutions of sectorial plates based on the Kirchhoff (or classical) thin plate theory and the Mindlin plate theory. The Kirchhoff plate theory neglects the effect of transverse shear deformation, and the Mindlin plate theory allows for this effect which becomes significant when dealing with thick plates and sandwich plates. The considered sectorial plates have simply supported radial edges while the circular curved edge may be either simply supported, clamped, or free. The availability of such relationships allow easy conversion of the existing Kirchhoff sectorial plate solutions into the corresponding Mindlin solutions. The use of the relationships is illustrated using some sectorial plate examples and sample solutions obtained were checked with existing results and those computed from the finite element analysis software ABAQUS.

10.1 Introduction

This chapter focuses on the elastic bending problem of sectorial plates with simply supported radial edges while the circular edge may be either simply supported, clamped or free. The relationships between the bending solutions for such plates are derived herein. On the basis of existing exact Kirchhoff solutions for these plates, (see Timoshenko and Woinowsky-Krieger 1959 and Mansfield 1989), the relationships enable the easy deduction of the corresponding exact Mindlin plate solutions. These solutions and thus the relationships are verified by comparison with the existing results, e.g., the 3-D finite strip solutions of Cheung and Chan 1981, and the finite element analysis results obtained using the ABAQUS computer program.

10.2 Formulation

Consider a sectorial plate of radius a, thickness h, and with a subtended angle θ, as shown in Figure 10.1. The sectorial plate is simply supported along its two radial edges ($\theta = 0$, $\theta = \alpha$) while its circular edge can be either simply supported, clamped or free. The flexural rigidity of the isotropic plate is given by $D = Eh^3/[12(1-\nu^2)]$ where E is Young's modulus and ν Poisson's ratio. The shear modulus of the plate is $G = E/[2(1+\nu)]$. For analysis, it is expedient to adopt the polar coordinate system for such a plate shape with the origin of the coordinate system located at the vertex of the sectorial plate.

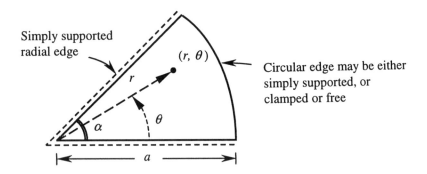

Figure 10.2.1. Geometry and the coordinate system used for a sectorial plate. The circular edge may be either simply supported, clamped, or free.

The form of the transverse loading on the plate is assumed to be the same for all sections parallel to the radial direction and is defined by

$$q(r,\theta) = \sum_{m=1}^{\infty} \bar{q}_m(r) \sin \mu\theta, \qquad \mu = \frac{m\pi}{\alpha} \qquad (10.2.1)$$

10.2.1 The Kirchhoff Plate Theory (CPT)

First consider the bending problem of a sectorial plate under the loading defined in Eq. (10.2.1) based on the Kirchhoff plate theory.

The equilibrium equations of the plate problem at hand are given by (Timoshenko and Woinowsky-Krieger, 1959)

$$\frac{\partial Q_r^K}{\partial r} + \frac{1}{r}\frac{\partial Q_\theta^K}{\partial \theta} + \frac{1}{r}Q_r^K + q = 0 \qquad (10.2.2a)$$

$$\frac{\partial M_{rr}^K}{\partial r} + \frac{1}{r}\frac{\partial M_{r\theta}^K}{\partial \theta} + \frac{M_{rr}^K - M_{\theta\theta}^K}{r} - Q_r^K = 0 \qquad (10.2.2b)$$

$$\frac{\partial M_{r\theta}^K}{\partial r} + \frac{1}{r}\frac{\partial M_{\theta\theta}^K}{\partial \theta} + \frac{2}{r}M_{r\theta}^K - Q_\theta^K = 0 \qquad (10.2.2c)$$

and the stress resultants are related to the displacement as follows

$$M_{rr}^K = -D\left[\frac{\partial^2 w_0^K}{\partial r^2} + \nu\left(\frac{1}{r}\frac{\partial w_0^K}{\partial r} + \frac{1}{r^2}\frac{\partial^2 w_0^K}{\partial \theta^2}\right)\right] \qquad (10.2.3a)$$

$$M_{\theta\theta}^K = -D\left[\nu\frac{\partial^2 w_0^K}{\partial r^2} + \left(\frac{1}{r}\frac{\partial w_0^K}{\partial r} + \frac{1}{r^2}\frac{\partial^2 w_0^K}{\partial \theta^2}\right)\right] \qquad (10.2.3b)$$

$$M_{r\theta}^K = -D(1-\nu)\frac{\partial}{\partial r}\left(\frac{1}{r}\frac{\partial w_0^K}{\partial \theta}\right) \qquad (10.2.3c)$$

$$Q_r^K = -D\frac{\partial}{\partial r}\left(\nabla^2 w_0^K\right) \qquad (10.2.3d)$$

$$Q_\theta^K = -D\frac{1}{r}\frac{\partial}{\partial \theta}\left(\nabla^2 w_0^K\right) \qquad (10.2.3e)$$

where the superscript 'K' denotes the Kirchhoff plate quantities and

$$\nabla^2 = \frac{\partial^2}{\partial r^2} + \frac{1}{r}\frac{\partial}{\partial r} + \frac{1}{r^2}\frac{\partial^2}{\partial \theta^2}$$

is the Laplacian operator in polar coordinates.

The above transverse deflection w_0^K in Eqs. (10.2.3a) to (10.2.3e) may be assumed to take the form

$$w_0^K(r,\theta) = \sum_{m=1}^{\infty} \bar{W}_m^K(r)\sin\mu\theta \qquad (10.2.4)$$

The bending solutions of the sectorial plate based on the Kirchhoff plate theory may be obtained by substituting Eq.(10.2.5) into Eqs. (10.2.3a-e) and then into Eqs. (10.2.2a), (10.2.2b), and (10.2.2c) and finally solving the governing equations together with the boundary conditions.

10.2.2 The Mindlin Plate Theory (FSDT)

When using the Mindlin plate theory, the equilibrium equations remain the same as those given in Eqs. (10.2.2) with 'K' replaced by 'M' but the Mindlin stress resultant-displacement relations are given by

$$M_{rr}^M = D \left[\frac{\partial \phi_r}{\partial r} + \frac{\nu}{r} \left(\phi_r + \frac{\partial \phi_\theta}{\partial \theta} \right) \right] \tag{10.2.5a}$$

$$M_{\theta\theta}^M = D \left[\nu \frac{\partial \phi_r}{\partial r} + \frac{1}{r} \left(\phi_r + \frac{\partial \phi_\theta}{\partial \theta} \right) \right] \tag{10.2.5b}$$

$$M_{r\theta}^M = \frac{1}{2} D(1-\nu) \left(\frac{1}{r} \frac{\partial \phi_y}{\partial \theta} - \frac{1}{r} \phi_\theta + \frac{\partial \phi_\theta}{\partial r} \right) \tag{10.2.5c}$$

$$Q_r^M = K_s G h \left(\phi_r + \frac{\partial w_0^M}{\partial r} \right) \tag{10.2.5d}$$

$$Q_\theta^M = K_s G h \left(\phi_\theta + \frac{1}{r} \frac{\partial w_0^M}{\partial \theta} \right) \tag{10.2.5e}$$

where ϕ_r and ϕ_θ are the Mindlin rotations about the circumferential and radial directions, respectively, the superscript 'M' denotes the Mindlin plate quantities, and K_s denotes the shear correction factor. Throughout this chapter, the shear correction factor is assumed to be 5/6.

Under the transverse loading (10.2.1), the transverse deflection and the rotations for the sectorial plate are given by (Mindlin, 1951)

$$w_0^M(r, \theta) = \sum_{m=1}^{\infty} \bar{W}_m^M(r) \sin \mu \theta \tag{10.2.6a}$$

$$\phi_r(r, \theta) = \sum_{m=1}^{\infty} \phi_{rm}(r) \sin \mu \theta \tag{10.2.6b}$$

$$\phi_\theta(r, \theta) = \sum_{m=1}^{\infty} \phi_{\theta m}(r) \cos \mu \theta \tag{10.2.6c}$$

Introducing the moment sum or *Marcus* moment

$$\mathcal{M} = \frac{M_{rr} + M_{\theta\theta}}{(1+\nu)} \tag{10.2.7}$$

and using Eqs. (10.2.5a)–(10.2.5c), one may express the force equilibrium equations (10.2.2a) and (10.2.2b) for the Mindlin plate theory as

$$Q_r^M = \frac{\partial \mathcal{M}^M}{\partial r} + \frac{1}{2r^2}(1-\nu)D\frac{\partial}{\partial \theta}\left[\frac{\partial \phi_r}{\partial \theta} - \frac{\partial}{\partial r}(r\phi_\theta)\right] \quad (10.2.8a)$$

$$Q_\theta^M = \frac{1}{r}\frac{\partial \mathcal{M}^M}{\partial \theta} + \frac{1}{2}(1-\nu)D\frac{\partial}{\partial r}\left[\frac{1}{r}\frac{\partial}{\partial r}(r\phi_\theta) - \frac{1}{r}\frac{\partial \phi_r}{\partial \theta}\right] \quad (10.2.8b)$$

10.2.3 Governing Equations

In view of Eqs. (10.2.3d), (10.2.3e) and (10.2.2a), the governing equation for the bending of the Kirchhoff sectorial plate can be expressed as

$$\nabla^2 \mathcal{M}^K = -q \quad (10.2.9)$$

and from Eqs. (10.2.8a), (10.2.8b) and (10.2.2a), one can similarly write the Mindlin governing equation as

$$\nabla^2 \mathcal{M}^M = -q \quad (10.2.10)$$

Alternatively, one can obtain the above governing equation based on the Mindlin plate theory by substituting the constitutive shear forces given by Eqs. (10.2.5d) and (10.2.5e) into Eq. (10.2.2a). This yields

$$K_s G h \left(\nabla^2 w_0^M + \frac{\mathcal{M}^M}{D}\right) = -q \quad (10.2.11)$$

In view of Eqs. (10.2.5d), (10.2.5e), (10.2.8a) and (10.2.8b) and eliminating $\mathcal{M}^M$ from the equations, one may deduce that

$$\nabla^2 \Omega = c^2 \Omega, \quad c^2 = \frac{2 K_s G h}{D(1-\nu)} = \frac{12 K_s}{h^2} \quad (10.2.12a)$$

where

$$\Omega = \frac{1}{r}\frac{\partial \phi_r}{\partial \theta} - \frac{1}{r}\frac{\partial}{\partial r}(r\phi_\theta) \quad (10.2.12b)$$

10.3 Exact Bending Relationships
10.3.1 General Relationships

Based on the concept of load equivalence, it follows from Eqs. (10.2.9) and (10.2.10) that

$$\nabla^2 \left(\mathcal{M}^M - \mathcal{M}^K \right) = 0 \tag{10.3.1}$$

Eq. (10.2.1) yields the following Kirchhoff–Mindlin Marcus moment relationship

$$\mathcal{M}^M = \mathcal{M}^K + \Xi \tag{10.3.2}$$

where Ξ is a harmonic function of (r, θ) which satisfies $\nabla^2 \Xi = 0$. For the sectorial plates considered here, the function Ξ is given by

$$\Xi = D \sum_{m=1}^{\infty} (C_{1m} r^\mu) \sin \mu\theta \tag{10.3.3}$$

where C_{1m} ($m = 1, 2, 3, ...$) are constants. Thus, Eq. (10.3.2) becomes

$$\mathcal{M}^M = \mathcal{M}^K + D \sum_{m=1}^{\infty} (C_{1m} r^\mu) \sin \mu\theta \tag{10.3.4}$$

The substitution of Eqs. (10.2.9) and (10.3.4) into Eq. (10.2.11) yields

$$\nabla^2 \left[w_0^M - w_0^K + \sum_{m=1}^{\infty} \left(\frac{C_{1m}}{4(\mu+1)} r^{\mu+2} \right) \sin \mu\theta \right] = \nabla^2 \left(\frac{\mathcal{M}^K}{K_s G h} \right) \tag{10.3.5}$$

In view of Eq. (10.3.5), one may deduce that the Kirchhoff–Mindlin deflection relationship is given by

$$w_0^M = w_0^K + \frac{\mathcal{M}^K}{K_s G h} + \sum_{m=1}^{\infty} \left[C_{2m} - \frac{C_{1m} r^2}{4(\mu+1)} \right] r^\mu \sin \mu\theta \tag{10.3.6}$$

where C_{1m} and C_{2m} are constants to be determined using the boundary conditions along the circular edge, $r = a$.

To obtain the Kirchhoff–Mindlin slope relationships, one has to solve Eq. (10.2.12). Noting the rotation functions given in Eqs. (10.2.6b) and (10.2.6c), the solution to Eq. (10.2.12) takes the form of

$$\Omega = \frac{1}{r} \frac{\partial \phi_r}{\partial \theta} - \frac{1}{r} \frac{\partial}{\partial r}(r\phi_\theta) = \sum_{m=1}^{\infty} R_m(r) \cos \mu\theta \tag{10.3.7}$$

Therefore by substituting Eq. (10.3.7) into Eq. (10.2.12) and using the method of separation of variables, Eq. (10.2.12) may be reduced to

$$r^2 \frac{d^2 R_m}{dr^2} + r \frac{dR_m}{dr} - \left[(cr)^2 + \mu^2\right] R_m = 0 \qquad (10.3.8)$$

Eq. (10.3.8) is the modified Bessel's equation of order μ and its solution is given by (Kresyzig, 1993)

$$R_m = C_{3m} I_\mu(cr) + C_{4m} K_\mu(cr) \qquad (10.3.9a)$$

where I_μ and K_μ are modified Bessel functions of the first and second kinds of order μ, respectively, while C_{3m} and C_{4m} are constants.

In view of Eqs. (10.3.7) and (10.3.9a), one obtains

$$\frac{1}{r}\frac{\partial \phi_y}{\partial \theta} - \frac{1}{r}\frac{\partial}{\partial r}(r\phi_\theta) = \sum_{m=1}^{\infty} \left[C_{3m} I_\mu(cr) + C_{4m} K_\mu(cr)\right] \cos \mu\theta \qquad (10.3.9b)$$

By substituting Eqs. (10.2.5d), (10.3.4) and (10.3.6) into Eq. (10.2.8a) and then combining with Eq. (10.3.9b), one arrives at

$$\phi_r = -\frac{\partial w_0^K}{\partial r} + \sum_{m=1}^{\infty} \left\{ \frac{(\mu+2)}{4(\mu+1)} C_{1m} r^{\mu+1} + \mu \left[\frac{D}{K_s G h} C_{1m} - C_{2m}\right] r^{\mu-1} \right.$$
$$\left. - \left(\frac{\mu}{c^2 r}\right)\left[C_{3m} I_\mu(cr) + C_{4m} K_\mu(cr)\right] \right\} \sin \mu\theta \qquad (10.3.10a)$$

Similarly, in view of Eqs. (10.2.5e), (10.2.7b), (10.3.4), (10.3.6) and (10.3.9b)

$$\phi_\theta = -\frac{1}{r}\frac{\partial w_0^K}{\partial \theta} + \sum_{m=1}^{\infty} \left\{ \frac{\mu}{4(\mu+1)} C_{1m} r^{\mu+1} + \mu \left[\frac{D}{K_s G h} C_{1m} - C_{2m}\right] r^{\mu-1} \right.$$
$$\left. - \frac{1}{c}\left[C_{3m} I'_\mu(cr) + C_{4m} K'_\mu(cr)\right] \right\} \cos \mu\theta \qquad (10.3.10b)$$

where the prime indicates partial differentiation with respect to r. For a finite Mindlin rotation ϕ_r and Kirchhoff slope $\partial w_0^K/\partial r$ along the radial direction, the term $\left(\frac{1}{r}\right) K_\mu(cr)$ which becomes singular as $r \to 0$, must be dropped from Eq. (10.3.10a). For its elimination, C_{4m} must be zero.

In view of the foregoing deflection and rotation expressions, the relationships between solutions of Mindlin and Kirchhoff sectorial plates may be summarized below.

Deflection relationship

$$w_0^M = w_0^K + \frac{\mathcal{M}^K}{K_sGh} + \sum_{m=1}^{\infty}\left[C_{2m} - \frac{C_{1m}r^2}{4(\mu+1)}\right]r^\mu \sin\mu\theta \qquad (10.3.11)$$

Rotation-slope relationships

$$\phi_r = \frac{\partial w_0^K}{\partial r} + \sum_{m=1}^{\infty}\left\{\frac{C_{1m}(\mu+2)}{4(\mu+1)}r^{\mu+1} + \mu\left[\frac{D}{K_sGh}C_{1m} - C_{2m}\right]r^{\mu-1}\right.$$
$$\left. - \left(\frac{\mu}{c^2 r}\right)C_{3m}I_\mu(cr)\right\}\sin\mu\theta \qquad (10.3.12a)$$

$$\phi_\theta = -\frac{1}{r}\frac{\partial w_0^K}{\partial \theta} + \sum_{m=1}^{\infty}\left\{\frac{\mu C_{1m}}{4(\mu+1)}r^{\mu+1} + \mu\left[\frac{D}{K_sGh}C_{1m} - C_{2m}\right]r^{\mu-1}\right.$$
$$\left. - \left(\frac{1}{c}\right)C_{3m}I'_\mu(cr)\right\}\cos\mu\theta \qquad (10.3.12b)$$

Moment relationships

$$M_{rr}^M = M_{rr}^K - D(\nu-1)\sum_{m=1}^{\infty}\left\{\frac{1}{4}(\mu+2)C_{1m}r^\mu + \mu(\mu-1)D_{1m}r^{\mu-2}\right.$$
$$\left. + \left(\frac{\mu}{cr}\right)C_{3m}\left[\frac{1}{cr}I_\mu(cr) - I'_\mu(cr)\right]\right\}\sin\mu\theta$$
$$+ \nu D\sum_{m=1}^{\infty}(C_{1m}r^\mu)\sin\mu\theta \qquad (10.3.13a)$$

$$M_{\theta\theta}^M = M_{\theta\theta}^K + D(\nu-1)\sum_{m=1}^{\infty}\left\{\frac{1}{4}(\mu+2)C_{1m}r^\mu + \mu(\mu-1)D_{1m}r^{\mu-2}\right.$$
$$\left. + \left(\frac{\mu}{cr}\right)C_{3m}\left[\frac{1}{cr}I_\mu(cr) - I'_\mu(cr)\right]\right\}\sin\mu\theta$$
$$+ D\sum_{m=1}^{\infty}(C_{1m}r^\mu)\sin\mu\theta \qquad (10.3.13b)$$

$$M_{r\theta}^M = M_{r\theta}^K + D(1-\nu)\sum_{m=1}^{\infty}\left\{\frac{\mu}{4}C_{1m}r^\mu + \mu(\mu-1)D_{1m}r^{\mu-2}\right.$$
$$\left. + \left[\frac{1}{cr}I'_\mu(cr) - \left(\frac{1}{2} + \left(\frac{\mu}{cr}\right)^2\right)I_\mu(cr)\right]C_{3m}\right\}\cos\mu\theta \qquad (10.3.13c)$$

$$D_{1m} = \left[\frac{D}{K_sGh}C_{1m} - C_{2m}\right] \qquad (10.3.13d)$$

Shear-force relationships

$$Q_r^M = Q_r^K + K_s Gh \sum_{m=1}^{\infty} \mu \left[\frac{D}{K_s Gh} C_{1m} r^{\mu-1} - \frac{1}{c^2 r} I_\mu(cr) C_{3m} \right] \sin \mu\theta \tag{10.3.14a}$$

$$Q_\theta^M = Q_\theta^K + K_s Gh \sum_{m=1}^{\infty} \left[\mu \frac{D}{K_s Gh} C_{1m} r^{\mu-1} - \frac{1}{c} I'_\mu(cr) C_{3m} \right] \cos \mu\theta \tag{10.3.14b}$$

The foregoing relationships contain a total of three unknown constants which are dependent on the three boundary conditions at the circular edge ($r = a$). In the sequel, these constants are evaluated for sectorial plates with various types of boundary conditions for its circular edge.

10.3.2 SSS Sectorial Plates

Consider the SSS sectorial plate where the circular edge is simply supported. The boundary conditions at $r = a$ are

$$w_0^M = w_0^K = 0, \quad M_{rr}^M = M_{rr}^K = 0, \quad \phi_\theta = 0 \tag{10.3.15}$$

By substituting Eq. (10.3.15) into Eqs. (10.3.11), (10.3.13a) and (10.3.12b), respectively, and solving for the constants, one obtains

$$C_{1m} = \frac{\mu \Omega_m}{C_{10}} \left[1 - \frac{\mu I_\mu(ca)}{ca I'_\mu(ca)} \right] \tag{10.3.16a}$$

$$C_{10} = a^{\mu+2} \left\{ \frac{2\mu+1}{2(\mu+1)} + \frac{2}{1-\nu} \frac{\mu}{(ca)^2} \left[\frac{\mu I_\mu(ca)}{ca I'_\mu(ca)} - 1 \right] + \frac{\nu}{1-\nu} \right\} \tag{10.3.16b}$$

$$C_{2m} = \frac{a^2}{4(\mu+1)} C_{1m} - \Omega_m a^{-\mu} \tag{10.3.16c}$$

$$C_{3m} = \frac{c\mu}{a I'_\mu(ca)} \left(\Omega_m + \frac{D}{K_s Gh} C_{1m} a^\mu \right) \tag{10.3.16d}$$

where

$$\Omega_m = \frac{\mathcal{M}_m^K|_{r=a}}{K_s Gh} \tag{10.3.16e}$$

10.3.3 SSC Sectorial Plates

Next, consider the SSC sectorial plate where the circular edge is clamped. The boundary conditions at $r = a$ are

$$w_0^M = w_0^K = 0, \quad \phi_r = \frac{\partial w_0^K}{\partial r} = 0, \quad \phi_\theta = 0 \qquad (10.3.17)$$

In view of Eqs. (10.3.17), (10.3.11) and (10.3.12), the constants are found to be

$$C_{1m} = \frac{\Omega_m \left[(ca)\frac{I'_\mu(ca)}{I_\mu(ca)} - 1\right]}{a^{\mu+2}\left\{\frac{\mu}{(ca)^2}\frac{2}{1-\nu} - \left(\frac{ca}{\mu}\right)\frac{I'_\mu(ca)}{I_\mu(ca)}\left[\frac{1}{2(\mu+1)} + \frac{\mu}{(ca)^2}\frac{2}{1-\nu}\right]\right\}} \qquad (10.3.18a)$$

$$C_{2m} = \frac{a^2}{4(\mu+1)}C_{1m} - \Omega_m a^{-\mu} \qquad (10.3.18b)$$

$$C_{3m} = \frac{c^2}{\mu I_\mu(ca)}\left\{C_{1m}a^\mu\left[\frac{a^2}{2(\mu+1)} + \mu\frac{D}{K_s Gh}\right] + \mu\Omega_m\right\} \qquad (10.3.18c)$$

where Ω_m is given by Eq. (10.3.16e).

10.3.4 SSF Sectorial Plates

Finally, consider the SSF sectorial plate where the circular edge is free. The boundary conditions at $r = a$ of a free edge are

$$M_{rr}^M = M_{rr}^K = 0, \quad Q_r^M = V_r^K = 0, \quad M_{r\theta}^M = 0 \qquad (10.3.19a)$$

where

$$V_r^K = Q_r^K + \frac{1}{r}\left(\frac{\partial M_{r\theta}^K}{\partial \theta}\right) \qquad (10.3.19b)$$

is the Kirchhoff effective shear force.

In view of Eqs. (10.3.19), (10.3.13a), (10.3.13c) and (10.3.14a), the constants are found to be

$$C_{1m} = \frac{\Gamma_m + \Psi_m\left(\frac{c}{\mu}\right)\left\{(\mu+1)\frac{I'_\mu(ca)}{I_\mu(ca)} - (ca)\left[\frac{1}{2} + \frac{\mu(\mu+1)}{(ca)^2}\right]\right\}}{a^\mu\left\{\frac{3+\nu}{2(1-\nu)} + \frac{\mu+1}{ca}\left[\frac{\mu}{ca} - \frac{I'_\mu(ca)}{I_\mu(ca)}\right]\frac{2}{1-\nu}\right\}} \qquad (10.3.20a)$$

$$C_{2m} = \frac{\Gamma_m + \Psi_m\left(\frac{c}{\mu}\right)\left\{\frac{I'_\mu(ca)}{I_\mu(ca)} - (ca)\left[\frac{1}{2} + \left(\frac{\mu}{ca}\right)^2\right]\right\} + C_{1m}a^\mu\mu_0}{a^{\mu-2}[\mu(\mu-1)]}$$

$$\qquad (10.3.20b)$$

$$C_{3m} = \frac{c^2 a}{I_\mu(ca)}\left[\frac{\Psi_m}{\mu} + \frac{D}{K_s Gh}C_{1m}a^{\mu-1}\right] \qquad (10.3.20c)$$

where

$$\mu_0 = \left\{\frac{\mu}{4} - \left[\frac{\mu}{(ca)^2} + \frac{1}{2} - \frac{I'_\mu(ca)}{caI_\mu(ca)}\right]\frac{2}{1-\nu}\right\}$$

$$\Gamma_m = \left(\frac{1}{D(1-\nu)}\right) M^K_{r\theta m}|_{r=a}$$

$$\Psi_m = \left(\frac{1}{K_s Gh}\right) Q^K_{rm}|_{r=a} \quad (10.3.21)$$

10.4 Examples

The foregoing relationships can be used to furnish the Mindlin deflection, rotations and stress-resultants of Mindlin sectorial plates upon supplying the corresponding Kirchhoff plate solutions. This is illustrated below using the examples of sectorial plates under a uniformly distributed load q_0.

Based on the Kirchhoff plate theory, the transverse deflection of a sectorial plate under a uniform load q_0 is given by (Mansfield 1989)

$$w_0^K = \frac{1}{D}\sum_{m=1}^{\infty}\left[\frac{\bar{q}_m r^4}{(16-\mu^2)(4-\mu^2)} + A_m r^\mu + B_m r^{\mu+2}\right]\sin\mu\theta \quad (10.4.1)$$

where

$$\bar{q}_m = \frac{2q_0}{m\pi}[1-(-1)^m] \quad (10.4.2)$$

which is zero for even m.

10.4.1 SSS Plates

For SSS plates, we have

$$A_m = \frac{\bar{q}_m a^{4-\mu}(\mu+5+\nu)}{2(16-\mu^2)(2+\mu)(2\mu+1+\nu)} \quad (10.4.3a)$$

$$B_m = -\frac{\bar{q}_m a^{2-\mu}(\mu+3+\nu)}{2(4-\mu^2)(4+\mu)(2\mu+1+\nu)} \quad (10.4.3b)$$

The expressions for A_m and B_m agree with those given by Timoshenko and Woinowsky-Krieger (1959).

10.4.2 SSC Plates

For SSC plates, we have

$$A_m = \frac{\bar{q}_m a^{4-\mu}}{2(16-\mu^2)(2+\mu)} \quad (10.4.4a)$$

$$B_m = -\frac{\bar{q}_m a^{2-\mu}}{2(4-\mu^2)(4+\mu)} \quad (10.4.4b)$$

and the Kirchhoff plate solution is given by Eq. (10.4.1).

10.4.3 SSF Plates

For SSF plates, the constants are given by

$$A_m = \frac{\bar{q}_m a^{4-\mu}[(\mu-4)A_0 A_1 + 2\mu(3+\nu)A_2]}{2\mu^2(16-\mu^2)(4-\mu^2)(1-\mu)(1-\nu)(3+\nu)} \quad (10.4.5a)$$

$$B_m = -\frac{\bar{q}_m a^{2-\mu}[8+5\mu+\nu\mu(1+\mu)]}{2\mu(4-\mu^2)(4+\mu)(1+\mu)(3+\nu)} \quad (10.4.5b)$$

$$A_0 = 8 + \mu(5+\nu) + \nu\mu^2, \quad A_1 = \mu(1-\nu) + 2(1+\nu)$$
$$A_2 = 4(3+\nu) - \nu\mu^2 \quad (10.4.5c)$$

Note that the expressions for A_m and B_m for the SSC and SSF sectorial plates are not available in the open literature at the time of this writing. Substituting Eqs. (10.4.1) and (10.4.2) into Eq. (10.2.3a) and (10.2.3b) and then combining them together we obtain the Kirchhoff Marcus moment

$$\mathcal{M}^K = \frac{M_{rr}^K + M_{\theta\theta}^K}{1+\nu} = -\sum_{m=1}^{\infty}\left[\frac{\bar{q}_m r^2}{(4-\mu^2)} + 4(\mu+1)B_m r^\mu\right]\sin\mu\theta \quad (10.4.6)$$

By substituting the Kirchhoff solutions given by Eqs. (10.4.1), (10.4.2) and (10.4.6) into the relationships [Eqs. (10.3.11)–(10.3.14)], one can readily obtain the corresponding Mindlin plate results. In the next section, we consider several examples. The results are verified using the finite element method.

10.4.4 Numerical Results

To verify the correctness of these relationships, Abaqus (1997) finite element program was used for the bending analysis. Two different shell elements, namely, S8R5 shell elements for thin plates and S8R elements for shear deformable thick plates have been used in the modeling of the sectorial plates. Three different - increasingly refined - uniform meshes designated Type A, Type B and Type C were used. Table 10.4.1 shows the results obtained from the convergence study for the deflection parameter $\bar{w} = 10^3(w_0 D)/(q_0 a^4)$ and radial moment parameter $\bar{M}_{rr} = 10^2 \bar{M}_{rr}/(q_0 a^2)$ at $r = 0.75a$, $\theta = \alpha/6$ for a uniformly loaded SSS sectorial plate with a subtended angle of $\pi/3$. It can be established from Table 10.4.1 that Mesh C will suffice in providing converged results and thus will be used to generate all the numerical solutions in this study.

Table 10.4.1. Convergence tests for SSS Sectorial plates ($\alpha = \pi/3$).

Mesh type (Elements)	Deflection $\bar{w}(0.75a, \pi/6)$			Radial Moment $\bar{M}_{rr}(0.75a, \pi/6)$		
	h/a			h/a		
	0.001	0.1	0.2	0.001	0.1	0.2
Type A (75)	0.9247	1.0215	1.3101	2.4433	2.4554	2.4582
Type B (300)	0.9248	1.0212	1.3097	2.4282	2.4320	2.4422
Type C (675)	0.9248	1.0210	1.3097	2.4284	2.4322	2.4423

It is worth noting that there has been little work done on the bending of thick sectorial plates. In the open literature, we find that Cheung and Chan (1981) used the three-dimensional finite strip method for the analysis of such plates. In Tables 10.4.2 and 10.4.3, the results

$$\bar{w} = 10^3 \frac{w_0 D}{q_0 a^4}, \quad \bar{M}_{rr} = 10^2 \frac{M_{rr}}{q_0 a^4}, \quad \bar{M}_{\theta\theta} = 10^2 \frac{M_{\theta\theta}}{q_0 a^2} \quad (10.4.7)$$

of Cheung and Chan (1981) are compared with the present solutions furnished by the relationships [Eqs. (10.3.11), (10.3.13a), (10.3.13b) and (10.3.16)] and the numerical results generated using the finite element program Abaqus (1997). From Table 10.4.2, it can be observed that the present Mindlin plate deflections furnished by the exact relationship are in excellent agreement with the Abaqus results. Although the present and the Abaqus bending moments do not exhibit the same degree of agreement, the difference in results is within 3%. When compared to the 3-D results furnished by Cheung and Chan (1981), one can observe that the deflections obtained by Cheung and Chan are consistently slightly lower than the present results evaluated from the relationships. For bending moments as shown in Table 10.4.3, the differences are within 10% of each other. Table 10.4.4 presents the deflection parameters $\bar{w} = 10^3(w_0 D)/(q_0 a^4)$ of SSC and SSF sectorial plates obtained from the deflection relationships, [Eqs. (10.3.11), (10.3.17) and (10.3.18)] and from Abaqus (1997). The results are in very good agreement with each other.

From the bending results, one can observe that as the plate thickness increases, the thick plate solutions deviate significantly from the thin plate results, especially for the transverse deflection. For example, in the case of thick SSS sectorial plates ($h/a = 0.2$), its maximum deflection can be lower by 40% when the analysis is based on the Kirchhoff (classical thin) plate theory, and for SSC sectorial plates, the difference can be as much as 75%! This shows the significant effect of transverse shear strains on the bending behavior of thick plates. The effect of shear deformation is to increase the deflection.

Table 10.4.2. Comparison of deflection $\bar{w} = 10^3 w_0 D/(q_0 a^4)$ for SSS sectorial plates with $\alpha = \pi/3$.

h/a	Thin plate[†] Results	Cheung & Chan (1981)	FEM (Abaqus)	Present[*] results
0.001	0.9975	0.9840	0.9974	0.9975
0.067	-	1.0264	1.0430	1.0430
0.100	-	1.0839	1.1000	1.1000
0.133	-	1.1573	1.1797	1.1797
0.200	-	1.3547	1.4076	1.4076

[†]Timoshenko and Woinowsky-Krieger (1970).

[*]From Eqs. (10.3.11) and (10.3.16).

Table 10.4.3. Comparison of bending moments, $\bar{M}_{rr} = 10^2 M_{rr}/(q_0 a^4)$ and $\bar{M}_{\theta\theta} = 10^2 M_{\theta\theta}/(q_0 a^2)$, for SSS sectorial plates with $\alpha = \pi/3$.

	Radial moment $\bar{M}_{rr}(0.75a, \pi/6)$				Circumferential moment $\bar{M}_{\theta\theta}(0.75a, \pi/6)$			
h/a	Thin[†] Plate	Cheung & Chan	Abaqus	Present* Results	Thin[†] Plate	Cheung &Chan	Abaqus	Present** Results
0.001	2.4260	2.4944	2.4284	2.4260	2.1316	2.1452	2.1342	2.1316
0.067	-	2.5175	2.4302	2.4276	-	2.1490	2.1332	2.1305
0.100	-	2.5029	2.4322	2.4296	-	2.1873	2.1318	2.1291
0.133	-	2.5533	2.4349	2.4324	-	2.2350	2.1299	2.1272
0.200	-	2.6384	2.4423	2.4398	-	2.2678	2.1247	2.1219

[†]Timoshenko and Woinowsky-Krieger (1970).
*From Eqs. (10.3.13a) and (10.3.16).
**From Eqs. (10.3.13b) and (10.3.16).

Table 10.4.4. Deflection parameters of SSC and SSF sectorial plates.

	SSC plate, $\alpha = \pi/3$ $\bar{w}(0.75a, \pi/6)$			SSF plate, $\alpha = \pi$ $\bar{w}(a, \pi/4)$		
h/a	Thin Plate[†]	Abaqus*	Present Results**	Thin Plate[†]	Abaqus*	Present Results[‡]
0.001	0.4674	0.4673	0.4674	63.2800	63.2800	63.2800
0.1	-	0.5872	0.5872	-	64.6480	64.6474
0.2	-	0.9262	0.9262	-	66.9260	66.9254

[†]Timoshenko and Woinowsky-Krieger (1959).
*Obtained using Mesh Type C with 675 shells elements.
**From Eqs. (10.3.11) & (10.3.18).
[‡]From Eqs. (10.3.11) and (10.3.20).

10.5 Conclusions

In this chapter, the exact relationships between the bending solutions of sectorial plates based on the Kirchhoff plate theory and the Mindlin thick plate theory are presented. These relationships, unavailable in the literature, allow engineers to obtain accurate Mindlin plate results upon supplying the corresponding Kirchhoff plate solutions, which are readily available for most problems in the open literature. These relationships can also help to elucidate the effect of transverse shear deformation on the flexural behavior of thick sectorial plates. Moreover, the Mindlin plate solutions furnished by the bending relationships can serve as benchmark results to check the accuracy of numerical methods and software developed for thick plate analysis.

The general relationships included in this chapter for sectorial plates are also valid for *annular* sectorial plates with outer radius a and inner radius b. The edges $\theta = 0$ and $\theta = \alpha$ are assumed to be simply supported while the other two edges, i.e. $r = b$ and $r = a$, may be each free, simply supported, or clamped. The relationships given in Problems 10.3 and 10.4 also hold for these plates. However, the expressions for plates with specific boundary conditions are too long, and interested readers may consult the paper by Lim and Wang (2000).

Problems

10.1 Verify Eqs. (10.2.8a,b).

10.2 Verify Eqs. (10.2.12a,b).

10.3 Write Eq. (10.3.2) as

$$\mathcal{M}^M = \mathcal{M}^K + D\nabla^2 \Phi \qquad (i)$$

where $\Phi(r, \theta)$ is a biharmonic function (i.e., $\nabla^4 \Phi = 0$).

(a) Use Eqs. (i) and (10.2.9) in Eq. (10.2.11) to show that

$$w_0^M = w_0^K + \frac{\mathcal{M}^K}{K_s G h} - \Phi + \Psi \qquad (ii)$$

where Ψ is a harmonic function satisfying the equation $\nabla^2 \Psi = 0$.

(b) Use Eqs. (10.2.5d,e), (10.2.8a,b), (i), and (ii) to show that

$$\phi_r^M = -\frac{\partial w_0^K}{\partial r} + \frac{\partial}{\partial r}\left[\frac{D}{K_s Gh}\left(\nabla^2 \Phi\right) + \Phi - \Psi\right] + \frac{1}{rc^2}\frac{\partial \Omega}{\partial \theta} \quad (iii)$$

$$\phi_\theta^M = -\frac{1}{r}\frac{\partial w_0^K}{\partial \theta} + \frac{1}{r}\frac{\partial}{\partial \theta}\left[\frac{D}{K_s Gh}\left(\nabla^2 \Phi\right) + \Phi - \Psi\right] - \frac{1}{c^2}\frac{\partial \Omega}{\partial r} \quad (iv)$$

10.4 Use Eqs. (i)-(iv) of Problem 10.3 and (10.2.3a-e) in Eq. (10.2.5a-c) to establish the following relationships:

$$M_{rr}^M = M_{rr}^K - D(1-\nu)\frac{1}{r}\frac{\partial}{\partial \theta}\left(\frac{1}{r}\frac{\partial \Lambda}{\partial \theta} - \frac{1}{c^2}\frac{\partial \Omega}{\partial r}\right) + D\nabla^2 \Phi \quad (i)$$

$$M_{\theta\theta}^M = M_{\theta\theta}^K - D(1-\nu)\frac{\partial}{\partial r}\left(\frac{\partial \Lambda}{\partial r} + \frac{1}{rc^2}\frac{\partial \Omega}{\partial \theta}\right) + D\nabla^2 \Phi \quad (ii)$$

$$M_{r\theta}^M = M_{r\theta}^K + D(1-\nu)\left[\frac{\partial}{\partial r}\left(\frac{1}{r}\frac{\partial \Lambda}{\partial \theta}\right)\right.$$

$$\left. + \frac{1}{2c^2}\left(\frac{1}{r}\frac{\partial \Omega}{\partial r} + \frac{1}{r^2}\frac{\partial^2 \Omega}{\partial \theta^2} - \frac{\partial^2 \Omega}{\partial r^2}\right)\right] \quad (iii)$$

$$Q_r^M = Q_r^K + D\frac{\partial}{\partial r}\left(\nabla^2 \Phi\right) + \frac{D(1-\nu)}{2}\frac{1}{r}\frac{\partial \Omega}{\partial \theta} \quad (iv)$$

$$Q_\theta^M = Q_\theta^K + \frac{D}{r}\frac{\partial}{\partial \theta}\left(\nabla^2 \Phi\right) - \frac{D(1-\nu)}{2}\frac{\partial \Omega}{\partial r} \quad (v)$$

where

$$\Lambda = \frac{D}{K_s Gh}\left(\nabla^2 \Phi\right) + \Phi - \Psi \quad (vi)$$

and Ω is the solution of Eq. (10.2.12a,b).

10.5 Verify Eqs. (10.3.10a,b).

Chapter 11

Buckling Relationships

This chapter presents exact relationships between the buckling loads of the classical Kirchhoff plate theory, the Mindlin plate theory and the Reddy plate theory for simply supported polygonal plates, and circular and sectorial plates subjected to hydrostatic in-plane loads. The buckling load relationships enable one to obtain the solutions of the shear deformable plate theories from the known Kirchhoff plate theory for the same problem. As examples, some buckling loads for rectangular plates, regular polygonal plates, and circular and sectorial plates are determined using these relationships.

11.1 Polygonal Plates

11.1.1 Governing Equations

Here we consider buckling of polygonal plates (with straight edges) under uniform in-plane compressive load N (measured per unit length) on all edges as shown in Figure 11.1.1. The potential energy V of the in-plane load N must be added to the strain energy U of the plate to form the total potential energy functional Π

$$\Pi = U + V \qquad (11.1.1)$$

The strain energy functionals for the Kirchhoff (CPT), Mindlin (FSDT), and Reddy (TSDT) plate theories were presented in Chapter 6. The potential energy of the in-plane load N is given by

$$V = -\int_{\Omega_0} N \left[\left(\frac{\partial w}{\partial x}\right)^2 + \left(\frac{\partial w}{\partial y}\right)^2 \right] dx dy \qquad (11.1.2)$$

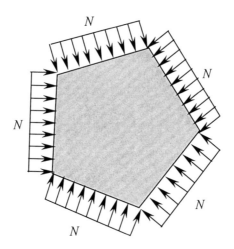

Figure 11.1.1. Polygonal plate under uniform compressive load N.

Using the principle of minimum potential energy, the minimization of the total potential energy functional Π with respect to the generalized displacements yields the equations for buckling. These equations are summarized below for the three theories.

CPT:

$$\frac{\partial M_{xx}^K}{\partial x} + \frac{\partial M_{xy}^K}{\partial y} - Q_x^K = 0 \qquad (11.1.3a)$$

$$\frac{\partial M_{xy}^K}{\partial x} + \frac{\partial M_{yy}^K}{\partial y} - Q_y^K = 0 \qquad (11.1.3b)$$

$$\frac{\partial Q_x^K}{\partial x} + \frac{\partial Q_y^K}{\partial y} = N^K \nabla^2 w^K \qquad (11.1.3c)$$

FSDT:

$$\frac{\partial M_{xx}^M}{\partial x} + \frac{\partial M_{xy}^M}{\partial y} - Q_x^M = 0 \qquad (11.1.4a)$$

$$\frac{\partial M_{xy}^M}{\partial x} + \frac{\partial M_{yy}^M}{\partial y} - Q_y^M = 0 \qquad (11.1.4b)$$

$$\frac{\partial Q_x^M}{\partial x} + \frac{Q_y^M}{\partial y} = N^M \nabla^2 w^M \qquad (11.1.4c)$$

TSDT:

$$\frac{\partial \hat{M}_{xx}}{\partial x} + \frac{\partial \hat{M}_{xy}}{\partial y} - \hat{Q}_x = 0 \qquad (11.1.5a)$$

$$\frac{\partial \hat{M}_{xy}}{\partial x} + \frac{\partial \hat{M}_{yy}}{\partial y} - \hat{Q}_y = 0 \qquad (11.1.5b)$$

$$\frac{\partial \hat{Q}_x}{\partial x} + \frac{\partial \hat{Q}_y}{\partial y} + \alpha \left(\frac{\partial^2 P_{xx}}{\partial x^2} + 2\frac{\partial^2 P_{xy}}{\partial x \partial y} + \frac{\partial^2 P_{yy}}{\partial y^2} \right) = N^R \nabla^2 w^R \qquad (11.1.5c)$$

where

$$\hat{M}_{\xi\eta} = M_{\xi\eta} - \alpha P_{\xi\eta} \qquad (11.1.6a)$$

$$\hat{Q}_\xi = Q_\xi - \beta R_\xi \qquad (11.1.6b)$$

and $\xi, \eta = x, y$, and α and β are the parameters introduced in the TSDT

$$\alpha = \frac{4}{3h^2}, \quad \beta = \frac{4}{h^2} \qquad (11.1.7)$$

The Laplace operator ∇^2 in the rectangular Cartesian coordinate system is

$$\nabla^2 = \frac{\partial^2}{\partial x^2} + \frac{\partial^2}{\partial y^2} \qquad (11.1.8)$$

and the definitions of the stress resultants of the various theories are given in Chapter 6.

The relationships between the force and moment resultants (Ms and Qs) and the generalized displacements (w, ϕ_x, ϕ_y) for various theories were presented in Chapter 6 [see Eqs. (6.2.22a-c) for the Kirchhoff plate theory, Eqs. (6.3.13a-e) for the Mindlin theory, and Eqs. (6.4.11a-j) for the Reddy theory]. These are restated here for ready reference.

CPT:

$$M_{xx}^K = -D \left(\frac{\partial^2 w^K}{\partial x^2} + \nu \frac{\partial^2 w^K}{\partial y^2} \right) \qquad (11.1.9a)$$

$$M_{yy}^K = -D \left(\nu \frac{\partial^2 w^K}{\partial x^2} + \frac{\partial^2 w^K}{\partial y^2} \right) \qquad (11.1.9b)$$

$$M_{xy}^K = -D(1-\nu) \frac{\partial^2 w^K}{\partial x \partial y} \qquad (11.1.9c)$$

FSDT:

$$M_{xx}^M = D\left(\frac{\partial \phi_x^M}{\partial x} + \nu \frac{\partial \phi_y^M}{\partial y}\right) \tag{11.1.10a}$$

$$M_{yy}^M = D\left(\nu \frac{\partial \phi_x^M}{\partial x} + \frac{\partial \phi_y^M}{\partial y}\right) \tag{11.1.10b}$$

$$M_{xy}^M = \frac{1}{2}D(1-\nu)\left(\frac{\partial \phi_x^M}{\partial y} + \frac{\partial \phi_y^M}{\partial x}\right) \tag{11.1.10c}$$

$$Q_x^M = K_s Gh\left(\phi_x^M + \frac{\partial w^M}{\partial x}\right) \tag{11.1.10d}$$

$$Q_y^M = K_s Gh\left(\phi_y^M + \frac{\partial w^M}{\partial y}\right) \tag{11.1.10e}$$

TSDT:

$$M_{xx}^R = \frac{4D}{5}\left(\frac{\partial \phi_x^R}{\partial x} + \nu \frac{\partial \phi_y}{\partial y}\right) - \frac{D}{5}\left(\frac{\partial^2 w^R}{\partial x^2} + \nu \frac{\partial^2 w^R}{\partial y^2}\right) \tag{11.1.11a}$$

$$P_{xx} = \frac{4h^2 D}{35}\left(\frac{\partial \phi_x^R}{\partial x} + \nu \frac{\partial \phi_y^R}{\partial y}\right) - \frac{h^2 D}{28}\left(\frac{\partial^2 w^R}{\partial x^2} + \nu \frac{\partial^2 w^R}{\partial y^2}\right) \tag{11.1.11b}$$

$$M_{yy}^R = \frac{4D}{5}\left(\nu \frac{\partial \phi_x^R}{\partial x} + \frac{\partial \phi_y^R}{\partial y}\right) - \frac{D}{5}\left(\nu \frac{\partial^2 w^R}{\partial x^2} + \frac{\partial^2 w^R}{\partial y^2}\right) \tag{11.1.11c}$$

$$P_{yy} = \frac{4h^2 D}{35}\left(\nu \frac{\partial \phi_x^R}{\partial x} + \frac{\partial \phi_y^R}{\partial y}\right) - \frac{h^2 D}{28}\left(\nu \frac{\partial^2 w^R}{\partial y^2} + \frac{\partial^2 w^R}{\partial y^2}\right) \tag{11.1.11d}$$

$$M_{xy}^R = \left(\frac{1-\nu}{2}\right)\left[\frac{4D}{5}\left(\frac{\partial \phi_x^R}{\partial y} + \frac{\partial \phi_y^R}{\partial x}\right) - \frac{D}{5}\left(2\frac{\partial^2 w^R}{\partial x \partial y}\right)\right] \tag{11.1.11e}$$

$$P_{xy} = \left(\frac{1-\nu}{2}\right)\left[\frac{4h^2 D}{35}\left(\frac{\partial \phi_x^R}{\partial y} + \frac{\partial \phi_y^R}{\partial x}\right) - \frac{h^2 D}{28}\left(2\frac{\partial^2 w^R}{\partial x \partial y}\right)\right] \tag{11.1.11f}$$

$$Q_x = \frac{2hG}{3}\left(\phi_x^R + \frac{\partial w^R}{\partial x}\right) \tag{11.1.11g}$$

$$R_x = \frac{h^3 G}{30}\left(\phi_x^R + \frac{\partial w^R}{\partial x}\right) \tag{11.1.11h}$$

$$Q_y = \frac{2hG}{3}\left(\phi_y^R + \frac{\partial w^R}{\partial y}\right) \tag{11.1.11i}$$

$$R_y = \frac{h^3 G}{30}\left(\phi_y^R + \frac{\partial w^R}{\partial y}\right) \tag{11.1.11j}$$

where D is the flexural rigidity of the plate

$$D = \frac{Eh^3}{12(1-\nu^2)} \quad (11.1.12)$$

h being the thickness, E the Young's modulus, and ν Poisson's ratio of the plate.

11.1.2 Relationships Between CPT and FSDT

In view of Eqs. (11.1.10a-e) and (11.1.4a-c), we obtain [see Eqs. (7.2.2a,b)]

$$\nabla^2 \mathcal{M}^K = N^K \nabla^2 w^K, \qquad \nabla^2 w_0^K = -\frac{\mathcal{M}^K}{D} \quad (11.1.13a,b)$$

$$\nabla^2 \mathcal{M}^M = N^M \nabla^2 w^M, \quad \nabla^2 \left(w_0^M - \frac{\mathcal{M}^M}{K_s Gh} \right) = -\frac{\mathcal{M}^M}{D} \quad (11.1.14a,b)$$

where $\mathcal{M}^K$ and $\mathcal{M}^M$ are the moment sums

$$\mathcal{M}^K = \frac{M_{xx}^K + M_{yy}^K}{1+\nu} = -D \left(\frac{\partial^2 w_0^K}{\partial x^2} + \frac{\partial^2 w_0^K}{\partial y^2} \right) = -D \nabla^2 w_0^K \quad (11.1.15a)$$

$$\mathcal{M}^M = \frac{M_{xx}^M + M_{yy}^M}{1+\nu} = D \left(\frac{\partial \phi_x}{\partial x} + \frac{\partial \phi_y}{\partial y} \right) \quad (11.1.15b)$$

Equations (11.1.13a,b) can be combined into the single equation

$$\left(\nabla^2 + \frac{N^K}{D} \right) \nabla^2 w^K = 0 \quad (11.1.16)$$

Similarly, Eqs. (11.1.14a,b) yield

$$\left(\nabla^2 + \lambda^M \right) \nabla^2 w^M = 0 \quad (11.1.17)$$

where

$$\lambda^M = \frac{N^M}{D \left[1 - \left(\frac{N^M}{K_s Gh} \right) \right]} \quad (11.1.18)$$

For simply supported, isotropic polygonal plate the following boundary conditions hold:

$$w^K = 0, \quad \mathcal{M}^K = \nabla^2 w^K = 0 \text{ for the CPT} \quad (11.1.19)$$
$$w^M = 0, \quad \mathcal{M}^M = \nabla^2 w^M = 0 \text{ for the FSDT} \quad (11.1.20)$$

Comparing Eqs. (11.1.16) and (11.1.17), and in view of the boundary conditions (11.1.19) and (11.1.20), it follows that

$$\lambda^M = \frac{N^K}{D} \qquad (11.1.21)$$

or

$$N^M = \frac{N^K}{\left(1 + \frac{N^K}{K_s G h}\right)} \qquad (11.1.22)$$

which provides a relationship between the buckling loads N^K and N^M of a simply supported Kirchhoff plate and a simply supported Mindlin plate. Note that Eq. (11.1.22) is similar in form to Eq. (4.2.14) for columns, and the effect of shear deformation is to reduce the buckling load.

11.1.3 Relationships Between CPT and TSDT

By differentiating Eq. (11.1.5a) with respect to x and Eq. (11.1.5b) with respect to y, and adding them and using Eq. (11.1.5c), we arrive at the governing buckling equation

$$\frac{\partial^2 M_{xx}^R}{\partial x^2} + 2\frac{\partial^2 M_{xy}^R}{\partial x \partial y} + \frac{\partial^2 M_{yy}^R}{\partial y^2} = N^R \nabla^2 w^R \qquad (11.1.23)$$

Next, we introduce the moment sum

$$\mathcal{M}^R = \frac{M_{xx}^R + M_{yy}^R}{1+\nu} = \frac{4D}{5}\left(\frac{\partial \phi_x^R}{\partial x} + \frac{\partial \phi_y^R}{\partial y}\right) - \frac{D}{5}\nabla^2 w^R \qquad (11.1.24)$$

where the moment-deflection relations (11.1.11a,c) are used in arriving at the last result.

Using Eqs. (11.1.11a,c,e) in Eq. (11.1.23) and noting the definition (11.1.24), we obtain

$$\nabla^2 \mathcal{M}^R = N^R \nabla^2 w^R \qquad (11.1.25)$$

Next, from Eqs. (11.1.5c) and (11.1.11b,d,f-j), we have

$$\frac{8Gh}{15}\left(\frac{\partial \phi_x^R}{\partial x} + \frac{\partial \phi_y^R}{\partial y}\right) = \left(N^R - \frac{8Gh}{15}\right)\nabla^2 w^R - \frac{4}{21}\nabla^2 \mathcal{M}^R + \frac{D}{105}\nabla^4 w^R$$

$$(11.1.26)$$

Substituting for $\phi_{x,x} + \phi_{y,y}$ from Eq. (11.1.26) into Eq. (11.1.24), we obtain

$$\mathcal{M}^R = \frac{3D}{2Gh}\left(N^R - \frac{8Gh}{15}\right)\nabla^2 w^R - \frac{2D}{7Gh}\nabla^2\mathcal{M}^R + \frac{D}{70}\frac{D}{Gh}\nabla^4 w^R - \frac{D}{5}\nabla^2 w^R \quad (11.1.27)$$

Finally, using Eq. (11.1.25) one can eliminate $\nabla^2 w^R$ from Eq. (11.1.27)

$$\nabla^4\mathcal{M}^R - \frac{70Gh}{D}\left(1 - \frac{17}{14Gh}N^R\right)\nabla^2\mathcal{M}^R - \frac{70}{D}\frac{Gh}{D}\mathcal{M}^R = 0 \quad (11.1.28)$$

which can be expressed as

$$\left(\nabla^2 + \lambda_1^R\right)\left(\nabla^2 + \lambda_2^R\right)\mathcal{M}^R = 0 \quad (11.1.29)$$

where $(j = 1, 2)$

$$\lambda_j^R = -\xi_1 + (-1)^j\sqrt{\xi_1^2 + \xi_2} \quad (11.1.30a)$$

$$\xi_1 = \frac{35Gh}{D}\left(1 - \frac{17}{14Gh}N^R\right), \quad \xi_2 = \frac{70}{D}\frac{Gh}{D} \quad (11.1.30b)$$

Since λ_1^R is negative, it does not lead to a feasible buckling solution. Thus, the buckling equation of Reddy polygonal plate is governed by

$$\left(\nabla^2 + \lambda_2^R\right)\mathcal{M}^R = 0 \quad (11.1.31)$$

For polygonal plates with simply supported edges, the TSDT requires the specification of the following boundary conditions:

$$w^R = 0, \quad \mathcal{M}^R = 0 \quad (11.1.32)$$

In view of Eqs. (11.1.31), (11.1.32), (11.1.16), and (11.1.19), it follows that

$$\lambda^R = \frac{N^K}{D} \quad (11.1.33)$$

or, in view of the expressions for λ^R and λ^K, we obtain

$$N^R = \frac{N^K\left(1 + \frac{N^K}{70Gh}\right)}{1 + \frac{N^K}{\frac{14}{17}Gh}} \quad (11.1.34)$$

Figure 11.1.2 shows a comparison of the buckling loads predicted by the two theories for a simply supported, isotropic ($\nu = 0.3$) polygonal plate. Both the FSDT and TSDT predict virtually the same buckling loads.

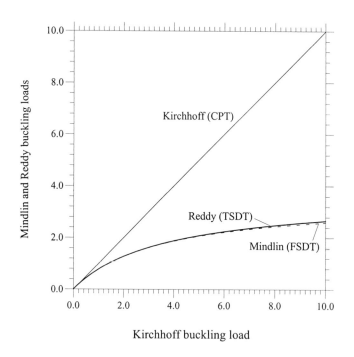

Figure 11.1.2. Comparison of the buckling loads of a simply supported plate as predicted by the Mindlin and Reddy plate theories.

It should be remarked that the relationships developed in this section are valid only for simply supported polygonal plates under uniform inplane forces (i.e., the same uniform load applied on all sides). For example, the relationships in Eqs. (11.1.22) and (11.1.34) do not hold for a simply supported rectangular plate subjected to biaxial loads (see Figure 11.1.3)

$$N_{xx} = N, \quad N_{yy} = \gamma N \tag{11.1.35}$$

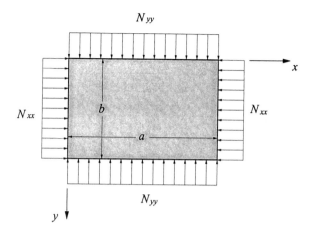

Figure 11.1.3. Rectangular plate under biaxial compression.

For this case, a relationship between the Kirchhoff and Mindlin plate can be derived using the solutions of the Kirchhoff plate theory and the Mindlin plate theory (see Timoshenko and Gere 1961, Herrmann and Armenakas 1960, and Reddy 1997a, 1999a)

$$N^K(m) = \frac{D\pi^2}{s^2 b^2} \frac{(s^2 + m^2)^2}{(\gamma s^2 + m^2)} \qquad (11.1.36)$$

$$N^M(m) = \frac{D\pi^2}{s^2 b^2} \frac{(s^2 + m^2)^2}{(\gamma s^2 + m^2)} \left[\frac{1}{1 + \kappa \pi^2 (1 + \frac{m^2}{s^2})} \right] \qquad (11.1.37)$$

where m is the number of half waves in the x–direction, and

$$s = \frac{a}{b}, \quad \kappa = \frac{h^2}{6 b^2 K_s (1 - \nu)} \qquad (11.1.38)$$

Although N^M can be expressed in terms of N^K as

$$N^M(m) = N^K(m) \left[\frac{1}{1 + \kappa \pi^2 (1 + \frac{m^2}{s^2})} \right] \qquad (11.1.39)$$

they do not necessarily correspond, in general, to the same number of half waves m. This is because $N^M(m)$ contains an additional factor involving m. In cases in which both theories yield the critical buckling load (i.e. the minimum buckling load) for the same half wave number m, it is possible to arrive at the following relationship

$$N^M = \frac{N^K}{1 + \frac{N^K}{2K_sGh}\left[1 + \sqrt{1 - \frac{4\pi^2(1-\gamma)D}{N^Kb^2}}\right]} \quad (11.1.40)$$

Note that Eq. (11.1.40) is independent of the aspect ratio s and the half-wave number m. When $\gamma = 1$ (i.e. uniform compression), the relationship in Eq. (11.1.40) reduces to the one in Eq. (11.1.22).

For buckling of rectangular plates under uniform in-plane shear load, Wang, Xiang, and Kitipornchai (1994) developed an approximate relationship in the same form as in Eq. (11.1.22):

$$N^M = \frac{N^K}{(1 + f\frac{N^K}{K_sGh})} \quad (11.1.41)$$

The preceding form of the formula must be used with the values of the modification factor f given in Table 11.1.1 for various boundary conditions. These values were generated using a curve-fitting exercise. A detailed comparison study between the solutions furnished by the formula and the solutions obtained from the Rayleigh–Ritz method assured that the maximum difference is 2.5% for $h/b \leq 0.15$. Comprehensive sets of thin plate solutions for N^K are given in Table 11.1.2 for ready use in conjunction with the formula (11.1.41).

Table 11.1.1. Modification factor f for various boundary conditions of rectangular plates under uniform shear load.

Boundary conditions	Modification factor f
SSSS	0.72
CCCC	0.94/[1+0.05(b/a)]
CCCS	0.85
CCSS	0.82
CSCS	0.74/[1-0.07(b/a)]
SCSC	0.99/[1+0.17(b/a)]

Table 11.1.2. Critical shear load factors $\lambda = N_{xy}b^2/(\pi^2 D)$ for thin (Kirchhoff) rectangular plates.

a/b	SSSS	CCCC	CCCS	CCSS	SCSC	SCSC
0.5	26.18	40.99	40.38	33.01	40.02	26.84
0.6	18.95	30.70	29.61	24.57	28.83	20.20
0.7	14.73	23.76	23.29	18.88	22.51	16.61
0.8	12.13	19.29	18.58	15.38	18.15	14.56
0.9	10.46	16.47	15.47	13.17	14.86	13.33
1.0	9.324	14.64	13.38	11.72	12.57	12.57
1.1	8.540	13.44	11.96	10.75	10.95	12.08
1.2	7.983	12.64	10.96	10.09	9.778	11.75
1.3	7.581	12.10	10.27	9.640	8.923	11.51
1.4	7.287	11.73	9.771	9.324	8.290	11.14
1.5	7.070	11.46	9.417	9.100	7.816	10.78
1.6	6.907	11.25	9.161	8.935	7.459	10.51
1.7	6.784	10.94	8.973	8.803	7.189	10.32
1.8	6.688	10.64	8.830	8.642	6.984	10.18
1.9	6.611	10.42	8.711	8.431	6.829	10.08
2.0	6.546	10.25	8.534	8.254	6.710	10.01
2.5	6.033	9.859	7.890	7.806	6.305	9.642
3.0	5.840	9.535	7.695	7.609	5.928	9.482
3.5	5.734	9.401	7.487	7.453	5.793	9.339
4.0	5.625	9.298	7.412	7.378	5.685	9.273
4.5	5.582	9.264	7.323	7.306	5.600	9.225
5.0	5.531	9.225	7.288	7.272	5.568	9.244

11.2 Circular Plates

11.2.1 Governing Equations

Consider an elastic, isotropic circular plate of radius R, uniform thickness h, Young's modulus E, shear modulus G and Poisson's ratio ν subjected to a uniform radial load N (see Figure 11.2.1). The governing equations of the classical plate theory (CPT), first-order shear deformation theory (FSDT), and third-order shear deformation theory (TSDT) for axisymmetric buckling of isotropic circular plates are summarized below.

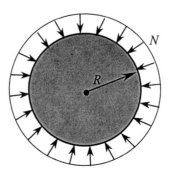

Figure 11.2.1. Circular plate under uniform compression.

CPT:

$$\frac{d}{dr}\left(rQ_r^K\right) = rN^K\nabla^2 w^R, \quad rQ_r^K \equiv \frac{d}{dr}\left(rM_{rr}^K\right) - M_{\theta\theta}^K \qquad (11.2.1)$$

$$M_{rr}^K = -D\left(\frac{d^2 w^K}{dr^2} + \frac{\nu}{r}\frac{dw^K}{dr}\right), \quad M_{\theta\theta}^K = -D\left(\nu\frac{d^2 w^K}{dr^2} + \frac{1}{r}\frac{dw^K}{dr}\right) \qquad (11.2.2)$$

FSDT:

$$\frac{d}{dr}\left(rQ_r^M\right) = rN^M\nabla^2 w^R, \quad rQ_r^M = \frac{d}{dr}\left(rM_{rr}^M\right) - M_{\theta\theta}^M \qquad (11.2.3)$$

$$M_{rr}^M = D\left(\frac{d\phi_r^M}{dr} + \frac{\nu}{r}\phi_r^M\right), \quad M_{\theta\theta}^M = D\left(\nu\frac{d\phi_r^M}{dr} + \frac{1}{r}\phi_r^M\right) \qquad (11.2.4a)$$

$$Q_r^M = K_s Gh\left(\phi_r^M + \frac{dw^M}{dr}\right) \qquad (11.2.4b)$$

TSDT:

$$\left(Q_r^R - \beta R_r^R\right) = \frac{d}{dr}\left(rM_{rr}^R - \alpha r P_{rr}^R\right) - \left(M_{\theta\theta}^R - \alpha P_{\theta\theta}^R\right) \qquad (11.2.5a)$$

$$\frac{d}{dr}\left(rQ_r^R - \beta r R_r^R\right) + \alpha\left[\frac{d^2}{dr^2}\left(rP_{rr}^R\right) - \frac{dP_{\theta\theta}^R}{dr}\right] = rN^R\nabla^2 w^R \qquad (11.2.5b)$$

$$M_{rr}^R = \frac{4D}{5}\left(\frac{d\phi_r^R}{dr} + \frac{\nu}{r}\phi_r^R\right) - \frac{D}{5}\left(\frac{d^2w^R}{dr^2} + \frac{\nu}{r}\frac{dw^R}{dr}\right) \qquad (11.2.6a)$$

$$M_{\theta\theta}^R = \frac{4D}{5}\left(\nu\frac{d\phi_r^R}{dr} + \frac{1}{r}\phi_r^R\right) - \frac{D}{5}\left(\nu\frac{d^2w^R}{dr^2} + \frac{1}{r}\frac{dw^R}{dr}\right) \qquad (11.2.6b)$$

$$P_{rr}^R = \frac{4h^2D}{35}\left(\frac{d\phi_r^R}{dr} + \frac{\nu}{r}\phi_r^R\right) - \frac{h^2D}{28}\left(\frac{d^2w^R}{dr^2} + \frac{\nu}{r}\frac{dw^R}{dr}\right) \qquad (11.2.6c)$$

$$P_{\theta\theta}^R = \frac{4h^2D}{35}\left(\nu\frac{d\phi_r^R}{dr} + \frac{1}{r}\phi_r^R\right) - \frac{h^2D}{28}\left(\nu\frac{d^2w^R}{dr^2} + \frac{1}{r}\frac{dw^R}{dr}\right) \qquad (11.2.6d)$$

$$Q_r^R = \frac{2Gh}{3}\left(\phi_r^R + \frac{dw^R}{dr}\right), \quad R_r^R = \frac{Gh^3}{30}\left(\phi_r^R + \frac{dw^R}{dr}\right) \qquad (11.2.6e)$$

Here the Laplace operator ∇^2 is understood to be in polar coordinates given by

$$\nabla^2 = \frac{d^2}{dr^2} + \frac{1}{r}\frac{d}{dr} \qquad (11.2.7)$$

11.2.2 Relationship Between CPT and FSDT

Equations (11.2.1) and (11.2.2) of the CPT and Eqs. (11.2.3) and (11.2.4a,b) of the FSDT can be reduced to

$$\frac{d^3\psi}{dr^3} + \frac{2}{r}\frac{d^2\psi}{dr^2} + (\lambda_0 - \frac{1}{r^2})\frac{d\psi}{dr} + \frac{1}{r}(\lambda_0 + \frac{1}{r^2})\psi = 0 \qquad (11.2.8)$$

where

$$\psi = \begin{cases} -\frac{dw}{dr}, & \text{for CPT} \\ \phi^M, & \text{for FSDT} \end{cases} \qquad (11.2.9)$$

$$\lambda_0 = \begin{cases} \frac{N^K}{D}, & \text{for CPT} \\ \frac{N^M}{1 - \frac{N^M}{K_sGh}}, & \text{for FSDT} \end{cases} \qquad (11.2.10)$$

Equation ψ is subject to the boundary conditions

At $r = R$:

$\psi = 0$ for clamped plates

$\dfrac{d\psi}{dr} + \dfrac{\nu}{r}\psi = 0$ for simply supported plates

$\dfrac{d\psi}{dr} + \dfrac{\nu}{r}\psi = k_2 \psi$ for rotational elastic restraint

At $r = 0$:

$\psi = 0$ for all boundary conditions (11.2.11)

where k_2 is the rotational spring constant. In view of the similarity of the governing equations and boundary conditions, we obtain

$$N^M = \dfrac{N^K}{\left(1 + \dfrac{N^K}{K_s G h}\right)} \qquad (11.2.13)$$

A relationship similar to Eq. (11.2.13) was obtained by Hong, Wang, and Tan (1993) for circular plates allowing for in-plane pre-buckling deformation.

11.2.3 Relationship Between CPT and TSDT

Introducing the higher order moment sum $\mathcal{P}^R$ as

$$\mathcal{P}^R = \dfrac{P_{rr}^R + P_{\theta\theta}^R}{1+\nu} = \dfrac{4h^2 D}{35} \dfrac{1}{r}\dfrac{d}{dr}(r\phi_r^R) - \dfrac{h^2 d}{28}\nabla^2 w^R \qquad (11.2.14)$$

we can write Eq. (11.2.5b) as

$$\dfrac{d}{dr}\left(rQ_r^R - r\dfrac{4}{h^2}R_r^R\right) + \dfrac{4}{3h^2}r\nabla^2 \mathcal{P}^R = rN^R \nabla^2 w^R \qquad (11.2.15)$$

The substitution of Eq. (11.2.5a) into Eq. (11.2.15) leads to

$$\mathcal{M}^R = N^R \nabla^2 w^R \qquad (11.2.16)$$

where the moment sum $\mathcal{M}^R$ is defined as

$$\mathcal{M}^R = \dfrac{M_{rr}^R + M_{\theta\theta}^R}{1+\nu} = \dfrac{4D}{5}\dfrac{1}{r}\dfrac{d}{dr}(r\phi_r^R) - \dfrac{D}{5}\nabla^2 w^R \qquad (11.2.17)$$

By substituting Eqs. (11.2.6e) and (11.2.14) into Eq. (11.2.15), one obtains

$$\frac{8Gh}{15}\left[\frac{1}{r}\frac{d}{dr}(r\phi_r^R) + \nabla^2 w^R\right] + \frac{16D}{105}\nabla^2\left[\frac{1}{r}\frac{d}{dr}(r\phi_r^R)\right]\frac{D}{21}\nabla^4 w^R = N^R\nabla^2 w^R \quad (11.2.18)$$

From Eq. (11.2.17), we have the relation

$$\nabla^2\left[\frac{1}{r}\frac{d}{dr}(r\phi_r^R)\right] = \frac{5}{4D}\nabla^2\mathcal{M}^R + \frac{1}{4}\nabla^4 w^R \quad (11.2.19)$$

In view of Eqs. (11.2.16), (11.2.17) and (11.2.19), we may express Eq. (11.2.18) as

$$\nabla^4\mathcal{M}^R - \left(\frac{420(1-\nu)}{h^2} - \frac{85}{D}N^R\right)\nabla^2\mathcal{M}^R - \left(\frac{420(1-\nu)}{h^2}N^R\right)\mathcal{M}^R = 0 \quad (11.2.20)$$

Equation (11.2.20) can be expressed in the form

$$(\nabla^2 + \lambda_1^R)(\nabla^2 + \lambda_2^R)\mathcal{M}^R = 0 \quad (11.2.21a)$$

or

$$(\nabla^2 + \lambda_1^R)(\nabla^2 + \lambda_2^R)\nabla^2 w^R = 0 \quad (11.2.21b)$$

where

$$\lambda_{1,2}^R = -\xi_1 \pm \sqrt{\xi_1^2 + \xi_2} \quad (11.2.22a)$$

$$\xi_1 = \frac{210(1-\nu)}{h^2} - \frac{85}{2D}N^R, \quad \xi_2 = \frac{410(1-\nu)}{h^2}N^R \quad (11.2.22b)$$

The general solution to Eq. (11.2.21) is of the form

$$w^R(r) = C_1 + C_2 \ln r + C_3 J_0(\sqrt{\lambda_1^R}r) + C_4 K_0(\sqrt{\lambda_1^R}r)$$
$$+ C_5 J_0(\sqrt{\lambda_2^R}r) + C_6 K_0(\sqrt{\lambda_2^R}r) \quad (11.2.23)$$

where J_0 and K_0 are the Bessel functions, and C_i are constants to be determined using the boundary conditions. We have

At $r = R$:
$$w^R = \frac{dw^R}{dr} = \phi_r^R = 0 \text{ for clamped plates}$$
$$w^R = M_{rr}^R = P_{rr}^R = 0 \text{ for simply supported plates}$$
$$w^R = P_{rr}^R = 0, \quad M_{rr}^R = k_2\phi_r^R \text{ for rotational elastic restraint}$$

At $r = 0$:
$$\frac{dw^R}{dr} = 0 \quad \frac{d}{dr}\left(rM_{rr}^R - \alpha r P_{rr}^R\right) - \left(M_{\theta\theta}^R - \alpha P_{\theta\theta}^R\right) = 0$$
$$\text{for all boundary conditions} \quad (11.2.24)$$

where k_2 is the rotational spring constant. For example, boundary conditions for the clamped plate yield $C_2 = C_4 = C_6 = 0$ and

$$\begin{bmatrix} 1 & 1 & 1 \\ 0 & J_0'(\sqrt{\lambda_1^R}R) & J_0'(\sqrt{\lambda_2^R}R) \\ 0 & -(\lambda_1^R)^2 J_0'(\sqrt{\lambda_1^R}R) & -(\lambda_2^R)^2 J_0'(\sqrt{\lambda_2^R}R) \end{bmatrix} \begin{Bmatrix} C_1 \\ C_3 \\ C_5 \end{Bmatrix} = \begin{Bmatrix} 0 \\ 0 \\ 0 \end{Bmatrix} \quad (11.2.25)$$

or

$$J_1(\sqrt{\lambda_1^R}R)\, J_1(\sqrt{\lambda_2^R}R) = 0 \quad (11.2.26)$$

The same type of equation holds for the Kirchhoff plate theory with $\lambda_1^R = \lambda_2^R = \lambda^K$. Hence, by analogy, we have

$$N^R = \frac{N^K \left(1 + \frac{N^K}{70Gh}\right)}{1 + \frac{N^K}{\frac{14}{17}Gh}} \quad (11.2.27)$$

The foregoing relationship given in Eq. (11.2.27) is valid for circular plates with any homogeneous edge condition such as (i) simply supported edges, (ii) clamped edges, (iii) simply supported edges with elastic rotational restraints and (iv) free edges with the centre clamped. Cases (i) and (iv) produce identical buckling solutions. Now, the Kirchhoff buckling solution for these plate cases may be unified and expressed as

$$\sqrt{\frac{N^R R^2}{D}} J_0\left(\sqrt{\frac{N^K r^2}{D}}\right) + \left[\frac{k_2 R}{D} - (1-\nu)\right] J_1\left(\sqrt{\frac{N^K R^2}{D}}\right) = 0 \quad (11.2.28)$$

where $J_0(\cdot)$ and $J_1(\cdot)$ are Bessel functions of the first kind of order 0 and 1, respectively, and k_2 is the rotational spring stiffness with extreme values covering the two ideal edges of simply supported ($k_2 = 0$) and clamped ($k_2 = \infty$).

11.2.4 Numerical Results

Table 11.2.1 presents the Kirchhoff, Mindlin, Reddy and Ye's buckling factors NR^2/D for circular plates with various values of the thickness to radius ratio h/R, elastic rotational restraint parameter $k_2 R/D$ and Poisson's ratio $\nu = 0.3$. Note that the Kirchhoff buckling factor is independent of h/R due to the neglect of transverse shear deformation. Both the Mindlin and Reddy results are very close to each other but are somewhat lower than the three-dimensional elasticity solution of Ye (1995). Ye (1995) derived the buckling load of circular plates from three-dimensional elasticity considerations. The analysis is based on a recursive formulation that results in the need to solve for only the roots of a 2 × 2 determinant for the buckling load.

Table 11.2.1. Comparison of buckling load factors for circular plate based on different theories.

$\frac{h}{R}$	$\frac{k_2 R}{D}$	CPT	FSDT	TSDT	Ye (1995)
	0	4.1978	4.1853	4.1853	
0.05	1	6.3532	6.3245	6.3245	
	10	12.173	12.068	12.068	
	∞	14.682	14.530	14.530	14.552
	0	4.1978	4.1481	4.1481	
0.10	1	6.3532	6.2399	6.2400	
	10	12.173	11.764	11.764	
	∞	14.682	14.091	14.091	14.177
	0	4.1978	4.0056	4.0057	
0.20	1	6.3532	5.9231	5.9235	
	10	12.173	10.686	10.688	
	∞	14.682	12.572	12.576	12.824
	0	4.1978	3.7888	3.7893	
0.30	1	6.3532	5.4610	5.4625	
	10	12.173	9.2710	9.2792	
	∞	14.682	10.658	10.671	11.024

11.3. Sectorial Mindlin Plates

11.3.1 Governing Equations

The buckling load relationship derived in the last section also applies to sectorial plates with simply supported edges and may be applied to sectorial plates with simply supported radial edges and either a clamped or a free circular edge. The availability of this relationship allows easy and accurate deduction of buckling loads of the Mindlin plates from their corresponding Kirchhoff plate solutions.

Consider an elastic, isotropic, sectorial plate with uniform thickness h, radius R, included angle α, Young's modulus E, Poisson's ratio ν, and shear modulus $G = E/[2(1+\nu)]$. The sectorial plate is simply supported along the radial edges defined by $\theta = 0$ and $\theta = \alpha$ and its circular edge as well. The plate is subjected to uniform in-plane compressive load N, as shown in Figure 11.3.1.

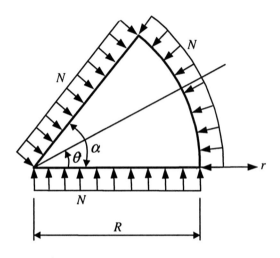

Figure 11.3.1. Buckling of a sectorial plate under compressive force N.

According to the Mindlin plate theory, the equations for buckling in polar coordinates (r, θ) are given by

$$\frac{\partial M_{rr}^M}{\partial r} + \frac{1}{r}\frac{\partial M_{r\theta}^M}{\partial \theta} + \frac{M_{rr}^M - M_{\theta\theta}^M}{r} - Q_r^M = 0 \qquad (11.3.1)$$

$$\frac{\partial M_{r\theta}^M}{\partial r} + \frac{1}{r}\frac{\partial M_{\theta\theta}^M}{\partial \theta} + \frac{2M_{r\theta}^M}{r} - Q_\theta^M = 0 \qquad (11.3.2)$$

$$\frac{\partial Q_r^M}{\partial r} + \frac{1}{r}\frac{\partial Q_\theta^M}{\partial \theta} + \frac{Q_r^M}{r} - N^M \nabla^2 w^M = 0 \qquad (11.3.3)$$

where

$$\nabla^2 = \frac{\partial^2}{\partial r^2} + \frac{1}{r}\frac{\partial}{\partial r} + \frac{1}{r^2}\frac{\partial^2}{\partial \theta^2}$$

is the Laplacian operator, w^M is the transverse displacement, and the bending moments per unit length $M_{rr}^M, M_{\theta\theta}^M, M_{r\theta}^M$ and shear forces per unit length Q_r^M, Q_θ^M are given by

$$M_{rr}^M = D\left[\frac{\partial \phi_r^M}{\partial r} + \frac{\nu}{r}\left(\phi_r^M + \frac{\partial \phi_\theta^M}{\partial \theta}\right)\right] \qquad (11.3.4)$$

$$M_{\theta\theta}^M = D\left[\nu\frac{\partial \phi_r^M}{\partial r} + \frac{1}{r}\left(\phi_r + \frac{\partial \phi_\theta^M}{\partial \theta}\right)\right] \qquad (11.3.5)$$

$$M_{r\theta}^M = \frac{D(1-\nu)}{2}\left(\frac{\partial \phi_\theta^M}{\partial r} - \frac{1}{r}\phi_\theta^M + \frac{1}{r}\frac{\partial \phi_r^M}{\partial \theta}\right) \qquad (11.3.6)$$

$$Q_r^M = K_s Gh\left(\phi_r^M + \frac{\partial w^M}{\partial r}\right) \qquad (11.3.7)$$

$$Q_\theta^M = K_s Gh\left(\phi_\theta^M + \frac{1}{r}\frac{\partial w^M}{\partial \theta}\right) \qquad (11.3.8)$$

where ϕ_r^M, ϕ_θ^M are the bending rotations, K_s is the shear correction factor and $D = Eh^3/[12(1-\nu^2)]$ the flexural rigidity of the plate.

By substituting the shear forces from Eqs. (11.3.1) and (11.3.2) into Eq. (11.3.3), and taking note of Eqs. (11.3.4)-(11.3.6), one can rewrite Eq. (11.3.3) as

$$N^M \nabla^2 w^M = \nabla^2 \mathcal{M}^M \qquad (11.3.9)$$

where $\mathcal{M}^M$ is the moment sum defined as

$$\mathcal{M}^M = \frac{M_{rr}^M + M_{\theta\theta}^M}{1+\nu} = D\left(\frac{\partial \phi_r^M}{\partial r} + \frac{1}{r}\phi_r^M + \frac{1}{r}\frac{\partial \phi_\theta^M}{\partial \theta}\right) \qquad (11.3.10)$$

In view of Eqs. (11.3.7), (11.3.8) and (11.3.10), one can also express Eq. (11.3.3) as

$$K_s Gh\left(\nabla^2 w^M + \frac{\mathcal{M}^M}{D}\right) - N^M \nabla^2 w^M = 0 \qquad (11.3.11)$$

Substituting the moment sum given by Eq. (11.3.11) into Eq. (11.3.9) furnishes

$$\nabla^2\left(\nabla^2 + \frac{N^M}{1 - \frac{N^M}{K_s Gh}}\right) w^M = 0 \qquad (11.3.12)$$

The boundary conditions for the simply supported circular edge of the sectorial Mindlin plate are

$$w^M(R,\theta) = 0, \quad \phi_\theta^M(R,\theta) = 0, \quad M_{rr}^M(R,\theta) = 0, \qquad (11.3.13)$$

and for the simply supported radial edges

$$w^M(r,0) = w^M(r,\alpha) = 0 \tag{11.3.14a}$$
$$\phi_r^M(r,0) = \phi_r^M(r,\alpha) = 0 \tag{11.3.14b}$$
$$M_{\theta\theta}^M(r,\theta) = M_{\theta\theta}^M(r,\alpha) = 0 \tag{11.3.14c}$$

The displacement functions for the considered sectorial plates may be assumed to take the following forms:

$$w^M(r,\theta) = W^M(r)\sin\frac{2\pi n\theta}{\alpha} \tag{11.3.15}$$

$$\phi_r^M(r,\theta) = \Phi_r^M(r)\sin\frac{2\pi n\theta}{\alpha} \tag{11.3.16}$$

$$\phi_\theta^M(r,\theta) = \Phi_\theta^M(r)\cos\frac{2\pi n\theta}{\alpha} \tag{11.3.17}$$

where n is the number of circumferential nodal diameters. In view of Eqs. (11.3.15)-(11.3.17) and Eqs. (11.3.3), (11.3.7), (11.3.8), (11.3.10), (11.3.11) and (11.3.13), the boundary conditions for the curved edge may be expressed as

$$w^M(R,\theta) = 0 \tag{11.3.18a}$$

and

$$\nabla^2 w^M(R,\theta) = \frac{M^M(R,\theta)}{CD} = \frac{1}{C}\left(\frac{d\Phi_r^M}{dr} + \frac{1}{r}\Phi_r^M\right)_{r=R}\sin\frac{2\pi n\theta}{\alpha}$$
$$= -\left(\frac{d^2 W^M}{dr^2} + \frac{1}{r}\frac{dW^M}{dr}\right)_{r=R}\sin\frac{2\pi n\theta}{\alpha} \tag{11.3.18b}$$

where $C = -1 + [N^M/(K_s G h)]$. Also, the boundary conditions given in Eq. (11.3.14) for the radial edges may be expressed as

$$w^M(r,0) = w^M(r,\alpha) = 0, \quad \nabla^2 w^M(r,0) = \nabla^2 w(r,\alpha) = -\frac{M^M}{D} = 0 \tag{11.3.19}$$

Now, let us consider the same buckling problem using the Kirchhoff plate theory. Based on this theory, the governing equation is given by

$$\nabla^2\left(\nabla^2 + \frac{N^K}{D}\right)w^K = 0 \tag{11.3.20}$$

where N^K and w^K are the buckling load and the transverse displacement of the CPT, respectively.

The boundary conditions for the simply supported circular edge of the sectorial plate in CPT are

$$w^K(R,\theta) = 0, \quad M_r^K(R,\theta) = 0, \qquad (11.3.21)$$

and for the simply supported, radial edges

$$w^K(r,0) = w^K(r,\alpha) = 0, \quad M_\theta^K(r,0) = M_\theta^K(r,\alpha) = 0 \qquad (11.3.22)$$

Using the function

$$w^K(r,\theta) = W^K(r) \sin \frac{2\pi n\theta}{\alpha} \qquad (11.3.23)$$

the boundary conditions given by Eq. (11.3.22) may be written as

$$w^K(R,\theta) = 0 \qquad (11.3.24a)$$

and

$$\nabla^2 w^K(R,\theta) = \left(\frac{d^2 W^K}{dr^2} + \frac{1}{r}\frac{dW^K}{dr}\right)_{r=R} \sin \frac{2\pi n\theta}{\alpha} \qquad (11.3.24b)$$

and Eq. (11.3.22) may be expressed as

$$w^K(r,0) = w^K(r,\alpha) = 0, \quad \nabla^2 w^K(r,0) = \nabla^2 w^K(r,\alpha) = 0 \qquad (11.3.25)$$

11.3.2 Buckling Load Relationship

In view of the governing equations (11.3.12) and (11.3.20), the boundary equations (11.3.18), (11.3.19) and (11.3.24), (11.3.25), one may deduce that

$$\frac{N^M}{1 - \frac{N^M}{K_s G h}} = N^K \quad \text{or} \quad N^M = \frac{N^K}{1 + \frac{N^K}{K_s G h}} \qquad (11.3.26)$$

This relationship, however, is not valid if the curved edge of the sectorial plate is either clamped or free because the expressions for their boundary conditions according to the Mindlin and Kirchhoff plate theories do not match exactly.

Table 11.3.1 shows the comparison studies between the buckling load factors obtained by (a) the Ritz method and (b) the buckling load relationship in Eq. (11.3.26). For details of the Ritz method, the reader may refer to the papers by Wang et al. (1994) and Xiang et al. (1993). From the table, it can be seen that the buckling results are in excellent good agreement, thus verifying the derived buckling load relationship. The small differences in results are due to round-off errors in the numerical calculations.

Table 11.3.1. Comparison of buckling load factors NR^2/D of simply supported sectorial plates ($\nu = 0.3$, $K_s = 5/6$).

α	h/r	Ritz	Eq. (11.3.26)
$\pi/6$	.001	96.92	-
	0.10	75.89	75.90
	0.20	45.97	45.98
$\pi/4$	.001	55.75	-
	0.10	48.08	48.09
	0.20	34.03	34.05
$\pi/3$	.001	38.85	-
	0.10	34.96	34.97
	0.20	26.88	26.90
$\pi/2$	.001	24.50	-
	0.10	22.89	22.90
	0.20	19.12	19.14
$3\pi/4$	.001	16.44	-
	0.10	15.65	15.70
	0.20	13.78	13.84
π	.001	12.69	-
	0.10	12.23	12.25
	0.20	11.06	11.08

When applied to sectorial plates with simply supported radial edges and either a clamped or a free circular edge, the relationship gives higher results. The relationship may, however, be made to give good results by introducing a modification factor f to the relationship as shown below

$$N^M = \frac{N^K}{1 + f\left(\frac{N^K}{K_s G h}\right)} \quad (11.3.27)$$

where the proposed factor is given by

$$f = \left(a_1 + a_2 \frac{h}{R}\right) \frac{h}{R} \quad (11.3.28)$$

with

$$a_1 = 16.39 + 0.14\alpha, \quad a_2 = -55.44 - 0.48\alpha \quad (11.3.29a)$$

Table 11.3.2. Comparison of buckling load factors NR^2/D of sectorial plates with simply supported radial edges and clamped circular edge ($\nu = 0.3$, $K_s = 5/6$).

α	h/R	Ritz	Eq. (11.3.27)
	.001	122.90	-
$\pi/6$	0.10	89.28	88.89
	0.20	49.72	49.24
	.001	76.94	-
$\pi/4$	0.10	62.10	62.05
	0.20	39.69	39.69
	.001	57.82	-
$\pi/3$	0.10	48.81	48.97
	0.20	33.70	33.88
	.001	40.71	-
$\pi/2$	0.10	36.12	36.10
	0.20	27.08	27.14
	.001	31.00	-
$3\pi/4$	0.10	28.18	28.23
	0.20	22.37	22.41
	.001	26.23	-
π	0.10	24.24	24.21
	0.20	19.83	19.78

for sectorial plates with clamped circular edge, and

$$a_1 = 18.9 + 7.78\alpha + 24.15\alpha^2, \quad a_2 = -59.28 - 53.94\alpha - 89.57\alpha^2 \tag{11.3.29b}$$

for sectorial plates with free circular edge. The above modification factors were obtained via regression analysis. Tables 11.3.2 and 11.3.3 show the accuracy of the modified relationship for sectorial plates with clamped circular edge and with free circular edge, respectively.

Table 11.3.3. Comparison of buckling load factors NR^2/D of sectorial plates with simply supported radial edges and free circular edge ($\nu = 0.3$, $K_s = 5/6$).

α	h/R	Ritz method	Eq. (11.3.27)
$\pi/6$	.001	35.77	-
	0.10	30.13	30.11
	0.20	22.61	22.54
$\pi/4$	.001	14.94	-
	0.10	13.53	13.54
	0.20	11.52	11.57
$\pi/3$	.001	7.73	-
	0.10	7.22	7.22
	0.20	6.51	6.50
$\pi/2$	.001	2.72	-
	0.10	2.61	2.61
	0.20	2.45	2.45

Problems

11.1 The equation governing the buckling of a biaxially loaded Kirchhoff plate is given by

$$D\left(\frac{\partial^4 w^K}{\partial x^4} + 2\frac{\partial^4 w^K}{\partial x^2 \partial y^2} + \frac{\partial^4 w^K}{\partial y^4}\right) = \hat{N}_{xx}\frac{\partial^2 w^K}{\partial x^2} + \hat{N}_{yy}\frac{\partial^2 w^K}{\partial y^2} \quad (i)$$

where $\hat{N}_{xx} < 0$ and $\hat{N}_{yy} < 0$ are the in-plane compressive forces on the edges of a rectangular plate. Suppose that

$$\hat{N}_{xx} = -N_0, \quad \hat{N}_{yy} = -\gamma N_0, \quad \gamma = \frac{\hat{N}_{yy}}{\hat{N}_{xx}} \qquad (ii)$$

and the edges are simply supported. Use the Navier solution procedure (see Problem 7.1) with

$$w(x,y) = W_{mn} \sin \alpha_m x \, \sin \beta_n y, \quad \alpha_m = \frac{m\pi}{a}, \quad \beta_n = \frac{n\pi}{b} \qquad (iii)$$

where a and b are the plate dimensions, to obtain the buckling load

$$N_0(m,n) = \frac{\pi^2 D}{b^2} \frac{(s^2 m^2 + n^2)}{s^2 m^2 + \gamma n^2} \qquad (iv)$$

where $s = b/a$ is the plate aspect ratio.

11.2 Consider the buckling of uniformly compressed rectangular plates simply supported along two opposite edges perpendicular to the direction of compression (see Figure P11.2) and having various edge conditions along the other two sides. For the case of uniform compression along the x axis, we have $\hat{N}_{xx} = -N_0$ and $\hat{N}_{yy} = 0$, and Eq. (i) of Problem 11.1 reduces to

$$D\left(\frac{\partial^4 w}{\partial x^4} + 2\frac{\partial^4 w}{\partial x^2 \partial y^2} + \frac{\partial^4 w}{\partial y^4}\right) = -N_0 \frac{\partial^2 w}{\partial x^2} \qquad (i)$$

This equation must be solved for the buckling load N_0 and mode shape w for any given boundary conditions.

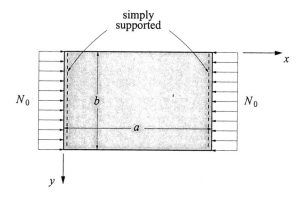

Figure P11.2

Assume the solution of Eq. (i) in the form

$$w(x,y) = W(y) \sin \alpha_m x, \quad \alpha_m = \frac{m\pi}{a} \quad (ii)$$

which satisfies the boundary conditions along the simply supported edges $x = 0, a$ of the plate. Use Eq. (ii) in Eq. (i) of Problem 11.1 and obtain

$$\left(\alpha_m^4 - \frac{N_0}{D}\alpha_m^2\right)W - 2\alpha_m^2 \frac{d^2W}{dy^2} + \frac{d^4W}{dy^4} = 0 \quad (iii)$$

Obtain the general solution of Eq. (iii) when

$$\frac{N_0}{D} > \alpha_m^2 \quad (iv)$$

In particular, show that

$$W(y) = C_1 \cosh \lambda_1 y + C_2 \sinh \lambda_1 y + C_3 \cos \lambda_2 y + C_4 \sin \lambda_2 y \quad (v)$$

where C_i ($i = 1, 2, 3, 4$) are constants, and

$$(\lambda_1)^2 = \sqrt{\alpha_m^2 \frac{N_0}{D} + \alpha_m^2}, \quad (\lambda_2)^2 = \sqrt{\alpha_m^2 \frac{N_0}{D} - \alpha_m^2} \quad (vi)$$

11.3 Consider the buckling of uniformly compressed rectangular plates with side $y = 0$ simply supported and side $y = b$ free (see Figure P11.3). The boundary conditions on the simply supported and free edges are

$$w = 0, \quad M_{yy} = -D\left(\nu\frac{\partial^2 w}{\partial x^2} + \frac{\partial^2 w}{\partial y^2}\right) = 0 \text{ at } y = 0 \quad (i)$$

$$M_{yy} = 0, \quad V_y = -D\left[\frac{\partial^3 w}{\partial y^3} + (1-\nu)\frac{\partial^3 w}{\partial x^2 \partial y}\right] = 0 \text{ at } y = b \quad (ii)$$

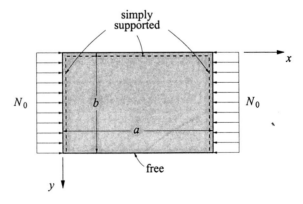

Figure P11.3

Use the above boundary conditions to show that the constants in Eq. (v) of Problem 11.2 are given by $C_1 = C_3 = 0$, and

$$\left(\nu\alpha_m^2 - \lambda_1^2\right)C_2 \sinh \lambda_1 b + \left(\nu\alpha_m^2 + \lambda_2^2\right)C_4 \sin \lambda_2 b = 0$$

$$\lambda_1\left[(1-\nu)\alpha_m^2 - \lambda_1^2\right]C_2 \cosh \lambda_1 b + \lambda_2\left[(1-\nu)\alpha_m^2 + \lambda_2^2\right]C_4 \cos \lambda_2 b = 0 \quad (iii)$$

Show that the characteristic equation associated with these equations is

$$\lambda_2 \Omega_1^2 \sinh \lambda_1 b \, \cos \lambda_2 b - \lambda_1 \Omega_2^2 \cosh \lambda_1 b \, \sin \lambda_2 b = 0 \quad (iv)$$

where Ω_1 and Ω_2 are defined by

$$\Omega_1 = \left(\lambda_1^2 - \nu\alpha_m^2\right), \quad \Omega_2 = \left(\lambda_2^2 + \nu\alpha_m^2\right) \quad (v)$$

11.4 Derive the relationship in Eq. (11.1.40) using Eqs. (11.1.36) and (11.1.37).

11.5 Derive the governing buckling equation in Eq. (11.2.8) using Eqs. (11.2.3) and (11.2.4a,b).

Chapter 12

Free Vibration Relationships

This chapter presents exact relationships between the natural frequencies of the classical Kirchhoff plate theory, the Mindlin plate theory and the Reddy plate theory for simply supported, polygonal isotropic plates, including rectangular plates. The relationship for the natural frequencies enables one to obtain the solutions of the shear deformable plate theories from the known Kirchhoff plate theory for the same problem. As examples, some vibration frequencies for rectangular and regular polygonal plates are determined using this relationship.

12.1 Introduction

To derive the equations of motion, Hamilton's principle is used. The Hamilton principle can be expressed as

$$\delta \int_{t_1}^{t_2} (U + V - K)\, dt = 0 \qquad (12.1.1)$$

where U denotes the strain energy, V the potential energy due to applied loads, K the kinetic energy, and t_1 and t_2 are initial and final times. The strain and potential energy functionals for the CPT, FSDT, and TSDT are given in Chapter 6 with all the displacement components being functions of time as well. The kinetic energy is given by

$$K = \frac{1}{2} \int_{\Omega_0} \left\{ \int_{-\frac{h}{2}}^{\frac{h}{2}} \rho \left[\left(\frac{\partial u_1}{\partial t}\right)^2 + \left(\frac{\partial u_2}{\partial t}\right)^2 + \left(\frac{\partial u_3}{\partial t}\right)^2 \right] dz \right\} dx\, dy \qquad (12.1.2)$$

where ρ is the mass density of the plate, h the plate thickness, and (u_1, u_2, u_3) are the displacements along the (x, y, z) coordinates. The assumed displacement expansions, for the pure bending case, in the Kirchhoff, Mindlin, and Reddy plate theories are given by Eqs. (6.1.1a-c), (6.1.2a-c), and (6.1.4a-c), respectively. Substituting the displacement expansions into Eq. (12.1.2) one can obtain the total kinetic energy in terms of the generalized displacements of each theory, as given below for uniform thickness homogeneous plates.

CPT:

$$K^K = \frac{1}{2}\int_{\Omega_0}\left\{I_0\left(\frac{\partial w_0^K}{\partial t}\right)^2 + I_2\left[\left(\frac{\partial^2 w_0^K}{\partial x \partial t}\right)^2 + \left(\frac{\partial^2 w_0^K}{\partial y \partial t}\right)^2\right]\right\}dxdy$$
(12.1.3)

FSDT:

$$K^M = \frac{1}{2}\int_{\Omega_0}\left\{I_0\left(\frac{\partial w_0^M}{\partial t}\right)^2 + I_2\left[\left(\frac{\partial \phi_x}{\partial t}\right)^2 + \left(\frac{\partial \phi_y}{\partial t}\right)^2\right]\right\}dxdy$$
(12.1.4)

TSDT:

$$\begin{aligned}K^R = \frac{1}{2}\int_{\Omega_0}&\left\{I_0\left(\frac{\partial w_0^R}{\partial t}\right)^2 + \bar{I}_2\left[\left(\frac{\partial \phi_x}{\partial t}\right)^2 + \left(\frac{\partial \phi_y}{\partial t}\right)^2\right]\right.\\&+ \alpha_0^2 I_6\left[\left(\frac{\partial^2 w_0^R}{\partial x \partial t}\right)^2 + \left(\frac{\partial^2 w_0^R}{\partial y \partial t}\right)^2\right]\\&\left.+ 2\alpha\bar{I}_4\left(\frac{\partial \phi_x}{\partial t}\frac{\partial^2 w_0^R}{\partial x \partial t} + \frac{\partial \phi_y}{\partial t}\frac{\partial^2 w_0^R}{\partial y \partial t}\right)\right\}dxdy\quad(12.1.5)\end{aligned}$$

where $\alpha = 4/3h^2$ and

$$I_i = \int_{-\frac{h}{2}}^{\frac{h}{2}} \rho(z)^i\, dz \qquad (12.1.6)$$

$$\bar{I}_2 = I_2 - 2\alpha I_4 + \alpha^2 I_6, \quad \bar{I}_4 = I_4 - \alpha I_6 \qquad (12.1.7)$$

or

$$I_0 = \rho h, \quad I_2 = \frac{\rho h^3}{12}, \quad I_4 = \frac{\rho h^5}{80}, \quad I_6 = \frac{\rho h^7}{448} \qquad (12.1.8a)$$

$$\bar{I}_2 = \frac{17}{315}\rho h^3, \quad \bar{I}_4 = \frac{1}{105}\rho h^5 \qquad (12.1.8b)$$

Employing Hamilton's principle (or the dynamic version of the principle of virtual displacements), one can derive the equations of motion associated with the theories [see Reddy (1999) for the details]. For free vibration, we assume periodic motion of the form

$$w_0(x,y,t) = w(x,y)e^{i\omega t} \qquad (12.1.9a)$$
$$\phi_x(x,y,t) = \varphi_x(x,y)e^{i\omega t} \qquad (12.1.9b)$$
$$\phi_y(x,y,t) = \varphi_y(x,y)e^{i\omega t} \qquad (12.1.9c)$$

where ω is the the angular frequency, and reduce the equations of motion to those governing free vibration. The equations of free vibration for various theories are summarized below. In the interest of simplicity, ϕ_x is used in place of φ_x, and ϕ_y is used in place of φ_y.

CPT:

$$\frac{\partial M_{xx}^K}{\partial x} + \frac{\partial M_{xy}^K}{\partial y} - Q_x^K = 0 \qquad (12.1.10a)$$

$$\frac{\partial M_{xy}^K}{\partial x} + \frac{\partial M_{yy}^K}{\partial y} - Q_y^K = 0 \qquad (12.1.10b)$$

$$\frac{\partial Q_x^K}{\partial x} + \frac{\partial Q_y^K}{\partial y} = -\rho h \omega_K^2 w^K \qquad (12.1.10c)$$

FSDT:

$$\frac{\partial M_{xx}^M}{\partial x} + \frac{\partial M_{xy}^M}{\partial y} - Q_x^M = -\frac{\rho h^3}{12}\omega_M^2 \phi_x^M \qquad (12.1.11a)$$

$$\frac{\partial M_{xy}^M}{\partial x} + \frac{\partial M_{yy}^M}{\partial y} - Q_y^M = -\frac{\rho h^3}{12}\omega_M^2 \phi_y^M \qquad (12.1.11b)$$

$$\frac{\partial Q_x^M}{\partial x} + \frac{\partial Q_y^M}{\partial y} = -\rho h \omega_M^2 w^M \qquad (12.1.11c)$$

TSDT:

$$\frac{\partial}{\partial x}\left(M_{xx}^R - \frac{4}{3h^2}P_{xx}^R\right) + \frac{\partial}{\partial y}\left(M_{xy}^R - \frac{4}{3h^2}P_{xy}^R\right) - \left(Q_x^R - \frac{4}{h^2}R_x^R\right)$$
$$= -\frac{17\rho h^3}{315}\omega_R^2 \phi_x^R + \frac{4\rho h^3}{315}\omega_R^2 \frac{\partial w^R}{\partial x} \qquad (12.1.12a)$$

$$\frac{\partial}{\partial x}\left(M_{xy}^R - \frac{4}{3h^2}P_{xy}^R\right) + \frac{\partial}{\partial y}\left(M_{yy}^R - \frac{4}{3h^2}P_{yy}^R\right) - \left(Q_y^R - \frac{4}{h^2}R_y^R\right)$$

$$= -\frac{17\rho h^3}{315}\omega_R^2\phi_y^R + \frac{4\rho h^3}{315}\omega_R^2\frac{\partial w^R}{\partial y} \quad (12.1.12b)$$

$$\frac{\partial}{\partial x}\left(Q_x^R - \frac{4}{h^2}R_x^R\right) + \frac{\partial}{\partial y}\left(Q_y^R - \frac{4}{h^2}R_y^R\right)$$
$$+ \frac{4}{3h^2}\left(\frac{\partial^2 P_{xx}^R}{\partial x^2} + 2\frac{\partial^2 P_{xy}^R}{\partial x\partial y} + \frac{\partial^2 P_{yy}^R}{\partial y^2}\right)$$
$$= \frac{\rho h^3}{252}\omega_R^2 \nabla^2 w^R - \rho h \omega_R^2 - \frac{4\rho h^3}{315}\omega_R^2\left(\frac{\partial \phi_x^R}{\partial x} + \frac{\partial \phi_y^R}{\partial y}\right) \quad (12.1.12c)$$

12.2 Relationship Between CPT and FSDT
12.2.1 General Relationship

One can derive an exact relationship between the natural frequencies of FSDT and those of the corresponding CPT. This relationship is, however, restricted to a class of polygonal plates in which all the straight edges are simply supported.

Substituting for Q_x^K and Q_y^K from Eqs. (12.1.10a,b) into Eq. (12.1.10c), we obtain

$$\nabla^4 w^K - \frac{\rho h}{D}\omega_K^2 w^K = 0 \quad (12.2.1)$$

Similarly, from Eqs. (12.1.11a-c), we obtain

$$\nabla^2 \mathcal{M}^M - \frac{K_s G h}{D}\left(\mathcal{M}^M + \nabla^2 w^M\right) = -\frac{\rho h^3}{12D}\omega_M^2 \mathcal{M}^M \quad (12.2.2)$$

$$K_s G h\left(\mathcal{M}^M + \nabla^2 w^M\right) = -\rho h \omega_M^2 w^M \quad (12.2.3)$$

where $\mathcal{M}^M$ is the moment sum

$$\mathcal{M}^M = \frac{M_{xx}^M + M_{yy}^M}{1+\nu} \quad (12.2.4)$$

Eliminating $\mathcal{M}^M$ from Eqs. (12.2.2) and (12.2.3), one obtains

$$\nabla^4 w^M + \left(\frac{\rho h}{K_s G h} + \frac{\rho h^3}{12D}\right)\omega_M^2 \nabla^2 w^M + \frac{\rho h}{D}\left(\frac{\rho h^3 \omega_M^2}{12K_s G h} - 1\right)\omega_M^2 w^M = 0$$
$$(12.2.5)$$

Equation (12.2.5) may be factored to give

$$\left(\nabla^2 + \lambda_1\right)\left(\nabla^2 + \lambda_2\right) w^M = 0 \qquad (12.2.6)$$

where

$$\lambda_j = \xi_1 + (-1)^j \sqrt{\xi_1^2 + \xi_2} \qquad (12.2.7)$$

$$\xi_1 = \frac{1}{2}\left(\frac{\rho h}{K_s G h} + \frac{\rho h^3}{12 D}\right) \omega_M^2, \quad \xi_2 = \frac{\rho h}{D} \omega_M^2 \qquad (12.2.8)$$

Alternatively, Eq. (12.2.6) may be written as two second order equations given by [see Pnueli (1975)]

$$\left(\nabla^2 + \lambda_i\right) w^M = \bar{w}, \quad \left(\nabla^2 + \lambda_j\right) \bar{w} = 0 \qquad (12.2.9)$$

where $i = 1$ if $j = 2$ and vice versa.

For the hard type of simply supported (S) polygonal plate, the boundary conditions are given by Eqs. (6.3.22). Since along the straight edge $\phi_s^M = 0$ implies that $\partial \phi_s^M / \partial s = 0$, then together with the condition $M_{nn}^M = 0$, one may deduce that $\partial \phi_n^M / \partial n = 0$. In view of this fact and Eq. (12.2.3), the boundary conditions may be given as

$$w^M = 0, \quad \mathcal{M}^M = 0, \quad \nabla^2 w^M = 0, \quad \bar{w} = 0 \qquad (12.2.10)$$

Note that the governing equation for Kirchhoff plates in vibration given by Eq. (12.2.1) may be obtained from Eq. (12.2.5) by setting $K_s \to \infty$ and omitting the rotary inertia term $\left[= \left(\rho h^3 \omega^2 \nabla^2 w\right)/(12D)\right]$. Similarly, Eq. (12.2.1) of Kirchhoff plate may be factored to give

$$\left(\nabla^4 - \lambda_K^2\right) w^K = \left(\nabla^2 + \lambda_K\right)\left(\nabla^2 - \lambda_K\right) w^K = 0 \qquad (12.2.11)$$

$$\lambda_K^2 = \frac{\rho h}{D} \left(\omega_K\right)^2 \qquad (12.2.12)$$

For a simply supported polygonal Kirchhoff plate, the deflection and the Kirchhoff Marcus moment are zero at the boundary, i.e.

$$w^K = 0, \quad \mathcal{M}^K = -D \nabla^2 w^K = 0 \qquad (12.2.13)$$

As pointed out by Conway (1960), and later proven by Pnueli (1975), the frequency solutions of the fourth order differential equation (12.2.11) and the boundary conditions given by Eq. (12.2.13) are the same as those given by solving simply the following second order differential equation

$$\left(\nabla^2 + \lambda_K\right) w^K = 0 \qquad (12.2.14)$$

and the boundary condition $w^K = 0$.

Owing to the mathematical similarity of Eqs. (12.2.9) and (12.2.10) with Eqs. (12.2.13) and (12.2.14), it follows that the aforementioned Mindlin plate vibration problem is analogous to the Kirchhoff plate vibration problem. Thus, for a given simply supported polygonal plate

$$\lambda_j = \lambda_K \qquad (12.2.15)$$

The substitution of Eqs. (12.2.8) and (12.2.12) into Eq. (12.2.15) furnishes the frequency relationship between the two kinds of plates

$$(\omega_M^2)_N = \frac{6K_s G}{\rho h^2} \left\{ \left[1 + \frac{h^2}{12}(\omega_K)_N \sqrt{\frac{\rho h}{D}} \left(1 + \frac{2}{K_s(1-\nu)}\right) \right] \right.$$
$$\left. - \sqrt{\left[1 + \frac{h^2}{12}(\omega_K)_N \sqrt{\frac{\rho h}{D}} \left(1 + \frac{2}{K_s(1-\nu)}\right)\right]^2 - \frac{\rho h^2}{3K_s G}(\omega_K^2)_N} \right\}$$
$$(12.2.16)$$

where $N = 1, 2, \ldots$, corresponds to the mode sequence number.

If the rotary inertia effect is neglected, it can be shown that the frequency relationship simplifies to

$$(\hat{\omega}_M^2)_N = \frac{(\omega_K^2)_N}{1 + \frac{(\omega_K^2)_N h^2}{6(1-\nu)K_s}\sqrt{\frac{\rho h}{D}}} \qquad (12.2.17)$$

where $\hat{\omega}_M$ is the frequency of Mindlin plate without the rotary inertia effect. This frequency value $\hat{\omega}_M$ is greater than its corresponding ω_M but smaller than ω_K.

12.2.2 Numerical Results

Graphical representations of the frequency relationships given by Eqs. (12.2.16) and (12.2.17) are shown in Figure 12.2.1, where $\nu = 0.3$ and $K_s = 5/6$ have been assumed. By nondimensionalizing the circular frequency using ρ, h, and D, the curves shown in Figure 12.2.1 become independent of the plate shape! Note that as one moves along the curves away from the origin, the plate gets thicker or the frequency value becomes higher. It is clear from the figure that when the frequencies are low (lower modes of frequency or thin plates), the FSDT solutions are close to the CPT solutions. When the plate thickness increases and for higher mode frequencies, the FSDT solutions decrease relative to the CPT solutions. The effect of rotary inertia is also shown in the same figure and it can be seen that this effect becomes significant for high frequency values. The effect of shear deformation is to reduce the magnitude of frequencies.

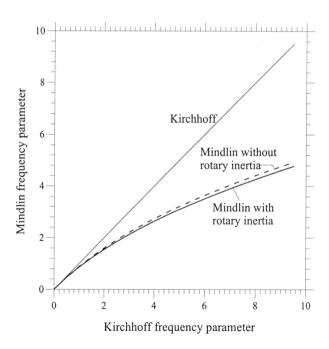

Figure 12.2.1. Frequency relationship between CPT and FSDT.

Although the relationship given in Eq. (12.2.16) is exact only for polygonal plates with straight edges, it has been shown by Wang (1994) that the relationship provides reasonably accurate frequencies for the FSDT from the CPT solutions even when the simply supported edges are curved. This relationship enables a quick deduction of simply supported Mindlin plate frequencies from the abundant Kirchhoff plate vibration solutions. It may be used as a basic form in which approximate formulas may be developed for predicting the FSDT frequencies for other boundary conditions. Moreover, the exact relationship provides a useful means to check the validity, convergence and accuracy of numerical results and software.

Conway (1960) pointed out the analogies between the vibration problem of the CPT, the buckling problem of the CPT and the vibration problem of uniform prestressed membranes. Owing to these existing analogies, one may choose to substitute into Eq. (12.2.16) the buckling solution for the corresponding simply supported Kirchhoff plate under hydrostatic inplane load, or the frequency of the corresponding uniformly prestressed membrane, instead of the CPT frequency. In other words, it may take any one of the following expressions:

$$(\omega_K)_N = \frac{N_N}{\sqrt{\rho h D}} = \hat{\omega}_N \frac{\mu}{T}\sqrt{\frac{D}{\rho h}} \qquad (12.2.18)$$

where N_N is the buckling load for the N-th mode, $\hat{\omega}_N$ the N-th frequency of the vibrating prestressed membrane, μ the mass density per unit area of membrane and T the uniform tension per unit length of membrane.

Owing to the importance of having exact solutions for checking the convergence and accuracy of numerical results, highly accurate simply supported Kirchhoff (classical thin) plate frequencies for various polygonal shapes are presented in Tables 12.2.1–12.2.4. These Kirchhoff solutions when used together with Eq. (12.2.16) or Figure 12.2.1 provide benchmark Mindlin plate vibration results for analysts. The accurate results for simply supported circular plates, annular plates, sectorial plates and annular sectorial plates are also presented in Tables 12.2.4 to 12.2.7, respectively, as they may be used to generate the corresponding FSDT results quite accurately.

FREE VIBRATION RELATIONSHIPS

Table 12.2.1. Frequency parameter λ_K of triangular and rectangular Kirchhoff plates with simply supported edges ($\lambda_K = \hat{\omega}_n\sqrt{\rho h/D}$).

Triangle	d/a	b/a	Mode sequence number					
Liew (1993)			1	2	3	4	5	6
	1/4	2/5	23.75	40.80	60.54	70.33	83.39	101.8
		1/2	27.12	49.47	75.18	78.27	107.4	117.4
		2/3	33.11	65.26	88.68	106.1	141.2	156.4
		1.0	46.70	100.2	117.2	171.8	197.0	220.4
		$2/\sqrt{3}$	53.78	115.9	134.7	198.1	229.2	252.0
		2.0	101.5	195.8	275.0	315.6	427.7	464.2
	1/2	2/5	23.61	40.70	60.55	69.78	83.42	101.5
		1/2	26.91	49.33	76.29	76.30	108.0	116.4
		2/3	32.72	65.22	87.38	106.0	142.5	154.8
		1.0	45.83	102.8	111.0	177.3	199.5	203.4
		$2/\sqrt{3}$	52.64	122.8	122.8	210.5	228.1	228.1
		2.0	98.57	197.4	256.2	335.4	394.8	492.5

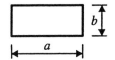

Rectangle [see Leissa(1969) and Reddy (1999)]

$\lambda_K = \left(\frac{m\pi}{a}\right)^2 + \left(\frac{n\pi}{b}\right)^2$

where m and n are the number of half waves

Note that Mindlin (1951) pointed out that for an isotropic plate, the shear correction factor K_s depends on Poisson's ratio ν and it may vary from $K_s = 0.76$ for $\nu = 0$ to $K_s = 0.91$ for $\nu = 0.5$. Following Mindlin's suggestion of equating the angular frequency of the first antisymmetric mode of thickness-shear vibration according to the exact three-dimensional theory to the corresponding frequency according to his theory, it can be shown that the shear correction factor is given by the following cubic equation:

$$(K_s)^3 - 8(K_s)^2 + \frac{8(2-\nu)K_s}{1-\nu} - \frac{8}{1-\nu} = 0 \qquad (12.2.19)$$

For example, if $\nu = 0.3$, then $K_s = 0.86$ and if $\nu = 0.176$, then $K_s = \pi^2/12$.

Table 12.2.2. Frequency parameter λ_K of parallelogram Kirchhoff plates with simply supported edges [$\lambda_K = \hat{\omega}_n\sqrt{\rho h/D}$; see Liew, Kitipornchai, and Wang (1993)].

Shape	a/b	β	Mode sequence number					
			1	2	3	4	5	6
	1.0	15°	20.87	48.20	56.12	79.05	104.0	108.9
		30°	24.96	52.63	71.87	83.86	122.8	122.8
		45°	35.33	66.27	100.5	108.4	140.8	168.3
		60°	66.30	105.0	148.7	196.4	213.8	250.7
	1.5	15°	15.10	28.51	46.96	49.76	61.70	75.80
		30°	18.17	32.49	53.48	58.02	76.05	78.61
		45°	25.96	42.39	64.80	84.18	93.31	107.5
		60°	48.98	70.51	96.99	127.3	162.3	171.1
	2.0	15°	13.11	20.66	33.08	44.75	50.24	52.49
		30°	15.90	23.95	36.82	52.64	56.63	63.26
		45°	23.01	32.20	46.21	63.50	82.08	83.00
		60°	44.00	56.03	72.79	92.80	117.4	151.7

Table 12.2.3. Frequency parameter λ_K of symmetrical trapezoidal Kirchhoff plates with simply supported edges [$\lambda_K b^2 = \hat{\omega}_n b^2 \sqrt{\rho h/D}$; see Liew and Lim (1993)].

Shape	a/b	c/b	Mode sequence number					
			1	2	3	4	5	6
	1.0	1/5	3.336	4.595	6.860	10.19	10.23	11.53
		2/5	2.198	3.479	5.499	5.789	7.737	9.027
		3/5	1.654	3.066	3.728	5.394	6.037	6.156
		4/5	1.356	2.833	2.879	4.560	5.086	5.192
	1.5	1/5	6.158	7.269	9.507	12.85	17.30	20.42
		2/5	3.703	5.175	7.390	9.272	10.63	14.22
		3/5	2.636	4.313	5.971	6.575	9.787	10.05
		4/5	2.089	3.802	4.494	6.121	7.257	8.000
	2.0	1/5	9.919	10.76	13.17	16.51	20.99	26.74
		2/5	5.351	7.575	9.633	12.83	13.69	17.07
		3/5	3.680	5.973	8.255	8.398	11.41	14.48
		4/5	2.856	5.053	6.187	7.452	10.32	10.64

Table 12.2.4. Frequencies of regular polygonal and circular Kirchhoff plates with simply supported edges.

Shape	Frequency parameter $\lambda_K b^2 = \hat{\omega}_n b^2 \sqrt{\rho h/D}$						
	No. of	Mode sequence number					
	sides	1	2	3	4	5	6
	3	52.6	122	122	210	228	228
	4	19.7	49.3	49.3	79.0	98.7	98.7
	6	28.9	73.0	73.2	130	130	151
	8	22.2	57.8	57.8	102	102	117

$$\frac{J_{n+1}(\sqrt{\lambda_K}R)}{J_n(\sqrt{\lambda_K}R)} + \frac{I_{n+1}(\sqrt{\lambda_K}R)}{I_n(\sqrt{\lambda_K}R)} = \frac{2\lambda_K R}{(1-\nu)}$$

where J_n and I_n are the Bessel functions and the modified Bessel functions, respectively, of the first kind of order n [see Leissa (1969)]

Table 12.2.5. Frequencies of annular Kirchhoff plates with simply supported edges [see Vogel and Skinner (1965)].

Shape	Frequency parameter $\lambda_K b^2 = \hat{\omega}_n b^2 \sqrt{\rho h/D}$						
	Number of nodal diameters m and nodal circles n, (m,n)						
	a/b	(0,0)	(1,0)	(2,0)	(3,0)	(0,1)	(1,1)
	0.1	14.5	16.7	25.9	40.0	51.7	56.5
	0.3	21.1	23.3	30.2	42.0	81.8	84.6
	0.5	40.0	41.8	47.1	56.0	159	161
	0.7	110	112	116	122	439	441
	0.9	988	988	993	998	3948	3948

Table 12.2.6. Frequencies of sectorial Kirchhoff plates with simply supported edges [see Xiang, Liew, Kitipornchai (1993)].

Shape	Frequency parameter $\lambda_K b^2 = \hat{\omega}_n b^2 \sqrt{\rho h/D}$						
		Mode sequence number					
	β	1	2	3	4	5	6
	30°	97.82	183.7	277.9	288.2	412.0	431.3
	45°	56.67	121.5	148.5	205.6	256.3	277.9
	60°	39.78	94.38	97.82	168.5	177.4	183.7
	90°	25.43	56.67	69.95	97.82	121.5	134.1

Table 12.2.7. Frequencies of annular sectorial Kirchhoff plates with simply supported edges [see Xiang, Liew, Kitipornchai (1993)].

Shape		Frequency parameter $\lambda_K b^2 = \hat{\omega}_n b^2 \sqrt{\rho h/D}$						
			Mode sequence number					
	a/b	β	1	2	3	4	5	6
	0.2	30°	97.82	183.8	277.9	288.5	413.9	431.3
		45°	56.73	122.3	148.5	210.1	256.3	277.9
		60°	40.16	97.48	97.82	177.4	179.9	183.8
		90°	27.17	56.73	78.68	97.82	122.3	148.5
	0.4	30°	98.75	195.5	277.9	336.3	430.5	529.4
		45°	60.31	148.2	148.7	260.4	277.8	286.1
		60°	46.28	98.75	131.7	177.5	195.5	269.7
		90°	36.19	60.31	98.75	120.0	148.2	148.7
	0.5	30°	103.3	228.6	278.2	427.0	438.8	539.8
		45°	68.34	150.8	189.7	278.2	283.5	387.8
		60°	55.97	103.3	176.1	178.8	228.6	278.3
		90°	47.15	68.34	103.3	150.7	166.5	189.7

On the other hand, comparing the Mindlin plate equations for the constitutive shear forces with the ones proposed by Reissner (1945), who assumed a parabolic variation of the shear stress distribution, the implicit shear correction factor of Reissner takes the value of

$$K_s = \frac{5}{6} \qquad (12.2.20)$$

Based on an analytical vibration solution of three-dimensional, simply supported, rectangular, isotropic plate, Wittrick (1987) performed a calibration of the Mindlin shear correction factor. He proposed that the shear correction factor be given by

$$K_s = \frac{5}{6-\nu} \qquad (12.2.21)$$

Wittrick's shear correction factor gives a value of 0.877 for $\nu = 0.3$, which corresponds closely to the value of 0.88 observed earlier by Srinivas et al. (1970) and Dawe (1978). It appears that the Wittrick shear correction factor is the best to date as it has a simple form and allows for the effect of Poisson's ratio.

12.3 Relationship Between CPT and TSDT

By differentiating Eq. (12.1.12a) with respect to x and Eq. (12.1.12b) with respect to y, summing them up and using Eq. (12.1.12c), the governing equation of motion for Reddy plates in vibration may be expressed as

$$\frac{\partial^2 M_{xx}^R}{\partial x^2} + \frac{\partial^2 M_{yy}^R}{\partial y^2} + 2\frac{\partial^2 M_{xy}^R}{\partial x \partial y}$$
$$= \frac{\rho h^3}{60} \omega_R^2 \nabla^2 w^R - \rho h \omega_R^2 w^R - \frac{\rho h^3}{15} \omega_R^2 \left(\frac{\partial \phi_x^R}{\partial x} + \frac{\partial \phi_y^R}{\partial y} \right) \qquad (12.3.1)$$

Next, we introduce the moment sum $\mathcal{M}^R$

$$\mathcal{M}^R = \frac{M_{xx}^R + M_{yy}^R}{1+\nu} = \frac{4D}{5}\left(\frac{\partial \phi_x^R}{\partial x} + \frac{\partial \phi_y^R}{\partial y} \right) - \frac{D}{5} \nabla^2 w^R \qquad (12.3.2)$$

In view of this moment sum and the moment expressions in Eqs. (6.4.11), equation (12.3.2) may be written as

$$\nabla^2 \mathcal{M}^R = \frac{\rho h^3}{60} w_R^2 \nabla^2 w^R - \rho h w_R^2 w^R - \frac{\rho h^3}{15} w_R^2 \left(\frac{\partial \phi_x^R}{\partial x} + \frac{\partial \phi_y^R}{\partial y} \right) \quad (12.3.3)$$

The substitution of Eqs. (6.4.11) and (12.3.3) into Eq. (12.1.12c) leads to

$$\mathcal{K} \left(\frac{\partial \phi_x^R}{\partial x} + \frac{\partial \phi_y^R}{\partial y} \right) = \left(\frac{\rho h^3}{252} w_R^2 - \frac{8Gh}{15} \right) \nabla^2 w^R - \rho h w_R^2 w^R$$
$$- \frac{4}{21} \nabla^2 \mathcal{M}^R + \frac{D}{105} \nabla^4 w^R \quad (12.3.4)$$

where

$$\mathcal{K} = \frac{8Gh}{15} + \frac{4\rho h^3}{315} w_R^2 \quad (12.3.5)$$

By substituting Eq. (12.3.4) into Eq. (12.3.3), one obtains

$$\mathcal{J} \nabla^2 \mathcal{M}^R = -\rho h w_R^2 \left(1 - \frac{\rho h^3}{15\mathcal{K}} w_R^2 \right) w^R - \frac{D}{105} \frac{\rho h^3}{15\mathcal{K}} w_R^2 \nabla^4 w^R + \mathcal{L} \nabla^2 w^R$$
$$(12.3.6)$$

where

$$\mathcal{J} = 1 - \frac{4\rho h^3}{315\mathcal{K}} w_R^2 \quad (12.3.7)$$

$$\mathcal{L} = \left(\frac{\rho h^3}{60} w_R^2 \right) - \frac{\rho h^3}{15\mathcal{K}} w_R^2 \left(\frac{\rho h^3}{252} w_R^2 - \frac{8Gh}{15} \right) \quad (12.3.8)$$

Also by substituting Eq. (12.3.4) into Eq. (12.3.2), one obtains

$$\mathcal{M}^R = -\frac{4D}{5\mathcal{K}} \rho h w_R^2 w^R - \frac{16D}{105\mathcal{K}} \nabla^2 \mathcal{M}^R + \frac{4D}{5\mathcal{K}} \frac{D}{105} \nabla^4 w^R$$
$$+ \left[-\frac{D}{5} + \frac{4D}{5\mathcal{K}} \left(\frac{\rho h^3}{252} w_R^2 - \frac{8Gh}{15} \right) \right] \nabla^2 w^R \quad (12.3.9)$$

and noting Eq. (12.3.6), equation (12.3.9) can be expressed as

$$\mathcal{M}^R = \left[-\frac{4D}{5\mathcal{K}} \rho h w_R^2 + \frac{16D}{105\mathcal{K}\mathcal{J}} \rho h w_R^2 \left(1 - \frac{\rho h^3}{15\mathcal{K}} w_R^2 \right) \right] w^R$$
$$+ \left[-\frac{D}{5} + \frac{4D}{5\mathcal{K}} \left(\frac{\rho h^3}{252} w_R^2 - \frac{8Gh}{15} \right) - \frac{16D\mathcal{L}}{105\mathcal{K}\mathcal{J}} \right] \nabla^2 w^R$$
$$+ \left[\frac{4D}{5\mathcal{K}} \frac{D}{105} + \frac{16D}{105\mathcal{K}\mathcal{J}} \frac{D}{105} \frac{\rho h^3}{15\mathcal{K}} w_R^2 \right] \nabla^4 w^R \quad (12.3.10)$$

The substitution of Eq. (12.3.10) into Eq. (12.3.6) furnishes the following sixth-order governing differential equation in terms of w^R:

$$a_1 \nabla^6 w^R + a_2 \nabla^4 w^R + a_3 \nabla^2 w^R + a_4 w^R = 0 \qquad (12.3.11)$$

where

$$a_1 = \frac{4D}{5\mathcal{K}} \frac{D}{105} \left[1 + \frac{4}{21\mathcal{J}} \frac{\rho h^3}{15\mathcal{K}} \omega_R^2 \right] \qquad (12.3.12a)$$

$$a_2 = -\frac{D}{5} \left[1 - \frac{4}{\mathcal{K}} \left(\frac{\rho h^3}{252} \omega_R^2 - \frac{8Gh}{15} \right) + \frac{16\mathcal{L}}{21\mathcal{K}\mathcal{J}} - \frac{\rho h^3}{315\mathcal{K}\mathcal{J}} \omega_R^2 \right] \qquad (12.3.12b)$$

$$a_3 = -\frac{4D}{5\mathcal{K}} \rho h \omega_R^2 + \frac{16D}{105\mathcal{K}\mathcal{J}} \rho h \omega_R^2 \left(1 - \frac{\rho h^3}{15\mathcal{K}} \omega_R^2 \right) - \frac{\mathcal{L}}{\mathcal{J}} \qquad (12.3.12c)$$

$$a_4 = \frac{\rho h}{\mathcal{J}} \omega_R^2 \left(1 - \frac{\rho h^3}{15\mathcal{K}} \omega_R^2 \right) \qquad (12.3.12d)$$

The governing equation (12.3.11) may be factored to give

$$\left(\nabla^2 + \lambda_1 \right) \left(\nabla^2 + \lambda_2 \right) \left(\nabla^2 + \lambda_3 \right) w^R = 0 \qquad (12.3.13)$$

where

$$\lambda_1 = -2\sqrt{\Phi} \cos\left(\frac{\theta}{3}\right) + \frac{a_2}{3a_1} \qquad (12.3.14a)$$

$$\lambda_2 = -2\sqrt{\Phi} \cos\left(\frac{\theta + 2\pi}{3}\right) + \frac{a_2}{3a_1} \qquad (12.3.14b)$$

$$\lambda_3 = -2\sqrt{\Phi} \cos\left(\frac{\theta + 4\pi}{3}\right) + \frac{a_2}{3a_1} \qquad (12.3.14c)$$

$$\cos\theta = \frac{\Psi}{\sqrt{\Phi^3}}, \quad \Phi = -\frac{a_3}{3a_1} + \left(\frac{a_2}{3a_1}\right)^2 \qquad (12.3.15)$$

$$\Psi = \frac{a_2 a_3}{6a_1^2} - \frac{a_4}{2a_1} - \left(\frac{a_2}{3a_1}\right)^2 \qquad (12.3.16)$$

The following boundary conditions for a simply supported edge of a Reddy plate have been assumed (Reddy and Phan 1985)

$$w^R = 0, \quad \phi_s^R = 0, \quad M_{mm}^R = 0, \quad P_{nn}^R = 0 \qquad (12.3.17)$$

where the subscripts n, s denote, respectively, the normal and tangential directions to the edge. Since $w^R = 0$ implies that $\partial^2 w^R/\partial s^2 = 0$ and $\phi_s^R = 0$ implies that $\partial \phi_s^R/\partial s = 0$, then together with the conditions $M_{nn} = P_{nn} = 0$, we have $\partial^2 w^R/\partial n^2 = \partial \phi_s^R/\partial n = 0$ and $M_{ss}^R = P_{ss}^R = 0$. Also, it follows that $\partial^4 w^R/\partial s^4 = \partial^4 w^R/\partial n^4 = 0$. Thus, for a simply supported edge of a Reddy plate

$$w^R = 0, \quad \mathcal{M}^R = 0, \quad \nabla^2 w^R = 0, \quad \nabla^4 w^R = 0 \qquad (12.3.18)$$

In view of the boundary conditions given in Eq. (12.3.18), the sixth-order governing equation (12.3.13) of the Reddy plate may be written as three second-order differential equations given by (see Conway 1960, Pnueli 1975)

$$\left(\nabla^2 + \lambda_j\right) w^R = 0, \quad j = 1, 2, 3 \qquad (12.3.19)$$

with the boundary condition $w^R = 0$ along the edges. Note that although Pnueli (1975) proved that the frequency solutions of the fourth order differential plate equation are the same as those given by the second order differential equation for the case of straight, simply supported edges, the same (Pnueli's) proof together with the substitution of a variable (for example, let $v = \nabla^2 w^R + \lambda w^R$) can be used to reduce the sixth-order equation to a second-order equation.

In view of the mathematical similarity of Eqs. (12.3.18), (12.3.19), (12.2.13) and (12.2.14), it may be deduced that

$$\lambda_j = \lambda_K, \quad j = 1, 2, 3 \qquad (12.3.20)$$

Based on numerical tests, it was found that the first root $j = 1$ yields nonfeasible vibration solutions while the second root $j = 2$ of the Reddy plate solution gives the lowest frequency value when compared to the third root $j = 3$. Thus the relationship between the Reddy plate frequency ω_R and the Kirchhoff plate frequency ω_K is given by

$$\omega_K \sqrt{\frac{\rho h}{D}} = -2\sqrt{\Phi} \cos\left(\frac{\theta + 2\pi}{3}\right) + \frac{a_2}{3a_1} \qquad (12.3.21)$$

Upon supplying the Kirchhoff plate frequencies, the foregoing exact relationship (12.3.21) can be used to compute the Reddy plate frequencies. Note Eq. (12.3.21) is an explicit equation for ω_K as a function of ω_R, but it is a transcendental equation (12.3.21) of ω_R for given ω_K. The transcendental equation may be readily solved using the false position method.

Figure 12.3.1 shows a graphical representation of the relationship given by Eq. (12.3.21). The curve in Figure 12.3.1 applies to any polygonal shaped plate with straight, simply supported edges. It can be seen that as the frequency parameters increase (i.e., corresponding to increasing plate thickness or higher modes of vibration), the Reddy solutions decrease with respect to the Kirchhoff solutions due to the effects of transverse shear deformation and rotary inertia.

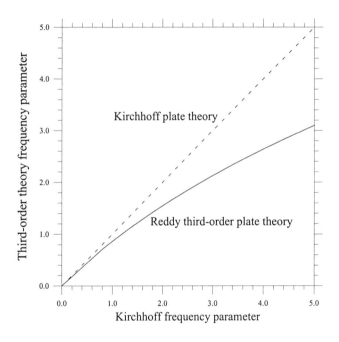

Figure 12.3.1. Frequency relationship between CPT and TSDT.

To illustrate the use of Eq. (12.3.21), the vibration frequencies of rectangular and regular polygonal plates are determined. Table 12.3.1 presents sample vibration frequencies of square and rectangular Reddy plates with simply supported edges. The vibration frequencies computed from the relationship given by Eq. (12.3.21) are in close agreement with those obtained earlier by Reddy and Phan (1985). It should be noted that Reddy and Phan (1985) determined their solutions by solving directly the governing equations of motion using trigonometric series. Table 12.3.2 gives the fundamental frequencies of regular polygonal plates with different thickness-to-side length ratios. Note that the Kirchhoff results for the regular polygonal plates used in determining the Reddy solutions are taken from Leissa's monograph (Leissa 1993).

Table 12.3.1. Natural frequencies $\bar{\omega} = \omega b^2 \sqrt{\rho h/D}$ of square plates ($b = a$) and rectangular plates ($b = \sqrt{2}a$) with $h/a = 0.1$ and $\nu = 0.3$.

Mode (m,n)	Thin plate	$a/b = 1$ Reddy plate Eq.(21)	Reddy &Phan	Mode (m,n)	Thin plate	$a/b = 1/\sqrt{2}$ Reddy plate Eq.(21)	Reddy &Phan
(1,1)	19.739	19.065	19.080	(1,1)	29.609	28.839	28.847
(1,2)	49.348	45.487	45.538	(1,2)	59.218	56.275	56.309
(2,2)	78.957	69.809	69.905	(2,1)	88.826	82.485	82.554
(1,3)	98.696	85.065	85.214	(1,3)	108.57	99.347	99.449
(2,3)	128.31	106.74	107.00	(2,2)	118.44	107.61	107.73
(1,4)	167.78	133.72	134.13	(2,3)	167.78	147.38	147.61
(3,3)	177.65	140.17	140.63	(1,4)	177.65	155.05	155.31
(2,4)	197.39	152.75	153.31	(3,1)	187.52	162.63	162.92
(3,4)	246.74	182.57	183.40	(3,2)	217.13	184.88	185.26
(1,5)	256.61	188.28	189.16	(2,4)	236.87	199.32	199.76
(2,5)	286.22	204.96	206.03	(3,3)	266.48	220.44	220.99
(4,4)	315.83	221.02	222.30	(1,5)	266.48	220.44	220.99
(3,5)	335.57	231.41	232.83	(2,5)	325.70	260.91	261.69

Table 12.3.2. Natural frequencies $\bar{\omega} = \omega a^2 \sqrt{\rho h/D}$ of regular polygonal plates with side length a and $\nu = 0.3$.

Polygon shape	Thin plate*	Reddy plate $h/a = 0.05$	$h/a = 0.10$	$h/a = 0.15$
Equilateral	52.638	51.414	48.284	44.294
Square	19.739	19.562	19.065	18.330
Pentagon	10.863	10.809	10.653	10.410
Hexagon	7.129	7.106	7.037	6.929
Octagon	3.624	3.618	3.600	3.571

*The CPT results are taken from Leissa's monograph (Leissa 1993).

12.4 Concluding Remarks

Presented herein are exact relationships between the natural frequencies of Mindlin plates, Reddy plates and Kirchhoff plates. The relationship is valid for any general polygonal plates with simply supported edges. Once the Kirchhoff vibration solutions are known, the Mindlin and Reddy solutions may be readily calculated from these relationships. As examples, vibration frequencies of rectangular and regular polygonal plates have been determined using the relationship. The relationships may also be used for symmetrically, isotropic laminated plates by modifying appropriately the stiffnesses. Unlike the FSDT-CPT frequency relationship, the TSDT-CPT frequency relationship does not need a shear correction factor. This feature is advantageous when considering laminated plates where the shear correction factor is not available.

Problems

12.1 The equation of motion governing the CPT is given by

$$D\left(\frac{\partial^4 w_0}{\partial x^4} + 2\frac{\partial^4 w_0}{\partial x^2 \partial y^2} + \frac{\partial^4 w_0}{\partial y^4}\right)$$
$$+ I_0 \frac{\partial^2 w_0}{\partial t^2} - I_2 \left(\frac{\partial^4 w_0}{\partial t^2 \partial x^2} + \frac{\partial^4 w_0}{\partial t^2 \partial y^2}\right) = 0 \quad (i)$$

where
$$I_0 = \rho_0 h, \quad I_2 = \frac{\rho_0 h^3}{12} \quad (ii)$$

For natural vibration, the solution is assumed to be periodic

$$w_0(x,y,t) = w(x,y) e^{i\omega t} \quad (iii)$$

where $i = \sqrt{-1}$ and ω is the frequency of natural vibration associated with mode shape w. Obtain the Navier solution to the resulting equation by assuming a solution of the form

$$w(x,y) = W_{mn} \sin\frac{m\pi x}{a} \sin\frac{n\pi y}{b} \quad (iv)$$

In particular, show that

$$\omega_{mn} = \frac{\pi^2}{b^2}\sqrt{\frac{D}{\tilde{I}_0}}\left[m^2\left(\frac{b}{a}\right)^2 + n^2\right] \quad (v)$$

where
$$\tilde{I}_0 = I_0 + I_2\left[\left(\frac{m\pi}{a}\right)^2 + \left(\frac{n\pi}{b}\right)^2\right] \qquad (vi)$$

12.2 Use the Lévy procedure described in Problems 11.2 and 11.3 to determine the characteristic (frequency) equation associated with the natural vibration of a rectangular plate with sides $x = 0, a$ and $y = 0$ simply supported and side $y = b$ free. The boundary conditions on edges $y = 0, b$ are, when neglecting the rotary inertia I_2 and $\omega^2 \geq \alpha_m^4 D/I_0$

$$w = 0, \quad M_{yy} = -D\left(\nu\frac{\partial^2 w}{\partial x^2} + \frac{\partial^2 w}{\partial y^2}\right) = 0 \text{ at } y = 0 \qquad (i)$$

$$M_{yy} = 0, \quad V_y = -D\left[\frac{\partial^3 w}{\partial y^3} + (1-\nu)\frac{\partial^3 w}{\partial x^2 \partial y}\right] = 0 \text{ at } y = b \qquad (ii)$$

where $\alpha_m = m\pi/a$. In particular, show that

$$\lambda_2 \Omega_1 \bar{\Omega}_2 \sinh \lambda_1 b \cos \lambda_2 b - \lambda_1 \Omega_2 \bar{\Omega}_1 \cosh \lambda_1 b \sin \lambda_2 b = 0 \qquad (iii)$$

where

$$(\lambda_1)^2 = \sqrt{\alpha_m^2 \frac{N_0}{D} + \alpha_m^2}, \quad (\lambda_2)^2 = \sqrt{\alpha_m^2 \frac{N_0}{D} - \alpha_m^2} \qquad (iv)$$

$$\Omega_1 = \left(\lambda_1^2 - \nu\alpha_m^2\right), \quad \Omega_2 = \left(\lambda_2^2 + \nu\alpha_m^2\right) \qquad (v)$$

$$\bar{\Omega}_1 = \left[\lambda_1^2 - (1-\nu)\alpha_m^2\right], \quad \bar{\Omega}_2 = \left[\lambda_2^2 + (1-\nu)\alpha_m^2\right] \qquad (vi)$$

12.3 Obtain the characteristic equation, using the Lévy solution procedure, for a rectangular plate with $x = 0, a$ and $y = b$ simply supported and side $y = 0$ clamped. The boundary conditions on edges $y = 0, b$ are

$$w = 0, \quad \frac{\partial w}{\partial y} = 0 \text{ at } y = 0 \qquad (i)$$

$$w = 0, \quad M_{yy} = -D\left(\nu\frac{\partial^2 w}{\partial x^2} + \frac{\partial^2 w}{\partial y^2}\right) = 0 \text{ at } y = b \qquad (ii)$$

These boundary conditions, with $\omega^2 \geq \alpha_m^4 D/I_0$, yield the frequency equation

$$\lambda_1 \cosh \lambda_1 b \sin \lambda_2 b - \lambda_2 \sinh \lambda_1 b \cos \lambda_2 b = 0 \qquad (iii)$$

where

$$(\lambda_1)^2 = \sqrt{\alpha_m^2 \frac{N_0}{D} + \alpha_m^2}, \quad (\lambda_2)^2 = \sqrt{\alpha_m^2 \frac{N_0}{D} - \alpha_m^2} \qquad (iv)$$

Chapter 13

Relationships for Inhomogeneous Plates

In this chapter, exact relationships are developed for (1) the deflection values of sandwich plates in terms of the corresponding Kirchhoff plates for simply supported polygonal plates under any transverse load or for simply supported and clamped circular plates under any axisymmetric load; (2) the bending solutions of the first-order plate theory (FSDT) for functionally graded circular plates in terms of the deflections of isotropic circular plates based on the classical plate theory (CPT); (3) the buckling load of sandwich plates based on the FSDT in terms of those of the Kirchhoff plates based on the CPT under uniform in-plane compressive load for simply supported general polygonal plates and simply supported and clamped circular plates; and (4) the vibration frequencies of smply supported sandwich polygonal plates in terms of those of polygonal Kirchhoff plates with the same boundary conditions and loads.

13.1 Deflection Relationships for Sandwich Plates

13.1.1 Introduction

Sandwich plates are made up of three layers, with the top and bottom layers (or facings) being thin although made from a high-strength material, and the thick middle layer (or core) being made from a relatively light and low-strength material. In all-steel sandwich panels, the two steel facings are spot-welded onto a core of stiffeners, which may consist of z-sections, tophats, channels, corrugated sheeting or honeycomb-type of construction. Owing to the thick core, the use of the Kirchhoff (classical thin) plate theory will lead to an underprediction of the deflections, since it does not allow for the effect of transverse shear deformation. For global responses such as maximum deflection, buckling loads, or vibration frequencies, the first-order shear deformation theory yields reasonably accurate solutions.

Here we present exact relationships between the deflection of sandwich Mindlin plates and their corresponding Kirchhoff plates. The sandwich plates considered here can either be (i) simply supported (S) plates of general polygonal shape and under any transverse loading or (ii) simply supported (S) and clamped (C) circular plates under axisymmetric loading. As the relationships are exact under the assumptions used in the plate theories, one may obtain exact deflection solutions of sandwich plates if the Kirchhoff plate solutions are also exact. The relationships should also be useful for the development of approximate formulas for plates with other shapes, boundary and loading conditions, and may serve to check numerical deflection values computed from sandwich plate analysis software.

13.1.2 Governing Equations of Kirchhoff Plates

The well-known governing equation for isotropic Kirchhoff plate bending problem is given by

$$D\nabla^4 w_0^K = q \qquad (13.1.1)$$

where $D = Eh^3/[12(1-\nu^2)]$ is the flexural rigidity of the plate, h the thickness, E the Young's modulus, ν Poisson's ratio, w_0^K the transverse deflection of the mid-plane, and q is the transverse load. The equation can be written in terms of the rectangular coordinates (x,y) or the radial coordinate r by appropriately selecting the biharmonic operator ∇^4 or the Laplace operator ∇^2

$$\nabla^2 = \frac{\partial^2}{\partial x^2} + \frac{\partial^2}{\partial y^2} \quad \text{for polygonal plates} \qquad (13.1.2)$$

$$\nabla^2 = \frac{\partial^2}{\partial r^2} + \frac{1}{r}\frac{\partial}{\partial r} \quad \text{for axisymmetric circular plates} \qquad (13.1.3)$$

Equation (13.1.1) can be written as a pair of Poisson equations [see Eqs. (7.2.1a,b)]

$$\nabla^2 \mathcal{M}^K = -q \qquad (13.1.4)$$

$$\nabla^2 w_0^K = -\frac{\mathcal{M}^K}{D} \qquad (13.1.5)$$

where the Marcus moment $\mathcal{M}^K$ is defined as [see Eq. (7.2.3a) and (9.2.1)]

$$\mathcal{M}^K = \frac{M_{xx}^K + M_{yy}^K}{1+\nu} = \frac{M_{rr}^K + M_{\theta\theta}^K}{1+\nu} \quad (13.1.6)$$

where (M_x, M_y) and (M_r, M_θ) are the bending moments in the Cartesian coordinate and the polar coordinates, respectively.

The boundary conditions associated with Eqs. (13.1.4) and (13.1.5) are given by

$$w_0^K = \mathcal{M}^K = 0 \quad (13.1.7)$$

in the case of polygonal Kirchhoff plates with straight simply supported edges, and

$$w_0^K = 0 \text{ for S and C plates} \quad (13.1.8)$$

$$\mathcal{M}^K = \begin{cases} -D(1-\nu)\left(\frac{1}{r}\frac{dw_0^K}{dr}\right) = D\frac{(1-\nu)}{\nu}\left(\frac{d^2 w_0^K}{dr^2}\right) & \text{for S plates} \\ -D\left(\frac{d^2 w_0^K}{dr^2}\right) & \text{for C plates} \end{cases} \quad (13.1.9)$$

at $r = R$ for circular plates with radius R. The expressions in Eqs. (13.1.8) and (13.1.9) for $\mathcal{M}^K$ were obtained using the boundary conditions $M_{rr} = 0$ for simply supported (S) plates and $dw_0^K/dr = 0$ for clamped (C) plates. Note that for a circular plate undergoing axisymmetric bending, the Marcus moment at the boundary takes on a constant value. This important feature will be used later in the derivation of the deflection relationship between circular sandwich plates and Kirchhoff plates.

13.1.3 Governing Equations for Sandwich Mindlin Plates

Consider a sandwich plate with isotropic core and facings, and core thickness h_c and the thickness of the facings h_f, as shown in Figure 13.1.1. Thus, the total thickness of the sandwich plate is $h = 2h_f + h_c$. The Young's modulus of elasticity E, Poisson's ν, and shear modulus $G = E/[2(1+\nu)]$ of the core and the facings will be identified with subscripts of c and f, respectively. The plate is either of general polygonal shape or of circular shape.

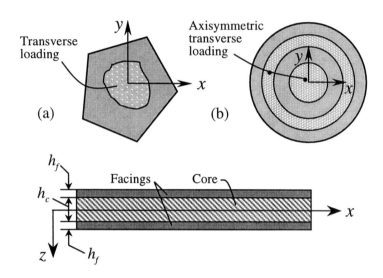

Figure 13.1.1. Geometry of a sandwich plate.

First consider the polygonal sandwich plate. Assuming that deformations are continuous across the thickness, the moment-displacement relations for sandwich plates are given by

$$M_{xx}^M = (D_c + D_f)\frac{\partial \phi_x^M}{\partial x} + (\nu_c D_c + \nu_f D_f)\frac{\partial \phi_y^M}{\partial y} \quad (13.1.10a)$$

$$M_{yy}^M = (\nu_c D_c + \nu_f D_f)\frac{\partial \phi_x^M}{\partial x} + (D_c + D_f)\frac{\partial \phi_y^M}{\partial y} \quad (13.1.10b)$$

$$M_{xy}^M = \frac{1}{2}[(1-\nu_c)D_c + (1-\nu_f)D_f]\left(\frac{\partial \phi_x^M}{\partial y} + \frac{\partial \phi_y^M}{\partial x}\right) \quad (13.1.10c)$$

$$Q_x^M = K_s(G_c h_c + 2G_f h_f)\left(\phi_x^M + \frac{\partial w_0^M}{\partial x}\right) \quad (13.1.10d)$$

$$Q_y^M = K_s(G_c h_c + 2G_f h_f)\left(\phi_y^M + \frac{\partial w_0^M}{\partial y}\right) \quad (13.1.10e)$$

where D_c and D_f are flexural rigidities of the core and facings, respectively

$$D_c = \frac{E_c h_c^3}{12(1-\nu_c^2)}, \quad D_f = \frac{2E_f h_f(\frac{3h_c^2}{4} + \frac{3h_c h_f}{2} + h_f^2)}{3(1-\nu_f^2)} \quad (13.1.11)$$

and (M_{xx}, M_{yy}) are the bending moments, M_{yx} is the twisting moment, (Q_x, Q_y) the transverse shearing forces (all measured per unit thickness), (ϕ_x, ϕ_y) the rotations of a transverse normal to the midsurface, and K_s the shear correction coefficient. The assumption of equal rotation for the core and the facings is valid, provided the sandwich plates are constructed from relatively thin facings.

The equilibrium equations of a Mindlin plate are given by Eqs. (6.3.9a-c). By substituting for the moments and shear forces from Eqs. (13.1.10a-e) into Eqs. (6.3.9a-c), we obtain

$$K_s(G_c h_c + 2G_f h_f)\left(\phi_x^M + \frac{\partial w_0^M}{\partial x}\right)$$
$$= (D_c + D_f)\frac{\partial^2 \phi_x^M}{\partial x^2} + (\nu_c D_c + \nu_f D_f)\frac{\partial^2 \phi_y^M}{\partial x \partial y}$$
$$+ \frac{1}{2}[(1-\nu_c)D_c + (1-\nu_f)D_f]\left(\frac{\partial^2 \phi_x^M}{\partial y^2} + \frac{\partial^2 \phi_y^M}{\partial x \partial y}\right) \quad (13.1.12a)$$

$$K_s(G_c h_c + 2G_f h_f)\left(\phi_y^M + \frac{\partial w_0^M}{\partial y}\right)$$
$$= (D_c + D_f)\frac{\partial^2 \phi_y^M}{\partial y^2} + (\nu_c D_c + \nu_f D_f)\frac{\partial^2 \phi_x^M}{\partial x \partial y}$$
$$+ \frac{1}{2}[(1-\nu_c)D_c + (1-\nu_f)D_f]\left(\frac{\partial^2 \phi_y^M}{\partial x^2} + \frac{\partial^2 \phi_x^M}{\partial x \partial y}\right) \quad (13.1.12b)$$

$$(G_c h_c + 2G_f h_f)\left(\mathcal{M}^M + \nabla^2 w_0^M\right) = -q \quad (13.1.12c)$$

where $\mathcal{M}^M$ is the moment sum of the Mindlin plate theory

$$\mathcal{M}^M = \frac{M_{xx}^M + M_{yy}^M}{(1+\nu_c)D_c + (1+\nu_f)D_f} = \frac{\partial \phi_x^M}{\partial x} + \frac{\partial \phi_y^M}{\partial y} \quad (13.1.13)$$

Equations (13.1.12a) and (13.1.12b) with respect to x and y, respectively, and adding the results and using Eq. (13.2.12c), we arrive at the result

$$\nabla^2 \mathcal{M}^M = -\frac{q}{D_c + D_f} \quad (13.1.14)$$

The substitution of (13.1.12c) and (13.1.4) into (13.1.14) furnishes

$$\nabla^2 \mathcal{M}^M = \nabla^2 \left[\nabla^2 \left(-w_0^M + \frac{\mathcal{M}^K}{K_s(G_c h_c + 2G_f h_f)} \right) \right] = -\frac{q}{D_c + D_f} \tag{13.1.15}$$

For a simply supported polygonal Mindlin plate, the boundary condition is such that

$$w_0^M = M_{nn}^M = \phi_s^M \equiv \phi_y^M n_x - \phi_x^M n_y = 0 \tag{13.1.16}$$

where n is the direction normal (in the xy-plane and with direction cosines n_x and n_y) to the straight simply supported edge, and s the direction tangential to the edge. Owing to the conditions $M_{nn} = \phi_s = 0$, we deduce that $\partial \psi_s / \partial s = 0$. Thus from Eq. (13.1.13), the moment sum has the boundary condition

$$\mathcal{M}^M = 0 \tag{13.1.17}$$

Equation (13.1.15) is also valid for axisymmetric bending of circular plates, except that the Laplace operator is in polar form (13.1.3), and

$$\mathcal{M}^K = -D \left(\frac{d^2 w_0^K}{dr^2} + \frac{1}{r} \frac{dw_0^K}{dr} \right) \tag{13.1.18}$$

$$\mathcal{M}^M = \frac{d\phi_r^M}{dr} + \frac{1}{r} \phi_r^M \tag{13.1.19}$$

where ϕ_r is the rotation. The boundary conditions for simply supported as well as clamped circular plates are given by

$$w_0^M = 0, \quad \mathcal{M}^M = \text{constant}, \ C \tag{13.1.20}$$

13.1.4 Relationship Between Sandwich and Kirchhoff Plates

In view of (13.1.5), (13.1.7), and (13.1.14)–(13.1.17), we have along a simply supported boundary

$$w_0^K = \mathcal{M}^K = \mathcal{M}^M = \nabla^2 w_0^K = \nabla^2 \left(-w_0^M + \frac{\mathcal{M}^K}{K_s(G_c h_c + 2G_f h_f)} \right) = 0 \tag{13.1.21}$$

Comparing (13.1.1), (13.1.15) and (13.1.21) and the boundary conditions in Eqs. (13.1.16) and (13.1.17), it can be readily deduced that

$$w_0^M = \frac{D}{D_c + D_f} w_0^K + \frac{\mathcal{M}^K}{K_S(G_c h_c + 2 G_f h_f)} \tag{13.1.22}$$

Owing to the fact that the Marcus moment does not vanish at the boundary for circular plates but takes on a constant value, Eq. (13.1.22) must be changed to include a constant C on one side of the equation. This constant C may be evaluated from the boundary conditions, $w_0^M = w_0^K = 0$. Thus, for a circular plate, the deflection relationship is given by

$$w_0^M = \frac{D}{D_c + D_f} w_0^K + \frac{\mathcal{M}^K - \mathcal{M}^K(R)}{K_S(G_c h_c + 2 G_f h_f)} \tag{13.1.23}$$

where the Marcus moment $\mathcal{M}^K(R)$ at the boundary is given by (13.1.9). Note that for axisymmetric loading $q(r)$, the Marcus moment can be obtained by integrating the loading function, i.e.

$$\mathcal{M}^K = -\int \frac{1}{r} dr \int q(r) r \, dr \tag{13.1.24}$$

It is also worth noting that the second term in (13.1.23) has the same value irrespective of whether the circular plate is simply supported or clamped at its edges. This conclusion can be readily proven using the fact that the difference between the deflections of a simply supported circular plate and that of its clamped counterpart is equal to the plate deflection due to a uniformly distributed boundary moment of the clamped plate.

Equation (13.1.22) furnishes an exact relationship between the deflection values of the simply supported sandwich plate and the corresponding simply supported Kirchhoff plate, while (13.1.23) gives the deflection relationship for circular plates. This means that the deflection of simply supported sandwich plates can be calculated upon supplying the deflection solution of the Kirchhoff plate and the Marcus moment $\mathcal{M}^K$, thus bypassing the necessity for a shear deformable sandwich plate bending analysis. Note that the deflection solutions of simply supported polygonal and axisymmetric Kirchhoff plates are available in the literature. If the deflection results are not available, thin plate bending analysis need only to be performed, and the sandwich plate solutions calculated from the derived relationships.

13.1.5 Examples

The use of the foregoing deflection relationships is illustrated using an equilateral triangular plate example and a circular plate example. First, consider a simply supported, equilateral triangular sandwich plate of side length $2L\sqrt{3}$ as shown in Figure 13.1.2 (see section 7.3.1). The plate is subjected to a uniformly distributed load of intensity q_0. The deflection and Marcus moment of this Kirchhoff plate are given by [see Eqs. (7.3.1) and (7.3.2); Timoshenko and Woinowsky-Krieger (1969) and Reddy (1999)]

$$w_0^K = \frac{q_0 L^4}{64 D}\left[\bar{x}^3 - 3\bar{y}^2\bar{x} - \left(\bar{x}^2 + \bar{y}^2\right) + \frac{4}{27}\right]\left(\frac{4}{9} - \bar{x}^2 - \bar{y}^2\right) \quad (13.1.25)$$

$$\mathcal{M}^K = -D\nabla^2 w_0^K = \frac{q_0 L^2}{4}\left[\bar{x}^3 - 3\bar{x}\bar{y}^2 - \left(\bar{x}^2 - \bar{y}^2\right) + \frac{4}{27}\right] \quad (13.1.26)$$

where $\bar{x} = x/L$ and $\bar{y} = y/L$.

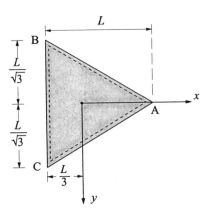

Figure 13.1.2. Simply supported equilateral triangular plate.

Using (13.1.22), the deflection surface of the uniformly loaded equilateral triangular sandwich plate is given by

$$w_0^M = \frac{q_0 L^4}{64(D_c + D_f)}\left[\bar{x}^3 - 3\bar{y}^2\bar{x} - \left(\bar{x}^2 + \bar{y}^2\right) + \frac{4}{27}\right]$$

$$\times \left[\frac{4}{9} - \bar{x}^2 - \bar{y}^2 + \frac{16(D_c + D_f)}{K_S(G_c h_c + 2G_f h_f)} \right] \quad (13.1.27)$$

Next, consider a circular sandwich plate under an axisymmetric linearly varying load. The load is zero at the centre and increases linearly to q_0 at the edge as shown in Figure 13.1.3. The plate can be either simply supported or clamped at the edge.

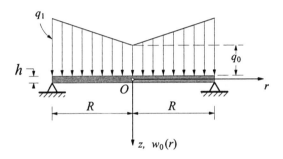

Figure 13.1.3. Circular plate under axisymmetric linearly varying load.

The deflection solutions for such loaded Kirchhoff circular plates are determined by Markus and are given by

$$w_0^K = \begin{cases} \frac{q_0 R^4}{450 D} \left[\frac{3(6+\nu)}{(1+\nu)} - \frac{5(4+\nu)}{(1+\nu)} \left(\frac{r}{R}\right)^2 + 2\left(\frac{r}{R}\right)^5 \right], & \text{for S plate} \\ \frac{q_0 R^4}{450 D} \left[3 - 5\left(\frac{r}{R}\right)^2 + 2\left(\frac{r}{R}\right)^5 \right], & \text{for C plate} \end{cases} \quad (13.1.28)$$

The Marcus moment is thus given by

$$\mathcal{M}^K = -D\nabla^2 w_0^K = \begin{cases} \frac{q_0 R^2}{45} \left[\frac{2(4+\nu)}{(1+\nu)} - 5\left(\frac{r}{R}\right)^3 \right], & \text{for S plate} \\ \frac{q_0 R^2}{45} \left[2 - 5\left(\frac{r}{R}\right)^3 \right], & \text{for C plate} \end{cases} \quad (13.1.29)$$

At the plate edge $r = R$, the Marcus moment is given by

$$\mathcal{M}^K(R) = \begin{cases} \frac{q_0 R^2}{45} \frac{(1-\nu)}{(1+\nu)}, & \text{for S plate} \\ -\frac{q_0 R^2}{15}, & \text{for C plate} \end{cases} \quad (13.1.30)$$

Using (13.1.23), the defelction of such loaded circular sandwich plate is given by

$$w_0^K = \begin{cases} \frac{q_0 R^4}{450(D_c+D_f)} \left[\frac{3(6+\nu)}{(1+\nu)} - \frac{5(4+\nu)}{(1+\nu)} \left(\frac{r}{R}\right)^2 + 2\left(\frac{r}{R}\right)^5 \right] \\ \quad + \frac{q_0 R^2}{15 K_S(G_c h_c + 2 G_f h_f)} \left[1 - \left(\frac{r}{R}\right)^3 \right], \quad \text{for S plate} \\ \\ \frac{q_0 R^4}{450(D_c+D_f)} \left[3 - 5\left(\frac{r}{R}\right)^2 + 2\left(\frac{r}{R}\right)^5 \right] \\ \quad + \frac{q_0 R^2}{15 K_S(G_c h_c + 2 G_f h_f)} \left[1 - \left(\frac{r}{R}\right)^3 \right], \quad \text{for C plate} \end{cases} \quad (13.1.31)$$

Note that the deflection component due to the shear deformation is the same for both simply supported and clamped plates as discussed earlier.

13.1.6 Relationship Between Sandwich and Solid Mindlin Plates

The foregoing derivation applies to solid Mindlin plates as well. By letting the thickness of the facings $h_f = 0$ and $h_c = h$ (i.e. $D_f = 0$, $D_c = D$), the sandwich plate becomes an isotropic, solid Mindlin plate of thickness h, Poisson's ratio ν, and Young's modulus E. Thus the key equations (13.1.22) and (13.1.23) reduce, respectively to

$$w_0^M = \begin{cases} w_0^K + \frac{\mathcal{M}^K}{K_S G h}, & \text{for simply supported polygonal plates} \\ w_0^K + \frac{\mathcal{M}^K - \mathcal{M}^K(R)}{K_S G h}, & \text{for circular plates} \end{cases}$$
$$(13.1.32)$$

where w_M is the deflection of the solid Mindlin plate.

In view of Eqs. (13.1.22), (13.1.23), and (13.1.32), the relationship between sandwich and solid Mindlin plate deflections is given by

$$\left[w_0^M - \left(\frac{D}{D_c + D_f} \right) w_0^K \right] (G_c h_c + 2 G_f h_f) = (w_0^M - w_0^K) G h \quad (13.1.33)$$

One may obtain the sandwich plate deflection from the Kirchhoff and solid Mindlin plate deflections using (13.1.33) without the need of the Marcus moment and vice versa.

13.2 Deflection Relationships for Functionally Graded Circular Plates

13.2.1 Introduction

Fiber-reinforced composites have a mismatch of mechanical properties across an interface due to two discrete materials bonded together. As a result, the constituents of fiber-matrix composites are prone to debonding at extremely high thermal loading. Further, cracks are likely to initiate at the interfaces and grow into weaker material sections. Additional problems include the presence of residual stresses due to the difference in coefficients of thermal expansion of the fiber and matrix in the composite materials. These problems can be avoided or reduced by gradually varying the volume fraction of the constituents rather than abruptly changing them across an interface. This gradation in properties of the material reduces thermal stresses, residual stresses, and stress concentration factors. Furthermore, the gradual change of mechanical properties can be tailored to different applications and working environments. Such materials, termed *functionally graded materials* (FGMs), were first introduced by a group of scientists in Sendai, Japan in 1984 [1,2].

Functionally graded materials are inhomogeneous materials in which the material properties are varied continuously from point to point. For example, a plate structure used as a thermal barrier may be graded through the plate thickness from ceramic on the face of the plate that is exposed to high temperature to metal on the other face. This is achieved by varying the volume fraction of the constituents *i.e.*, ceramic and metal in a predetermined manner. The ceramic constituent of the material provides the high temperature resistance due to its low thermal conductivity. The ductile metal constituent, on the other hand, prevents fracture caused by stresses due to high temperature gradient in a very short period of time. A mixture of the ceramic and a metal with a continuously varying volume fraction can be easily manufactured [2-6]. This eliminates interface problems and thus the stress distributions are smooth.

In this section, axisymmetric bending of through-the-thickness functionally graded circular plates are studied using the Mindlin plate theory, i.e., account for the transverse shear strains. Due to nonsymmetric grading of the material through the thickness, the bending-stretching coupling exists. General solution of the Mindlin plate problem for arbitrary variation of the constituents is derived in terms of the isotropic Kirchhoff plate solution. Particular solution are developed for a number of boundary conditions. The effect of material distribution through the thickness and boundary conditions on deflections and stresses are presented.

13.2.2. Formulation

Consider a functionally graded circular plate of total thickness h and subjected to axisymmetric transverse load q. The r−coordinate is taken radially outward from the center of the plate, z−coordinate along the thickness of the plate, and the θ−coordinate is taken along a circumference of the plate. Suppose that the grading of the material, applied loads, and boundary conditions are axisymmetric so that the displacement u_θ is identically zero and (u_r, u_z) are only functions of r and z. At the moment, we assume that $E = E(z)$ and $\nu = \nu(z)$, and their specific variation will be discussed in the sequel.

Because of the bending-stretching coupling present in functionally graded plates, we must include the in-plane displacement along with the bending deflection. Therefore, we will revisit the classical and first-order theories here.

The complete displacement field of the classical plate theory (CPT) is

$$u_r(r,z) = u_0(r) - z\frac{dw_0}{dr}$$
$$u_z(r,z) = w_0(r) \qquad (13.2.1)$$

where u_0 is the radial displacement and w_0 is the transverse deflection of the point $(r, 0)$ of a point on the mid-plane (i.e., $z = 0$) of the plate. The displacement field of the first order shear deformation plate theory (FST) is given by

$$u_r(r,z) = u_0(r) + z\phi(r)$$
$$u_z(r,z) = w_0(r) \qquad (13.2.2)$$

Both of the theories are governed by the equations

$$-\frac{d}{dr}(rN_{rr}) - N_{\theta\theta} = 0 \qquad (13.2.3)$$

$$-\frac{d}{dr}(rQ_r) = rq \tag{13.2.4}$$

$$\frac{d}{dr}(rM_{rr}) - M_{\theta\theta} = rQ_r \tag{13.2.5}$$

where N_{rr} and $N_{\theta\theta}$ are the radial and circumferential in-plane forces and M_{rr} and $M_{\theta\theta}$ are the radial and circumferential moments

$$(N_{rr}, N_{\theta\theta}) = \int_{-\frac{h}{2}}^{\frac{h}{2}} (\sigma_{rr}, \sigma_{\theta\theta}) dz \tag{13.2.6}$$

$$(M_{rr}, M_{\theta\theta}) = \int_{-\frac{h}{2}}^{\frac{h}{2}} (\sigma_{rr}, \sigma_{\theta\theta}) z \, dz \tag{13.2.7}$$

The plate constitutive equations of the two theories are given below.

CPT for isotropic plate:

$$N_{rr}^K = A_{11}\frac{du_0}{dr} + A_{12}\frac{u_0}{r} \tag{13.2.8}$$

$$N_{\theta\theta}^K = A_{12}\frac{du_0}{dr} + A_{11}\frac{u_0}{r} \tag{13.2.9}$$

$$M_{rr}^K = -D_{11}\frac{d^2 w_0^K}{dr^2} - D_{12}\frac{1}{r}\frac{dw_0^K}{dr} \tag{13.2.10}$$

$$M_{\theta\theta}^K = -D_{12}\frac{d^2 w_0^K}{dr^2} - D_{11}\frac{1}{r}\frac{dw_0^K}{dr} \tag{13.2.11}$$

$$Q_r^K = \frac{1}{r}\left[\frac{d}{dr}(rM_{rr}) - M_{\theta\theta}\right] \tag{13.2.12}$$

FST for functionally graded plate:

$$N_{rr}^F = A_{11}\frac{du_0^F}{dr} + A_{12}\frac{u_0^F}{r} + B_{11}\frac{d\phi}{dr} + B_{12}\frac{\phi}{r} \tag{13.2.13}$$

$$N_{\theta\theta}^F = A_{12}\frac{du_0^F}{dr} + A_{11}\frac{u_0^F}{r} + B_{12}\frac{d\phi}{dr} + B_{11}\frac{\phi}{r} \tag{13.2.14}$$

$$M_{rr}^F = B_{11}\frac{du_0^F}{dr} + B_{12}\frac{u_0^F}{r} + D_{11}\frac{d\phi}{dr} + D_{12}\frac{\phi}{r} \tag{13.2.15}$$

$$M_{\theta\theta}^F = B_{12}\frac{du_0^F}{dr} + B_{11}\frac{u_0^F}{r} + D_{12}\frac{d\phi}{dr} + D_{11}\frac{\phi}{r} \tag{13.2.16}$$

$$Q_r^F = A_{55}\left(\phi + \frac{dw_0^F}{dr}\right) \tag{13.2.17}$$

where superscript K denotes quantities in CPT and F denotes in FST. Of course, ϕ appears only in FST. The plate stiffnesses A_{ij}, B_{ij}, and D_{ij} are defined by

$$(A_{ij}, B_{ij}, D_{ij}) = \int_{-\frac{h}{2}}^{\frac{h}{2}} Q_{ij}(1, z, z^2) dz \quad (i, j = 1, 2) \quad (13.2.18a)$$

$$A_{55} = K_s \int_{-\frac{h}{2}}^{\frac{h}{2}} \frac{E}{2(1+\nu)} dz \quad (13.2.18b)$$

$$Q_{11} = Q_{22} = \frac{E}{(1-\nu^2)}, \quad Q_{12} = \nu Q_{11} \quad (13.2.18c)$$

where E is the modulus of elasticity, ν the Poisson ratio, and K_s the shear correction factor.

The strategy is to develop relations for the deflections, forces, and moments of functionally graded plates based on the first-order shear deformation theory (FSDT) in terms of the associated quantities of isotropic plates based on the classical plate theory. Then the relations developed are specialized for plates with various boundary conditions.

13.2.3. Relationships Between CPT and FSDT

From Eqs. (13.2.3), (13.2.13), and (13.2.14), we obtain

$$\begin{aligned} 0 &= -\frac{d}{dr}(rN_{rr}) - N_{\theta\theta} \\ &= A_{11}\left[\frac{d}{dr}\left(r\frac{du_0}{dr}\right) - \frac{u_0}{r}\right] + B_{11}\left[\frac{d}{dr}\left(r\frac{d\phi}{dr}\right) - \frac{\phi}{r}\right] \\ &= A_{11}\left[r\frac{d}{dr}\left(\frac{1}{r}\frac{d}{dr}(ru_0)\right)\right] + B_{11}\left[r\frac{d}{dr}\left(\frac{1}{r}\frac{d}{dr}(r\phi)\right)\right] \end{aligned}$$

or

$$\left[r\frac{d}{dr}\left(\frac{1}{r}\frac{d}{dr}(ru_0)\right)\right] = -\frac{B_{11}}{A_{11}}\left[r\frac{d}{dr}\left(\frac{1}{r}\frac{d}{dr}(r\phi)\right)\right]$$

Upon integration, we obtain

$$\frac{d}{dr}(ru_0) = -\frac{B_{11}}{A_{11}}\frac{d}{dr}(r\phi) + K_1 r \quad (13.2.19a)$$

and

$$u_0 = -\frac{B_{11}}{A_{11}}\phi + K_1 \frac{r}{2} + \frac{K_2}{r} \quad (13.2.19b)$$

from which we can compute

$$\frac{du_0}{dr} = -\frac{B_{11}}{A_{11}}\frac{d\phi}{dr} + \frac{K_1}{r} - \frac{K_2}{r^2} \qquad (13.2.20)$$

where K_1 and K_2 are integration constants. Using Eqs. (13.2.19) and (13.2.20), the forces and moments of Eqs. (13.2.13)–(13.2.16) can be expressed in terms of ϕ as

$$M_{rr}^F = \Omega_1\frac{d\phi}{dr} + \Omega_2\frac{\phi}{r} + \frac{1}{2}\Omega_3 K_1 + \frac{1}{r^2}\Omega_4 K_2 \qquad (13.2.21)$$

$$M_{\theta\theta}^F = \Omega_2\frac{d\phi}{dr} + \Omega_1\frac{\phi}{r} + \frac{1}{2}\Omega_3 K_1 - \frac{1}{r^2}\Omega_4 K_2 \qquad (13.2.22)$$

$$N_{rr}^F = \Omega_5\frac{\phi}{r} + \frac{1}{2}\Omega_6 K_1 + \frac{1}{r^2}\Omega_7 K_2 \qquad (13.2.23)$$

$$N_{\theta\theta}^F = \Omega_5\frac{d\phi}{dr} + \frac{1}{2}\Omega_6 K_1 - \frac{1}{r^2}\Omega_7 K_2 \qquad (13.2.24)$$

where

$$\Omega_1 = D_{11} - \frac{B_{11}^2}{A_{11}}, \quad \Omega_2 = D_{12} - \frac{B_{11}B_{12}}{A_{11}}$$

$$\Omega_3 = B_{11} + B_{12}, \quad \Omega_4 = B_{12} - B_{11}, \quad \Omega_5 = B_{12} - \frac{A_{12}B_{11}}{A_{11}}$$

$$\Omega_6 = A_{11} + A_{12}, \quad \Omega_7 = A_{12} - A_{11} \qquad (13.2.25)$$

Based on load equivalence, we have

$$\frac{d}{dr}\left(rQ_r^F\right) = \frac{d}{dr}\left(rQ_r^K\right) \qquad (13.2.26)$$

and after integration yields

$$rQ_r^F = rQ_r^K + C_1 \qquad (13.2.27)$$

where C_1 is a constant of integration. But from Eqs. (13.2.5), (13.2.21) and (13.2.22), we have

$$rQ_r^F = \frac{d}{dr}\left(rM_{rr}^F\right) - M_{\theta\theta}^F = \Omega_1\left[r\frac{d}{dr}\left(\frac{1}{r}\frac{d}{dr}(r\phi)\right)\right] \qquad (13.2.28)$$

Similarly,

$$rQ_r^K = \frac{d}{dr}\left(rM_{rr}^K\right) - M_{\theta\theta}^K = -D\left[r\frac{d}{dr}\left(\frac{1}{r}\frac{d}{dr}(r\frac{dw_0^K}{dr})\right)\right] \qquad (13.2.29)$$

Using relations (13.2.28) and (13.2.29) in (13.2.27), we obtain

$$\Omega_1 \left[r \frac{d}{dr} \left(\frac{1}{r} \frac{d}{dr} (r\phi) \right) \right] = -D \left[r \frac{d}{dr} \left(\frac{1}{r} \frac{d}{dr} (r \frac{dw_0^K}{dr}) \right) \right] + C_1 \quad (13.2.30)$$

Upon integrations, we have

$$\Omega_1 \frac{d}{dr} (r\phi) = -D \frac{d}{dr} \left(r \frac{dw_0^K}{dr} \right) + C_1 r \log r + C_2 r \quad (13.2.31a)$$

$$\Omega_1 \phi = -D \frac{dw_0^K}{dr} + \frac{1}{4} C_1 r (2 \log r - 1) + \frac{1}{2} C_2 r + \frac{1}{r} C_3 \quad (13.2.31b)$$

Next from Eqs. (13.2.17) and (13.2.28), we have

$$A_{55} \left(\phi + \frac{dw_0^F}{dr} \right) = \frac{d\mathcal{M}^K}{dr} + \frac{1}{r} C_1 \quad (13.2.32)$$

where $\mathcal{M}^K$ is the moment sum

$$\mathcal{M}^K = \frac{M_{rr}^K + M_{\theta\theta}^K}{(1+\nu)} \quad (13.2.33)$$

Substituting for ϕ from Eq. (13.2.31b) into Eq. (13.2.32), we obtain

$$A_{55} \left[-\frac{D}{\Omega_1} \frac{dw_0^K}{dr} + \frac{C_1}{4\Omega_1} r (2 \log r - 1) + \frac{C_2}{2\Omega_1} r + \frac{C_3}{\Omega_1 r} + \frac{dw_0^F}{dr} \right]$$
$$= \frac{d\mathcal{M}^K}{dr} + \frac{1}{r} C_1 \quad (13.2.34)$$

Integrating the above expression, we obtain

$$w_0^F = \frac{D}{\Omega_1} w_0^K + \frac{C_1}{\Omega_1} \left[\frac{r^2}{4} (1 - \log r) + \frac{\Omega_1}{A_{55}} \log r \right]$$
$$- \frac{C_2}{4\Omega_1} r^2 - \frac{C_3}{\Omega_1} \log r - C_4 + \frac{\mathcal{M}^K}{A_{55}} \quad (13.2.35)$$

From Eq. (13.2.31b), we have

$$\Omega_1 \frac{d\phi}{dr} = -D \frac{d^2 w_0^K}{dr^2} + \frac{C_1}{2} \left[\frac{1}{2} (2 \log r - 1) + 1 \right] + \frac{C_2}{r} - \frac{C_3}{r^2} \quad (13.2.36)$$

Substituting Eqs. (13.2.31b) and (13.2.36) into Eq. (13.2.21), we obtain

$$M_{rr}^F = -D\frac{d^2 w_0^K}{dr^2} + \frac{C_1}{2}\left[\frac{1}{2}(2\log r - 1) + 1\right] + \frac{C_2}{r} - \frac{C_3}{r^2}$$

$$+ \frac{\Omega_2}{r}\left[-\hat{D}\frac{dw_0^K}{dr} + \hat{C}_1\frac{r}{4}(2\log r - 1) + \hat{C}_2\frac{r}{2} + \hat{C}_3\frac{1}{r}\right] + \Omega_3\frac{K_1}{2} + \Omega_4\frac{K_2}{r^2}$$

or

$$M_{rr}^F = M_{rr}^K + \frac{D}{r}\frac{dw_0^K}{dr}\left(\nu - \hat{\Omega}_2\right) + \frac{C_1}{2}\left[\frac{1}{2}\left(1 - \hat{\Omega}_2\right) + \left(1 + \hat{\Omega}_2\right)\log r\right]$$

$$+ \frac{C_2}{2}\left(1 + \hat{\Omega}_2\right) - \frac{C_3}{r^2}\left(1 - \hat{\Omega}_2\right) + \Omega_3\frac{K_1}{2} + \Omega_4\frac{K_2}{r^2} \qquad (13.2.37)$$

where $\hat{D} = \frac{D}{\Omega_1}, \hat{C}_i = \frac{C_i}{\Omega_1}$ and so on. Similarly, we can write

$$M_{\theta\theta}^F = M_{\theta\theta}^K + D\frac{d^2 w_0^K}{dr^2}\left(\nu - \hat{\Omega}_2\right) + \frac{C_1}{2}\left[-\frac{1}{2}\left(1 - \hat{\Omega}_2\right) + \left(1 + \hat{\Omega}_2\right)\log r\right]$$

$$+ \frac{C_2}{2}\left(1 + \hat{\Omega}_2\right) + \frac{C_3}{r^2}\left(1 - \hat{\Omega}_2\right) + \Omega_3\frac{K_1}{2} - \Omega_4\frac{K_2}{r^2} \qquad (13.2.38)$$

Next we substitute (13.2.31b) into (13.2.23) and obtain

$$N_{rr}^F = \Omega_5\left[-\hat{D}\frac{1}{r}\frac{dw_0^K}{dr} + \hat{C}_1\frac{1}{4}(2\log r - 1) + \frac{1}{2}\hat{C}_2 + \frac{1}{r^2}\hat{C}_3\right]$$

$$+ \Omega_6\frac{K_1}{2} + \Omega_7\frac{K_2}{r^2} \qquad (13.2.39)$$

Substituting (13.2.36) into (13.2.24), we obtain

$$N_{\theta\theta}^F = \Omega_5\left[-\hat{D}\frac{1}{r}\frac{dw_0^K}{dr} + \hat{C}_1\frac{1}{4}(2\log r + 1) + \frac{1}{2}\hat{C}_2 - \frac{1}{r^2}\hat{C}_3\right]$$

$$+ \Omega_6\frac{K_1}{2} - \Omega_7\frac{K_2}{r^2} \qquad (13.2.40)$$

Define

$$\mathcal{M}^{\mathcal{F}} \equiv \frac{M_{rr}^F + M_{\theta\theta}^F}{(1 + \hat{\Omega}_2)}, \quad \hat{\Omega}_2 = \frac{\Omega_2}{\Omega_1} \qquad (13.2.41)$$

$$\mathcal{N}^F \equiv \left(N_{rr}^F + N_{\theta\theta}^F\right)\Omega_1 \qquad (13.2.42)$$

Then we have

$$\mathcal{M}^F = \mathcal{M}^K + C_1 \log r + C_2 + \left(\frac{\Omega_3}{(1+\hat{\Omega}_2)}\right) K_1 \qquad (13.2.43)$$

and

$$\mathcal{N}^F = \Omega_5 \left(\mathcal{M}^K + C_1 \log r + C_2\right) + \Omega_1 \Omega_6 K_1 \qquad (13.2.44)$$

This completes the development of equations for the deflections, forces, and moments of functionally graded plates based on the first-order shear deformation theory in terms of the associated quantities of isotropic plates based on the classical plate theory. Thus, it remains that we develop particular relationships for axisymmetric bending of plates with various boundary conditions.

13.2.4 Relationships for Various Boundary Conditions

Roller-supported circular plate

Consider a solid circular plate with a roller support at $r = a$, a being the radius of the plate. The boundary conditions are

$$\text{At } r = 0: \quad u = 0, \quad \phi = 0, \quad \frac{dw_0^K}{dr} = 0, \quad Q_r = 0 \quad (13.2.45a)$$

$$\text{At } r = a: \quad w = 0, \quad N_{rr} = 0, \quad M_{rr} = 0 \qquad (13.2.45b)$$

The above boundary conditions give

$$K_1 = -\frac{2\Omega_5}{a\Omega_6}\left(-\hat{D}\frac{dw_0^K(a)}{dr} + \frac{\hat{C}_2 a}{2}\right), \quad K_2 = 0 \qquad (13.2.46a)$$

$$C_2 = -\frac{2D}{a}\left(\frac{\nu - \hat{\Omega}_2 + \Omega_8}{1 + \hat{\Omega}_2 - \Omega_8}\right)\frac{dw_0^K(a)}{dr} \qquad (13.2.46b)$$

$$C_4 = \frac{\mathcal{M}^K(a)}{A_{55}} - \frac{\hat{C}_2 a^2}{4}, \quad C_1 = C_3 = 0 \qquad (13.2.46c)$$

where $\Omega_8 = \hat{\Omega}_3 \Omega_5 / \Omega_6$. Hence, we have the following relations between the deflections, forces, and moments of the two theories:

$$w_0^F(r) = \hat{D} w_0^K(r) + \frac{\mathcal{M}^K(r) - \mathcal{M}^K(a)}{A_{55}} + \frac{1}{4}\hat{C}_2(a^2 - r^2) \quad (13.2.47)$$

$$Q_r^F(r) = Q_r^K(r) \tag{13.2.48}$$

$$N_{rr}^F(r) = \Omega_5 \hat{D} \left(\frac{1}{a} \frac{dw_0^K(a)}{dr} - \frac{1}{r} \frac{dw_0^K}{dr} \right) \tag{13.2.49}$$

$$N_{rr}^F(r) = \Omega_5 \hat{D} \left(\frac{1}{a} \frac{dw_0^K(a)}{dr} - \frac{d^2 w_0^K}{dr^2} \right) \tag{13.2.50}$$

$$M_{rr}^F(r) = M_{rr}^K(r) + D \frac{1}{r} \frac{dw_0^K}{dr} \left(\nu - \hat{\Omega}_2 \right) + \frac{1}{2} C_2 \left(1 + \hat{\Omega}_2 \right)$$
$$+ \frac{1}{2} K_1 \Omega_3 \tag{13.2.51}$$

$$M_{\theta\theta}^F(r) = M_{\theta\theta}^K(r) + D \frac{d^2 w_0^K}{dr^2} \left(\nu - \hat{\Omega}_2 \right) + \frac{1}{2} C_2 \left(1 + \hat{\Omega}_2 \right)$$
$$+ \frac{1}{2} K_1 \Omega_3 \tag{13.2.52}$$

Hinged circular plate

Consider a solid circular plate with a hinged support at $r = a$. The boundary conditions are

At $r = 0$: $\quad u = 0, \quad \phi = 0, \quad \dfrac{dw_0^K}{dr} = 0, \quad Q_r = 0$ (13.2.53a)

At $r = a$: $\quad u = 0, \quad w = 0, \quad M_{rr} = 0$ (13.2.53b)

The boundary conditions give

$$K_1 = \frac{2B_{11}}{aA_{11}} \left(-\hat{D} \frac{dw_0^K(a)}{dr} + \frac{\hat{C}_2 a}{2} \right), \quad K_2 = 0 \tag{13.2.54a}$$

$$C_2 = -\frac{2D}{a} \left(\frac{\nu - \hat{\Omega}_2 - \Omega_9}{1 + \hat{\Omega}_2 + \Omega_9} \right) \frac{dw_0^K(a)}{dr} \tag{13.2.54b}$$

$$C_4 = \frac{M^K(a)}{A_{55}} - \frac{\hat{C}_2 a^2}{4}, \quad C_1 = C_3 = 0 \tag{13.2.54c}$$

where $\Omega_9 = \hat{\Omega}_3 B_{11}/A_{11}$. The relations follow the same form as those given in (13.2.47) to (13.2.52) but K_1, C_2, and C_4 take the expressions given in (13.2.54a) and (13.2.54b).

Clamped circular plate

Consider a solid circular plate with a clamped support at $r = a$. The boundary conditions are

$$\text{At } r = 0: \quad u = 0, \quad \phi = 0, \quad \frac{dw_0^K}{dr} = 0, \quad Q_r = 0 \quad (13.2.55a)$$

$$\text{At } r = a: \quad u = 0, \quad w = 0, \quad \phi = 0, \quad \frac{dw_0^K}{dr} = 0 \quad (13.2.55b)$$

The boundary conditions give

$$K_1 = K_2 = 0, \quad C_1 = C_2 = C_3 = 0, \quad C_4 = \frac{\mathcal{M}^K(a)}{A_{55}} \quad (13.2.56)$$

Hence, we have the following relations between the deflections, forces, and moments of the two theories:

$$w_0^F(r) = \hat{D} w_0^K(r) + \frac{\mathcal{M}^K(r) - \mathcal{M}^K(a)}{A_{55}} \quad (13.2.57)$$

$$Q_r^F(r) = Q_r^K(r) \quad (13.2.58)$$

$$N_{rr}^F(r) = -\Omega_5 \hat{D} \frac{1}{r} \frac{dw_0^K}{dr} \quad (13.2.59)$$

$$N_{\theta\theta}^F(r) = -\Omega_5 \hat{D} \frac{d^2 w_0^K}{dr^2} \quad (13.2.60)$$

$$M_{rr}^F(r) = M_{rr}^K(r) + D \frac{1}{r} \frac{dw_0^K}{dr} \left(\nu - \hat{\Omega}_2 \right) \quad (13.2.61)$$

$$M_{\theta\theta}^F(r) = M_{\theta\theta}^K(r) + D \frac{d^2 w_0^K}{dr^2} \left(\nu - \hat{\Omega}_2 \right) \quad (13.2.62)$$

Clamped-free annular plate

Consider an annular plate with a clamped support at the inner edge $r = b$ and free at the outer edge $r = a$. The boundary conditions are

$$\text{At } r = b: \quad u = 0, \quad w = 0, \quad \phi = 0, \quad \frac{dw_0^K}{dr} = 0 \quad (13.2.63a)$$

$$\text{At } r = a: \quad N_{rr} = 0, \quad M_{rr} = 0, \quad Q_r = 0 \quad (13.2.63b)$$

The boundary conditions give

$$K_1 = \Omega_5 \left(\frac{-2a\hat{D} \frac{dw_0^K(a)}{dr} + \hat{C}_2 a^2 + 2\hat{C}_3}{b^2 \Omega_7 - a^2 \Omega_6} \right), \quad K_2 = -\frac{b^2 K_1}{2} \quad (13.2.64a)$$

$$C_1 = 0, \quad C_2 = -2ad_1 D \frac{dw_0^K(a)}{dr}, \quad C_3 = -\frac{b^2 C_2}{2} \qquad (13.2.64\text{b})$$

$$C_4 = \frac{\mathcal{M}^K(b)}{A_{55}} - \frac{\hat{C}_2 b^2}{4}(1 - 2\log b) \qquad (13.2.64\text{c})$$

$$d_1 = \frac{\nu - \hat{\Omega}_2 - \hat{\Omega}_5 d_2}{a^2\left(1 + \hat{\Omega}_2\right) + b^2\left(1 - \hat{\Omega}_2\right) + \hat{\Omega}_5 d_2(a^2 - b^2)} \qquad (13.2.64\text{d})$$

$$d_2 = \left(\frac{a^2 \Omega_3 - b^2 \Omega_4}{b^2 \Omega_7 - a^2 \Omega_6}\right) \qquad (13.2.64\text{e})$$

Clamped-clamped annular plate

Consider an annular plate with clamped inner edge $r = b$ and outer edge $r = a$. The boundary conditions are

$$\text{At } r = b: \quad u = 0, \quad w = 0, \quad \phi = 0, \quad \frac{dw_0^K}{dr} = 0 \quad (13.2.65\text{a})$$

$$\text{At } r = a: \quad u = 0, \quad w = 0, \quad \phi = 0, \quad \frac{dw_0^K}{dr} = 0 \quad (13.2.65\text{b})$$

The boundary conditions give

$$K_1 = K_2 = 0, \quad C_1 = \frac{\Omega_1}{A_{55} C_5}\left(\mathcal{M}^K(b) - \mathcal{M}^K(a)\right) \qquad (13.2.66\text{a})$$

$$C_2 = \left(\frac{b^2 \log b - a^2 \log a}{a^2 - b^2} + \frac{1}{2}\right) C_1 \qquad (13.2.66\text{b})$$

$$C_3 = \left(\frac{a^2 b^2}{2(a^2 - b^2)} \log \frac{a}{b}\right) C_1 \qquad (13.2.66\text{c})$$

$$C_4 = \left[\frac{a^2 + b^2}{16} + \frac{\Omega_1}{2A_{55}} \log ab + \frac{a^2 b^2 \left(\log \frac{b}{a} - (\log ab)^2\right)}{4(b^2 - a^2)}\right] \hat{C}_1$$

$$\quad + \frac{\mathcal{M}^K(a) + \mathcal{M}^K(b)}{2A_{55}} \qquad (13.2.66\text{d})$$

$$C_5 = \frac{a^2 - b^2}{8} + \frac{\Omega_1}{A_{55}} \log \frac{a}{b} - \frac{a^2 b^2}{2(a^2 - b^2)}\left(\log \frac{a}{b}\right)^2 \qquad (13.2.66\text{e})$$

Clamped-roller supported annular plate

Lastly, consider an annular plate with clamped inner edge $r = b$ and roller supported at the outer edge $r = a$. The boundary conditions are

$$\text{At } r = b: \quad u = 0, \quad w = 0, \quad \phi = 0, \quad \frac{dw_0^K}{dr} = 0 \quad (13.2.67a)$$

$$\text{At } r = a: \quad w = 0, \quad N_{rr} = 0, \quad M_{rr} = 0 \quad (13.2.67b)$$

The boundary conditions yield

$$K_1 = \Omega_5 \left(\frac{-2a\hat{D}\frac{dw_0^K(a)}{dr} + \hat{C}_1 \frac{a^2}{2}(2\ln a - 1) + \hat{C}_2 a^2 + 2\hat{C}_3}{b^2 \Omega_7 - a^2 \Omega_6} \right) \quad (13.2.68a)$$

$$K_2 = -\frac{b^2}{2} K_1, \quad C_1 = \frac{f_1 e_4 - f_2 e_2}{e_1 e_4 - e_2 e_3}, \quad C_2 = \frac{f_1 - C_1 e_1}{e_2} \quad (13.2.68b)$$

$$C_3 = -\frac{C_1 b^2}{4}(2\ln b - 1) - \frac{C_2 b^2}{2} \quad (13.2.68c)$$

$$C_4 = \frac{\mathcal{M}^K(b) + \mathcal{M}^K(a)}{2A_{55}} + \frac{\hat{C}_1}{2} \left[\frac{b^2}{4}(1 - \ln b) + \frac{a^2}{4}(1 - \ln a) + \frac{\Omega_1}{A_{55}} \ln ab \right]$$
$$- \hat{C}_2 \frac{(a^2 + b^2)}{8} - \frac{1}{2}\hat{C}_3 \ln ab \quad (13.2.68d)$$

$$e_1 = \frac{b^2}{4}(1 - \ln b) - \frac{a^2}{4}(1 - \ln a) + \frac{b^2}{4}(2\ln b - 1)\ln \frac{b}{a}$$
$$+ \frac{\Omega_1}{A_{55}} \ln \frac{b}{a} \quad (13.2.68e)$$

$$e_2 = \frac{b^2}{2} \ln \frac{b}{a} - \frac{(b^2 - a^2)}{4} \quad (13.2.68f)$$

$$e_3 = \frac{1}{2}\left(1 + \hat{\Omega}_2\right)\ln a + \frac{1}{4}\left(1 - \hat{\Omega}_2\right) + \frac{b^2}{4a^2}\left(1 - \hat{\Omega}_2\right)$$

$$e_4 = \frac{1}{2}\left[\left(1 + \hat{\Omega}_2\right) + \frac{b^2}{a^2}\left(1 - \hat{\Omega}_2\right)\right] \quad (13.2.68g)$$

$$f_1 = \frac{\mathcal{M}^K(a) - \mathcal{M}^K(b)}{A_{55}} \Omega_1, \quad f_2 = \frac{D}{a}\left(\hat{\Omega}_2 - \nu\right)\frac{dw_0^K(a)}{dr} \quad (13.2.68h)$$

13.2.5 Illustrative Examples

To illustrate the usefulness of the relationships derived herein, we provide some examples. Consider the case of circular plates under uniformly distributed transverse load of intensity q_0. The classical plate solutions are given by (see [27,28])

$$w_0^K(r) = \begin{cases} \frac{q_0 a^4}{64D}\left[\left(\frac{r}{a}\right)^4 - 2\left(\frac{3+\nu}{1+\nu}\right)\left(\frac{r}{a}\right)^2 + \frac{5+\nu}{1+\nu}\right], & \text{for simple support} \\ \frac{q_0 a^4}{64D}\left[1 - \left(\frac{r}{a}\right)^2\right]^2, & \text{for clamped support} \end{cases}$$
(13.2.69)

Now consider a functionally graded plate whose modulus is assumed to be of the form

$$E^F(z) = E_m\left(\frac{h-2z}{2h}\right)^n + E_c\left[1 - \left(\frac{h-2z}{2h}\right)^n\right]$$
(13.2.70)

and $\nu^F = \nu$ (i.e., independent of z). Here E_c and E_m denote the moduli of two different constituents, namely ceramic and metal, n is the power-law exponent which is related to the volume fraction of ceramic and metal, and h denotes the total plate thickness.

We have

$$A_{ij} = \left(Q_{ij}^c - Q_{ij}^m\right)h\left(\frac{1}{1+n}\right) + Q_{ij}^m h$$

$$B_{ij} = \left(Q_{ij}^c - Q_{ij}^m\right)\frac{h^2}{2}\left(\frac{n}{(1+n)(2+n)}\right)$$

$$D_{ij} = \left(Q_{ij}^c - Q_{ij}^m\right)\frac{h^3}{12}\left(\frac{3(2+n+n^2)}{(1+n)(2+n)(3+n)}\right) + Q_{ij}^m \frac{h^3}{24} \quad (13.2.71)$$

or

$$A_{11} = A_{22} = \frac{h(E_m + nE_c)}{(1+n)(1-\nu^2)}$$

$$A_{55} = A_{44} = \frac{hK^2(E_m + nE_c)}{2(1+n)(1+\nu)}$$

$$B_{11} = B_{22} = \frac{nh^2(E_c - E_m)}{2(1+n)(2+n)(1-\nu^2)}$$

$$D_{11} = D_{22} = \frac{h^3\left[n(n^2+3n+8)E_c + 3(n^2+n+2)E_m\right]}{12(1+n)(2+n)(3+n)(1-\nu^2)}$$

$$A_{12} = \nu A_{11}, \quad B_{12} = \nu B_{11}, \quad D_{12} = \nu D_{11} \quad (13.2.72)$$

If we define

$$D_0 = \frac{E_c h^3}{12(1-\nu^2)}, \quad G = \frac{E_c}{2(1+\nu)}, \quad E_r = \frac{E_m}{E_c} \quad (13.2.73)$$

then the expressions in (13.2.72) can be written as

$$A_{11} = \frac{12(n+E_r)D}{h^2(1+n)}, \quad A_{55} = \frac{K_s Gh(n+E_r)}{(1+n)}, \quad B_{11} = \frac{6n(1-E_r)D}{h(1+n)(2+n)}$$

$$D_{11} = \frac{[n(n^2+3n+8)+3(n^2+n+2)E_r]D}{(1+n)(2+n)(3+n)} \tag{13.2.74}$$

The constants Ω_i have the following values: $\Omega_5 = \Omega_8 = 0$ and

$$\Omega_1 = D_{11} - \frac{B_{11}^2}{A_{11}}$$
$$= \frac{D[(n^4+4n^3+7n^2)+4E_r(n^3+4n^2+7n+3E_r)]}{(n+E_r)(3+n)(2+n)^2}$$

$$\Omega_2 = D_{12} - \frac{B_{12}B_{11}}{A_{11}} = \nu\Omega_1$$

$$\Omega_3 = B_{11} + B_{12} = \frac{6n(1-E_r)(1+\nu)D}{h(1+n)(2+n)}$$

$$\Omega_4 = B_{12} - B_{11} = -\frac{6n(1-E_r)(1-\nu)D}{h(1+n)(2+n)}$$

$$\Omega_6 = \frac{12(n+E_r)(1+\nu)D}{h^2(1+n)}$$

$$\Omega_7 = -\frac{12(n+E_r)(1-\nu)D}{h^2(1+n)}$$

$$\Omega_9 = \frac{3(1-E_r)^2 n^2(3+n)(1+\nu)}{(1+n)[7n^2+4n^3+n^4+4E_r(3E_r+7n+4n^2+n^3)]}$$

$$\tag{13.2.75}$$

By substituting the CPT solution given by Eq. (13.2.69) into (13.2.47) and (13.2.57), one obtains the deflection of the FGM plate as

$$\bar{w}_0^F = \frac{64 w_0^F D_c}{q_0 a^4}$$

$$= \frac{D_c}{\Omega_1}\left[\left(\frac{r}{a}\right)^4 - 2\left(\frac{3+\nu}{1+\nu}\right)\left(\frac{r}{a}\right)^2 + \frac{5+\nu}{1+\nu}\right]$$

$$+ \frac{8}{3K_s^2(1-\nu_c)}\left(\frac{h}{a}\right)^2\left[1-\left(\frac{r}{a}\right)^2\right]\left(\frac{1+n}{E_r+n}\right)$$

$$- \frac{4\eta D_c}{(1+\nu_c)\Omega_1}\left(\frac{\Omega_9}{1+\nu+\Omega_9}\right)\left[1-\left(\frac{r}{a}\right)^2\right] \tag{13.2.76}$$

where
$$\eta = \begin{cases} = 0, & \text{for roller supported edge} \\ = 1, & \text{for simply supported edge} \end{cases} \quad (13.2.77)$$

For clamped edge, we obtain

$$\bar{w}_0^F = \frac{64 w_0^F D_c}{q_0 a^4}$$
$$= \frac{D_c}{\Omega_1} \left[1 - \left(\frac{r}{a}\right)^2\right]^2 + \frac{8}{3K_s^2(1-\nu_c)} \left(\frac{h}{a}\right)^2 \left[1 - \left(\frac{r}{a}\right)^2\right]$$
$$\times \left(\frac{1+n}{E_r + n}\right) \quad (13.2.78)$$

In the above equations K_s denotes the shear correction factor.

Considering a Titanium-Zirconica FGM plate, *i.e.*, $\nu = 0.288$, $E_r = 0.396$, and taking $K_s = 5/6$, the maximum deflection parameters at the plate center ($r = 0$) are tabulated in Tables 13.2.1–13.2.3 for various values of n and h/a ratio. The deflection parameter increases with increasing h/a ratios but decreases with increasing values of n.

From Eqs. (13.2.48)–(13.2.52) and (13.2.59)–(13.2.62), the stress resultants for the FGM plates are (since $\Omega_5 = \Omega_8 = C_2 = K_1 = 0$)

$$N_{rr}^F = 0, \quad N_{\theta\theta}^F = 0, \quad M_{rr}^F = M_{rr}^K, \quad M_{\theta\theta}^F = M_{\theta\theta}^K \quad (13.2.79)$$

Table 13.2.1. Maximum deflection $\bar{w}_0^F$ of functionally graded roller-supported circular plates ($\nu = 0.288$, $E_r = 0.396$, $K_s = 5/6$).

	Thickness radius ratio, h/a				
n	0.0	0.05	0.1	0.15	0.2
0	10.368	10.396	10.481	10.623	10.822
2	5.700	5.714	5.756	5.826	5.925
4	5.210	5.223	5.261	5.325	5.414
6	4.958	4.970	5.007	5.069	5.155
8	4.800	4.812	4.848	4.909	4.993
10	4.692	4.704	4.739	4.799	4.882
15	4.527	4.538	4.573	4.632	4.714

(Table 13.2.1 is continued on the next page).

(Table 13.2.1 is continued from the previous page).

	\multicolumn{5}{c}{Thickness radius ratio, h/a}
n	0.0	0.05	0.1	0.15	0.2
20	4.434	4.446	4.480	4.538	4.619
25	4.375	4.386	4.421	4.478	4.559
30	4.334	4.345	4.379	4.437	4.517
35	4.303	4.315	4.349	4.406	4.486
40	4.280	4.291	4.326	4.383	4.463
50	4.247	4.258	4.293	4.349	4.429
10^2	4.178	4.189	4.223	4.280	4.359
10^3	4.113	4.124	4.158	4.214	4.293
10^4	4.106	4.118	4.151	4.208	4.286
10^5	4.106	4.117	4.151	4.207	4.285

Table 13.2.2. Maximum deflection $\bar{w}_0^F$ of functionally graded simply supported circular plates ($\nu = 0.288$, $E_r = 0.396$, $K_s = 5/6$).

	\multicolumn{5}{c}{Thickness radius ratio, h/a}
n	0.0	0.05	0.1	0.15	0.2
0	10.368	10.396	10.481	10.623	10.822
2	5.483	5.497	5.539	5.610	5.708
4	5.102	5.115	5.153	5.217	5.307
6	4.897	4.909	4.946	5.007	5.094
8	4.761	4.773	4.810	4.870	4.954
10	4.665	4.677	4.712	4.772	4.855
15	4.514	4.525	4.560	4.619	4.701
20	4.426	4.438	4.473	4.531	4.612
25	4.370	4.381	4.416	4.473	4.554
30	4.330	4.342	4.376	4.433	4.513
35	4.301	4.312	4.346	4.404	4.484
40	4.278	4.289	4.324	4.381	4.461
50	4.246	4.257	4.291	4.348	4.428
10^2	4.178	4.189	4.223	4.280	4.359
10^3	4.113	4.124	4.158	4.214	4.293
10^4	4.106	4.118	4.151	4.208	4.286
10^5	4.106	4.117	4.151	4.207	4.285

Table 13.2.3. Maximum deflection $\bar{w}_0^F$ of functionally graded clamped circular plates ($\nu = 0.288$, $E_r = 0.396$, $K_s = 5/6$).

	Thickness radius ratio, h/a				
n	0.0	0.05	0.1	0.15	0.2
0	2.525	2.554	2.639	2.781	2.979
2	1.388	1.402	1.444	1.515	1.613
4	1.269	1.282	1.320	1.384	1.473
6	1.208	1.220	1.257	1.318	1.404
8	1.169	1.181	1.217	1.278	1.362
10	1.143	1.155	1.190	1.250	1.333
15	1.103	1.114	1.149	1.208	1.289
20	1.080	1.092	1.126	1.184	1.265
25	1.066	1.077	1.112	1.169	1.250
30	1.056	1.067	1.101	1.159	1.239
35	1.048	1.060	1.094	1.151	1.231
40	1.043	1.054	1.088	1.145	1.225
50	1.034	1.046	1.080	1.137	1.216
10^2	1.018	1.029	1.063	1.119	1.199
10^3	1.002	1.013	1.047	1.103	1.182
10^4	1.000	1.011	1.045	1.101	1.180
10^5	1.000	1.011	1.045	1.101	1.180

13.3 Buckling Load Relationships for Sandwich Mindlin Plates

13.3.1 Governing Equations

Here we extend the buckling load relationships developed for isotropic polygonal plates in Chapter 11 to polygonal sandwich plates. The governing equations of the Kirchhoff and Mindlin plates are given by Eqs. (11.1.3a-c) and (11.1.4a-c), respectively. They are repeated for ready reference. The buckling equations for Kirchhoff plates are given by

$$\frac{\partial M_{xx}^K}{\partial x} + \frac{\partial M_{xy}^K}{\partial y} - Q_x^K = 0 \qquad (13.3.1a)$$

$$\frac{\partial M_{xy}^K}{\partial x} + \frac{\partial M_{yy}^K}{\partial y} - Q_y^K = 0 \qquad (13.3.1b)$$

$$\frac{\partial Q_x^K}{\partial x} + \frac{\partial Q_y^K}{\partial y} = N^K \nabla^2 w^K \qquad (13.3.1c)$$

Where the moment resultants (M_{xx}, M_{yy}, M_{xy}) are related to the deflection w_0 by Eqs. (6.2.22a-c).

In the case of Mindlin plates under hydrostatic in-plane loads, the buckling equations are

$$\frac{\partial M_{xx}^M}{\partial x} + \frac{\partial M_{xy}^M}{\partial y} - Q_x^M = 0 \qquad (13.3.2a)$$

$$\frac{\partial M_{xy}^M}{\partial x} + \frac{\partial M_{yy}^M}{\partial y} - Q_y^M = 0 \qquad (13.3.2b)$$

$$\frac{\partial Q_x^M}{\partial x} + \frac{Q_y^M}{\partial y} = N^M \nabla^2 w^M \qquad (13.2.2c)$$

The relationships between the force and moment resultants (Ms and Qs) and the generalized displacements (w_0, ϕ_x, ϕ_y) of sandwich plates are given by Eqs. (13.1.10a-e).

Substitution of Eqs. (13.1.10a-e) into Eqs. (13.3.2a-c) yields the result

$$K_s(G_c h_c + 2G_f h_f)\left(\phi_x^M + \frac{\partial w^M}{\partial x}\right)$$
$$= (D_c + D_f)\frac{\partial^2 \phi_x^M}{\partial x^2} + (\nu_c D_c + \nu_f D_f)\frac{\partial^2 \phi_y^M}{\partial x \partial y}$$
$$+ \frac{1}{2}[(1-\nu_c)D_c + (1-\nu_f)D_f]\left(\frac{\partial^2 \phi_x^M}{\partial y^2} + \frac{\partial^2 \phi_y^M}{\partial x \partial y}\right) \qquad (13.3.3a)$$

$$K_s(G_c h_c + 2G_f h_f)\left(\phi_y^M + \frac{\partial w^M}{\partial y}\right)$$
$$= (D_c + D_f)\frac{\partial^2 \phi_y^M}{\partial y^2} + (\nu_c D_c + \nu_f D_f)\frac{\partial^2 \phi_x^M}{\partial x \partial y}$$
$$+ \frac{1}{2}[(1-\nu_c)D_c + (1-\nu_f)D_f]\left(\frac{\partial^2 \phi_y^M}{\partial x^2} + \frac{\partial^2 \phi_x^M}{\partial x \partial y}\right) \qquad (13.3.3b)$$

$$(G_c h_c + 2G_f h_f)\left(\mathcal{M}^M + \nabla^2 w^M\right) = N^M \nabla^2 w^M \qquad (13.3.3c)$$

where $\mathcal{M}^M$ is the moment sum

$$\mathcal{M}^M = \frac{M_{xx}^M + M_{yy}^M}{(1+\nu_c)D_c + (1+\nu_f)D_f} = \frac{\partial \phi_x^M}{\partial x} + \frac{\partial \phi_y^M}{\partial y} \qquad (13.3.4)$$

Differentiating Eqs. (13.3.3a) and (13.3.3b) with respect to x and y, respectively, and adding the results and using Eq. (13.3.3c), we arrive at the result

$$\left(\nabla^2 + \lambda^2\right)\nabla^2 w^M = 0 \tag{13.3.5}$$

where

$$\lambda^2 = \frac{N^M}{(D_c + D_f)\left\{1 - [N^M/K_s(G_c h_c + 2G_f h_f)]\right\}} \tag{13.3.6}$$

The equation governing buckling of an isotropic Kirchhoff plate is given by [see Reddy (1999)]

$$\left(\nabla^2 + \frac{N^K}{D}\right)\nabla^2 w^K = 0 \tag{13.3.7}$$

where w^K is the deflection of the Kirchhoff plate and N^K is the associated buckling load.

13.3.2 Buckling Load Relationship

For simply supported, isotropic polygional plate the following boundary conditions hold on simply supported edges:

$$w^K = 0, \quad \mathcal{M}^K = \nabla^2 w^K = 0 \text{ for the Kirchhoff plate} \tag{13.3.8}$$
$$w^M = 0, \quad \mathcal{M}^M = \nabla^2 w^M = 0 \text{ for the Mindlin plate} \tag{13.3.9}$$

Comparing Eqs. (13.3.5) and (13.3.7), and in view of the boundary conditions (13.3.8) and (13.3.9), it follows that

$$\lambda^2 = \frac{N^K}{D} \tag{13.3.10}$$

or

$$N^M = \frac{(D_c + D_f)N^K}{D(1 + D_{eff}N^K)}, \quad D_{eff} = \frac{D_c + D_f}{DK_s(G_c h_c + 2G_f h_f)} \tag{13.3.11}$$

which provides a relationship between the buckling load N^K of a simply supported, solid polygonal Kirchhoff plate and that of a simply supported, sandwich Mindlin plate. Equation (13.3.11) is also valid for solid Mindlin plate. By setting $h_f = 0$ and omitting the subscript 'c' on quantities, we obtain the relationship between the buckling loads of solid polygonal Mindlin and Kirchhoff plates

$$N^S = \frac{N^K}{(1 + D_s N^K)}, \quad D_s = \frac{1}{K_s G h} \qquad (13.3.12)$$

Finally, in view of Eqs. (13.3.11) and (13.3.12), we have the relationship between the buckling loads of Mindlin sandwich and solid plates

$$\frac{N^M/(D_c + D_f)}{1 - N^M/[K_s(G_c h_c + 2G_f h_f)]} = \frac{N^S/D}{1 - N^S/(K_s G h)} \qquad (13.3.13)$$

13.4 Free Vibration Relationships for Sandwich Plates

13.4.1 Governing Equations

This section is concerned with the free vibration of general polygonal sandwich plates based on the first-order shear deformation plate theory. The polygonal plate considered is simply supported on all its straight edges. It will be shown herein that the vibration frequencies of such sandwich plates may be computed from their well-known Kirchhoff plate counterparts via an exact relationship. This exact relationship should be useful for the development of approximate vibration formulae for sandwich plates of other shapes, boundary and loading conditions.

Consider an arbitrary polygonal sandwich plate with simply supported edges (see Figure 13.1.1). On the basis of the first-order plate theory and upon assuming that the deformation are continuous through the plate thickness, the stress-displacement relations are given by Eqs. (13.1.1a-e).

The equations of motion of the plate are

$$\frac{\partial M_{xx}^M}{\partial x} + \frac{\partial M_{xy}^M}{\partial y} - Q_x^M = (\rho_c I_c + \rho_f I_f)\frac{\partial^2 \phi_x^M}{\partial t^2} \qquad (13.4.1)$$

$$\frac{\partial M_{xy}^M}{\partial x} + \frac{\partial M_{yy}^M}{\partial y} - Q_y^M = (\rho_c I_c + \rho_f I_f)\frac{\partial^2 \phi_y^M}{\partial t^2} \qquad (13.4.2)$$

$$\frac{\partial Q_x^M}{\partial x} + \frac{Q_y^M}{\partial y} = (\rho_c I_c + 2\rho_f I_f)\frac{\partial^2 w^M}{\partial t^2} \qquad (13.4.3)$$

where t is the time, ρ_c and ρ_f are the material densities of the core and facings, respectively, and

$$D_c = \frac{E_c I_c}{(1-\nu_c^2)}, \quad I_c = \frac{h_c^3}{12}$$
$$D_f = \frac{E_f I_f}{(1-\nu_f^2)}, \quad I_f = \frac{2}{3}h_f\left[h_f^2 + \frac{3}{4}h_c^2 + \frac{3}{2}h_c h_f\right] \quad (13.4.4)$$

Note that the terms on the right side of Eqs. (13.4.1) and (13.4.2) account for the rotary inertia effect. The bending moments (M_{xx}, M_{yy}, M_{xy}) and transverse shear forces (Q_x, Q_y) are known in terms of the generalized deflections through Eqs. (13.1.10a-e).

By substituting Eqs. (13.1.1a-e) into equations (13.1.1)–(13.1.3), the three plate equations for vibration can be written as

$$(D_c + D_f)\frac{\partial^2 \phi_x^M}{\partial x^2} + (\nu_c D_c + \nu_f D_f)\frac{\partial^2 \phi_y^M}{\partial x \partial y}$$
$$+ \frac{1}{2}[(1-\nu_c)D_c + (1-\nu_f)D_f]\left(\frac{\partial^2 \phi_x^M}{\partial y^2} + \frac{\partial^2 \phi_y^M}{\partial x \partial y}\right)$$
$$- K_s(G_c h_c + 2G_f h_f)\left(\phi_x^M + \frac{\partial w^M}{\partial x}\right) = (\rho_c I_c + \rho_f I_f)\frac{\partial^2 \phi_x^M}{\partial t^2}$$
$$(13.4.5)$$

$$(D_c + D_f)\frac{\partial^2 \phi_y^M}{\partial y^2} + (\nu_c D_c + \nu_f D_f)\frac{\partial^2 \phi_x^M}{\partial x \partial y}$$
$$+ \frac{1}{2}[(1-\nu_c)D_c + (1-\nu_f)D_f]\left(\frac{\partial^2 \phi_y^M}{\partial x^2} + \frac{\partial^2 \phi_x^M}{\partial x \partial y}\right)$$
$$- K_s(G_c h_c + 2G_f h_f)\left(\phi_y^M + \frac{\partial w^M}{\partial y}\right) = (\rho_c I_c + \rho_f I_f)\frac{\partial^2 \phi_y^M}{\partial t^2}$$
$$(13.4.6)$$

$$(G_c h_c + 2G_f h_f)\left(\mathcal{M}^M + \nabla^2 w^M\right) = (\rho_c I_c + 2\rho_f I_f)\frac{\partial^2 w^M}{\partial t^2} \quad (13.4.7)$$

where $\mathcal{M}^M$ is the moment sum [see Eq. (13.1.13)]

$$\mathcal{M}^M = \frac{M_{xx}^M + M_{yy}^M}{(1+\nu_c)D_c + (1+\nu_f)D_f} = \frac{\partial \phi_x^M}{\partial x} + \frac{\partial \phi_y^M}{\partial y} \quad (13.4.8)$$

For free vibration (i.e., harmonic motion), the displacement and rotations are assumed to be of the form

$$w^M(x,y,t) = \bar{w}^M(x,y)\, e^{i\omega_M t} \tag{13.4.9a}$$
$$\phi_x^M(x,y,t) = \bar{\phi}_x^M(x,y)\, e^{i\omega_M t} \tag{13.4.9b}$$
$$\phi_y^M(x,y,t) = \bar{\phi}_y^M(x,y)\, e^{i\omega_M t} \tag{13.4.9c}$$

where ω_M is the circular frequency of the plate and the over bar denotes that the amplitudes of natural vibration and they are functions of only the spatial co-ordinates x and y. In the interest of brevity, in the following discussion, the over bar on the variables will be omitted.

Substitution of Eqs. (13.4.9a-c) and differentiating Eqs. (13.4.6) and (13.4.7) with respect to x and y, respectively and adding them, and using Eq. (13.4.8), we obtain

$$(D_c + D_f)\nabla^2 \mathcal{M}^M - K_s(G_c h_c + 2G_f h_f)\left(\mathcal{M}^M + \nabla^2 w^M\right)$$
$$= -(\rho_c I_c + \rho_f I_f)\omega_M^2 \mathcal{M}^M \tag{13.4.10}$$

Moreover, the substitution of Eqs. (13.4.9a-c) into Eq. (13.4.8) yields

$$K_s(G_c h_c + 2G_f h_f)\left(\mathcal{M} + \nabla^2 w^M\right) = -(\rho_c h_c + 2\rho_f h_f)\omega_M^2 w^M \tag{13.4.11}$$

By eliminating $\mathcal{M}^M$ from Eqs. (13.4.10) and (13.4.11), one obtains

$$\nabla^2 w^M + \left[\frac{\rho_c h_c + 2\rho_f h_f}{K_S(G_c h_c + 2G_f h_f)} + \frac{\rho_c I_c + \rho_f I_f}{(D_c + D_f)}\right]\omega_M^2 \nabla^2 w^M$$
$$+ \frac{\rho_c h_c + 2\rho_f h_f}{(D_c + D_f)}\left[\frac{(\rho_c h_c + \rho_f h_f)\omega_M^2}{K_S(G_c h_c + 2G_f h_f)} - 1\right]\omega_M^2 w^M = 0 \tag{13.4.12}$$

which can be written as

$$(\nabla^2 + \lambda_{Mj})w_j^M = 0, \quad w^M = w_1^M + w_2^M, \quad j = 1, 2 \tag{13.4.13}$$

where

$$\lambda_{Mj} = \xi_1 + (-1)^j\sqrt{\xi_1^2 + \xi_2} \tag{13.4.14a}$$

$$\xi_1 = \frac{1}{2}\left[\frac{\rho_c h_c + 2\rho_f h_f}{K_S(G_c h_c + 2G_f h_f)} + \frac{\rho_c I_c + \rho_f I_f}{(D_c + D_f)}\right]\omega_M^2 \tag{13.4.14b}$$

$$\xi_2 = \left[\frac{\rho_c h_c + 2\rho_f h_f}{(D_c + D_f)}\right]\omega_M^2 \tag{13.4.14c}$$

The boundary conditions for simply supported polygonal sandwich plates, the boundary conditions are

$$w^M = \mathcal{M}^M = \nabla^2 w^M = 0 \qquad (13.4.15)$$

The equation governing free vibration of an isotropic Kirchhoff plate is given by [see Reddy (1999)]

$$\left(\nabla^4 - \lambda_K^2\right) w^K = (\nabla^2 + \lambda_K)(\nabla^2 - \lambda_K) w^K, \quad \lambda_K^2 = \frac{\rho h \omega_K^2}{D} \qquad (13.4.16)$$

where w^K is the deflection, h is the total thickness, ρ is the density, and ω_K is the natural frequency of the isotropic Kirchhoff plate. Since the equation $(\nabla^2 - \lambda_K^2) w^K = 0$ produces imaginary frequencies, the vibration of the Kirchhoff plate is thus governed by

$$(\nabla^2 + \lambda_K) w^K = 0 \qquad (13.4.17)$$

For a simple supported polygonal Kirchhoff plate, the deflection and the moment sum are zero at the boundary

$$w^K = 0 \quad \nabla^2 w^K = 0 \qquad (13.4.18)$$

13.4.2 Free Vibration Relationship

Comparing Eqs. (13.4.13) and (13.4.15) with (13.4.17) and (13.4.18), we note that the vibrating polygonal sandwich plate problem is analogous to the vibrating polygonal Kirchhoff plate problem. Thus, for a given simply supported polygonal plate, we have

$$\lambda^M = \lambda^K \qquad (13.4.19)$$

which yields the following relationship between the frequencies of the isotropic Kirchhoff polygonal plate and and the sandwich Mindlin polygonal plate:

$$\omega_M^2 = \left[(1 + a_1 \omega_K) - \sqrt{(1 + a_1 \omega_K)^2 - 4a_2 \omega_K^2}\right] \xi_1 \qquad (13.4.20)$$

$$a_1 = (1 + \xi_2)\xi_3 \sqrt{\frac{\rho h}{D}}, \quad a_2 = \xi_2 \xi_3^2 \qquad (13.4.21)$$

$$\xi_1 = \frac{1}{2} \frac{K_s (G_c h_c + 2G_f h_f)}{(\rho_c I_c + \rho_f I_f)} \qquad (13.4.22a)$$

$$\xi_2 = \left[\frac{(\rho_c h_c + 2\rho_f h_f)}{K_s (G_c h_c + 2G_f h_f)}\right] \left(\frac{D_c + D_f}{\rho_c I_c + \rho_f I_f}\right) \qquad (13.4.22b)$$

$$\xi_3 = \frac{\rho_c I_c + \rho_f I_f}{\rho_c h_c + 2\rho_f h_f} \qquad (13.4.22c)$$

Using Eq. (13.4.20), one can readily obtain the mth frequency of a simply supported, polygonal sandwich plate upon supplying the mth mode frequency of the corresponding Kirchhoff plate.

The foregoing derivation applies to a solid thick (Mindlin) plate solution as well. By letting the thickness of the facings, $h_f = 0$ and $h_c = h$ (i.e., $D_f = 0$, $D_c = D = Eh^3/[12(1-\nu^2)]$, $G_c = G = E/[2(1+\nu)]$, $\rho_c = \rho$), the sandwich plate becomes a solid Mindlin plate of thickness h, density ρ, shear modulus G, Poisson's ratio ν, and flexural rigidity D. Thus the Eq. (13.4.20) reduces to

$$\omega_S^2 = \frac{6K_sG}{\rho h^2}\left\{\left[1+\omega_K\frac{h^2}{12}\sqrt{\frac{\rho h}{D}}\left(1+\frac{2}{K_s(1-\nu)}\right)\right]\right.$$
$$\left. - \sqrt{\left[1+\omega_K\frac{h^2}{12}\sqrt{\frac{\rho h}{D}}\left(1+\frac{2}{K_S(1-\nu)}\right)\right]^2 - \frac{\rho h}{3K_SG}\omega_K^2}\right\}$$

(13.4.23)

where ω_S is the circular frequency of the solid Mindlin plate. This relationship (13.4.23) was derived by Wang (1994) for rectangular plates and shown to give accurate vibration frequencies for other simply supported plate shapes. A form similar to equation (13.4.23) was also derived by Irschik (1985). The latter expressed it in terms of the alternative form of the prestressed membrane vibration solution instead of the Kirchhoff plate solution. Note that Conway (1960) has established analogies between the buckling and vibration problem of polygonal Kirchhoff plates and the vibration problem of prestressed membranes. Thus, one may use either of these solutions for Eqs. (13.4.20) and (13.4.23). There are abundant buckling and vibration solutions of polygonal Kirchhoff plates and membranes available in the open literature.

13.5 Summary

In this chapter, an exact relationship between deflections of sandwich plates and the corresponding Kirchhoff plates is presented. The relationship is valid for simply supported polygonal plates under any transverse load or for simply supported and clamped circular plates under any axisymmetric load. Thus, for the bending problem of such plates, it suffices to perform only the classical thin (Kirchhoff) plate bending analysis, and the effect of shear deformation on the deflection can be readily computed. A more complicated shear deformable plate analysis may be avoided. It is clear that closed-form solutions can be obtained whenever there are closed-form solutions of the corresponding Kirchhoff plates, as illustrated by the two examples. A relationship between the deflection of sandwich plates and solid Mindlin plate is also given. Such relationships should be very useful to engineering designers and researchers working with sandwich plates.

Exact relationships between the bending solutions of the classical plate theory (CPT) and the first-order plate theory (FSDT) are also presented for functionally graded circular plates. Exact solutions of functionally graded plates using the first-order theory are presented in terms of the solutions of the classical plate theory for a number of boundary conditions. Numerical solutions of functionally graded plates under uniformly distributed load are presented as a function of the thickness-to-radius ratio and ratios of the volume fraction.

Next, exact relationships between the buckling load of sandwich plates and Kirchhoff plates under uniform in-plane compressive load are presented. The relationship applies for the simply supported general polygonal plate and simply supported and clamped circular plates. Exact solutions of sandwich plates can be obtained from existing exact Kirchhoff solutions. The more complicated buckling analysis of shear deformable polygonal sandwich plates can be avoided because of the relationship presented herein.

Lastly, an exact relationship has been presented between the vibration frequencies of sandwich plates and those of their Kirchhoff plate counterparts. The relationship applies for any general simply supported polygonal plate with straight edges. Exact vibration solutions for sandwich plates can be obtained by using existing exact Kirchhoff plate vibration solutions. Even buckling solutions for Kirchhoff plates under in-plane loads or vibration solutions for prestressed membranes can be used due to the analogies between these problems. The more complicated shear deformable buckling analysis for the considered sandwich plates may be bypassed because of the relationship derived. Vibration analysis of Kirchhoff plates can be readily performed, for example, by using the computerized Rayleigh–Ritz method and the sandwich plate solutions computer accordingly. This relationship between frequencies can also be used to check the convergence and accuracy of numerical shear deformable plate vibration solutions.

References

ABAQUS/Standard User's Manual (version 5.7), Vol. 1 - 3. Hibbitt, Karlsson & Sorensen, Inc., Providence, Rhode Island (1997).

Ambartsumyan, S. A., *Theory of Anisotropic Plates*, translated from Russian by T. Cheron, Technomic, Stamford, Connecticut (1969).

Anderson, R. A. "Flexural vibrations in uniform beams according to the Timoshenko theory", *J. Appl. Mech.*, **20**, 504-510 (1953).

Arnold, D. N. and Falk, R. S., "The boundary layer for the Reissner–Mindlin plate model", *SIAM J. Math. Anal.*, **21**(2), 281-312 (1990).

Averill, R. C. and Reddy, J. N., "On the behaviour of plate elements based on the first-order theory", *Engineering Computations*, **77**, 57-74 (1990).

Banerjee, J. R. and Williams, F. W., "The effect of shear deformation on the critical buckling of columns", *J. Sound and Vibration*, **174**, 607-616 (1994).

Barrett, K. E. and Ellis, S., "An exact theory of elastic plates", *Int. J. Solids and Structures*, **24**(9), 859-880 (1988).

Basset, A. B., "On the extension and flexure of cylindrical and spherical thin elastic shells", *Phil. Trans. Royal Soc.*, (London) Ser. A, **181**(6), 433-480 (1890).

Bert, C. W., "Simplified analysis static shear correction factors for beams of non-homogeneous cross section", *J. Composite Materials*, **7**, 525-529 (1973).

Bickford, W. B., "A consistent higher order beam theory", in *Developments in Theoretical and Applied Mechanics*, **11**, 137-150 (1982).

Cauchy, A. L., "Sur l'equilibre et le mouvement d'une plaque solide", *Exercies de Mathematique*, **3**, 328-355 (1828).

Cheng, Z. Q. and Kitipornchai, S., "Exact connection between deflections of the classical and shear deformation laminated plate theories", *J. Appl. Mech.*, **66**(1), 260-262 (1999).

Cheung, M. S. and Chan, M. Y. T., "Static and dynamic analysis of thin and thick sectorial plates by the finite strip method", *Computers & Structures*, **14**(1-2), 79-88 (1981).

Conway, H. D., "The bending of symmetrically loaded circular plates with variable thickness", *J. Appl. Mech.*, **15**, 1-6 (1948).

Conway, H. D., "Note on the bending of circular plates of variable thickness", *J. Appl. Mech.*, **16**, 209-210 (1949).

Conway, H. D., "Axially symmetric plates with linearly varying thickness", *J. Appl. Mech.*, **18**, 140-142 (1951).

Conway, H. D., "Closed form solutions for plates of variable thickness", *J. Appl. Mech.*, **20**, 564-565 (1953).

Conway, H. D., "Analogies between the buckling and vibration of polygonal plates and membranes", *Canadian Aeronautical J.*, **6**, 263 (1960).

Conway, H. D., "The bending, buckling and flexural vibration of simply supported polygonal plates by point-matching", *J. Appl. Mech.*, **28**, 288-291 (1961).

Conway, H. D. and Farnham, K. A., "The free flexural vibrations of triangular, rhombic and parallelogram plates and some analogies", *Int. J. Mechanical Science*, **7**, 811-816 (1965).

Cooke, D. W. and Levinson, M., "Thick rectangular plates- II, the generalized Lévy solution", *Int. J. Mechanical Sciences*, **25**, 207-215 (1983).

Cowper, G. R., "The shear coefficient in Timoshenko's beam theory", *J. Appl. Mech.*, **33**, 335-340 (1966).

Deverall, L. I. and Thorne, C. J., "Bending of thin ring sector plates", *J. Appl. Mech.*, **18**, 359-363 (1951).

Di Sciuva, M., "A refined transverse shear deformation theory for multilayered anisotropic plates", *Atti della Academia della Scienze di Torino*, **118**, 269-295 (1984).

Donnell, L. H., *Beams, Plates and Shells*, McGraw-Hill, New York, 269-295 (1976).

Dong, Y. F. and Teixeira de Freitas, J. A., "A quadrilateral hybrid stress element for Mindlin plates based on incompatible displacements", *Int. J. Numer. Meth. Engng.*, **37**, 279-296 (1994).

Dong, Y. F., Wu, C. C., and Teixeira de Freitas, J. A., "The hybrid stress model for Mindlin-Reissner plates based on a stress optimization condition", *Computers and Structures*, **46**, 877-897 (1993).

Eisenberger, M., "Derivation of shape functions for an exact 4-D.O.F. Timoshenko beam element", *Commun. Numerical Methods Engng.*, **10**, 673-681 (1994).

Engesser, F, "Die knickfestigkeit gerader stabe", *Zentralbl, Bauverwaltung* **11**, 483-486 (1891).

Flügge, W., *Viscoelasticity*, Springer-Verlag, New York (1975).

Friedman, Z. and Kosmatka, J. B., "An improved two-node Timoshenko beam finite element", *Computers & Structures*, **47**, 473-481 (1993).

Gere, J. M. and Weaver, Jr. W., *Analysis of Framed Structures* (see pp. 428-430), Van Nostrand, New York (1965).

Goodier, J. N., "On the problem of the beam and the plate in the theory of elasticity", *Transactions of the Royal Society of Canada*, **32**, 65-88 (1938).

Goodman, L. E. and Sutherland, J. G., Discussion of "Natural frequencies of continuous beams of uniform span length", by R. S. Ayre and L. S. Jacobsen, *J. Appl. Mech.*, **18**, 217-218 (1951).

Haggblad, B. and Bathe, K. J., "Specifications of boundary conditions for Reissner/Mindlin plate bending finite elements", *Int. J. Numer. Methods Engng*, **30**, 981-1011 (1990).

Hencky, H., "Uber die Berucksichtigung der Schubverzerrung in ebenen Platten", *Ing. Arch.*, **16**, 72-76 (1947).

Herrmann, G. and Armenakas, A. E., "Vibration and stability of plates under initial stress", *J. Engng. Mech., ASCE*, **86**, 65-94 (1960).

Heyliger, P. R. and Reddy, J. N., "A higher-order beam finite element for bending and vibration problems", *J. Sound and Vibration*, **126**(2), 309-326 (1988).

Hildebrande, F. B., Reissner, E. and Thomas, G. B., "Notes on the foundations of the theory of small displacements of orthotropic shells", NACA TN-1833, Washington, D.C. (1949).

Hinton, E. and Huang, H. C., "Shear forces and twisting moments in plates using Mindlin elements", *Engng. Computations*, **3**, 129-142 (1986).

Hinton, E. and Owen, D. R. J., *Finite Element Software for Plates and Shells*. Pineridge Press, Swansea, U.K. (1984).

Hong, G. M., Wang, C. M. and Tan, T. J., "Analytical buckling solutions for circular Mindlin plates: inclusion of in-plane prebuckling deformation", *Archive Appl. Mech.*, **63**, 534-542 (1993).

Huang, H. C., *Static and Dynamic Analyses of Plates and Shells*, Pineridge Press, Swansea, U.K. (1988).

Huang, H. C. and Hinton, E., "A nine node Lagrangian Mindlin plate element with enhanced shear interpolation", *Engng Computations*, **1**, 356-379 (1984).

Irschik, H., "Membrane-type eigenmotions of Mindlin plates", *Acta Mechanica*, **55**, 1-20 (1985).

Jemielita, G., "Techniczna Teoria Plyt Sredniej Grubbosci (Technical Theory of Plates with Moderate Thickness)", *Rozprawy Insynierskie (Engineering Transactions), Polska Akademia Nauk.*, **23**(3), 483-499 (1975).

Kant, T. and Hinton, E., "Mindlin plate analysis by segmentation method", *J. Engng. Mech., ASCE*, **109**(2), 537-556 (1983).

Karunasena, W., Wang, C. M. Kiripornchai, S. and Xiang, Y., "Exact solutions for axisymmetric bending of continuous annual plates", *Computers & Structures*, **63**(3), 455-464 (1997).

Kerr, A.D., "On the instability of circular plates", *J. Aero. Sci. (Readers' Forum)*, **29**(4), 486-487 (1962).

Khdeir, A.A., Librescu, L. and Reddy, J.N., "Analytical solution of a refined shear deformation theory for rectangular composite plates", *Int. J. Solids and Structures*, **23**(10), 1447-1463 (1987).

Kirchhoff, G., "Uber das Gleichgewicht und die Bewegung einer elastischen Scheibe", *J. Angew. Math.*, **40**, 51-88 (1850).

Kreyszig, E., *Advanced Engineering Mathematics*. 7th edition, John Wiley, Singapore (1993).

Krishna Murthy, A. V., "Vibrations of short beams", *AIAA J.*, **8**, 34-38 (1970a).

Krishna Murthy, A. V., "Analysis of short beams", *AIAA J.*, **8**, 2098-2100 (1970b).

Krishna Murthy, A. V., "Higher order theory for vibration of thick plates", *AIAA J.*, **15**(12), 1823-1824 (1977).

Krishna Murthy, A. V., "Flexure of composite plates", *Computers & Structures*, **7**(3), 161-177 (1987).

Lee, K. H. and Wang, C. M., "On the use of deflection components in Timoshenko beam theory", *J. Appl. Mech.*, **64**, 1006-1008 (1997).

Leissa, A. W., *Vibration of Plates*, Edition by Acoustical Society of America (originally issued by NASA, 1969) (1993).

Levinson, M., "An accurate, simple theory of the static and dynamics of elastic plates", *Mechanics Research Communications*, **7**(6), 343-350 (1980).

Levinson, M., "A new rectangular beam theory", *J. Sound and Vibration*, **74**, 81-87 (1981).

Lim, S. P., Lee, K. H., Chow, S. T., and Senthilnathan, N. R., "Linear and nonlinear bending of shear-deformable plates", *Computers & Structures*, **30**(4), 945-952 (1988).

Lim, C. W. and Wang, C. M., "Bending of annular sectorial Mindlin plates using Kirchhoff results", to appear (2000).

Lim, C. W., Wang, C. M., and Kitipornchai, S., "Timoshenko curved beam bending solutions in terms of Euler-Bernoulli solutions", *Archive of Applied Mechanics*, **67**(3), 179-190 (1997).

Lo, K. H., Christensen, R. M. and Wu, E. M., "A high-order theory of plate deformation: Part 1: Homogeneous plates", *J. Appl. Mech.*, **44**(4), 663-668 (1977).

Mansfield, E. H., *The Bending and Stretching of Plates* (2nd edition), Cambridge University Press, Cambridge (1989).

Marcus, H., *Die Theorie Elastischer Gewebe*, Springer–Verlag, Berlin, Germany (1932).

Márkus, G., *Theorie und Berechnung rotationssymmetrischer Bauwerke*, Werner–Verlag, Düsseldorf (1967).

McFarland, D., Smith, B. L., and Bernhart, W. D., *Analysis of Plates*, Spartan Books, Philadelphia, PA (1972).

Meek, J. L., *Matrix Structural Analysis*, McGraw–Hill, New York (1971).

Miklowitz, J., "Flexural wave solutions of coupled equations representing the more exact theory of bending", *J. Appl. Mech.*, **20**, 511-514 (1953).

Mindlin, R. D., "Influence of rotary inertia and shear on flexural motions of isotropic, elastic plates", *J. Appl. Mech.*, **18**, 31-38 (1951).

Mindlin, R. D. and Deresiewicz, H., "Timoshenko's shear coefficient for flexural vibrations of beams", *Proc. Second U.S. National Congress of Appl. Mech.*, 175-178 (1954).

Nickell, R. E. and Secor, G. A., "Convergence of consistently derived Timoshenko beam finite elements", *Int. J. Numer. Meth. Engng.*, **5**, 243-253 (1972).

Oden, J. T. and Reddy, J. N., *Variational Methods in Theoretical Mechanics*, 2nd Edition, Springer–Verlag, Berlin (1982).

Olhoff, N. and Akesson, B., "Minimum stiffness of optimally located supports for maximum value of column buckling loads", *Structural Optimization*, **3**, 163-175 (1991).

Panc, V., *Theories of Elastic Plates*, Noordhoff, The Netherlands (1975).

Phan, N. D. and Reddy, J. N., "Analysis of laminated composite plates using a higher-order shear deformation theory", *Int. J. Numer. Meth. Engng.*, **12**, 2201-2219 (1985).

Plantema, F. J., *Sandwich Construction: The Bending and Buckling of Sandwich Beams, Plates and Shells*, John Wiley, New York (1966).

Pnueli, D., "Lower bounds to the gravest and all higher frequencies of homogeneous vibrating plates of arbitrary shape", *J. Appl. Mech.*, **42**, 815-820 (1975).

Praveen, G. N. and Reddy, J. N., "Nonlinear transient thermomechanical analysis of functionally graded ceramic-metal plates", *Int. J. Solids and Structures*, **35**(33), 4457-4476 (1998).

Przemieniecki, J. S., *Theory of Matrix Structural Analysis*, McGraw–Hill, New York (1968).

Reddy, J. N., "A simple higher-order theory for laminated composite plates", *J. Appl. Mech.*, **51**, 745-752 (1984a).

Reddy, J. N., *Energy and Variational Methods in Applied Mechanics*, John Wiley, New York (1984b).

Reddy, J. N., "A general non-linear third-order theory of plates with moderate thickness", *Int. J. Non-Linear Mechanics*, **25**(6), 677-686 (1990b).

Reddy, J. N., *Applied Functional Analysis and Variational Methods in Engineering*, McGraw–Hill, New York, 1986; reprinted by Krieger, Melbourne, Florida (1992).

Reddy, J. N., *An Introduction to the Finite Element Method*, Second Edition, McGraw–Hill, New York (1993).

Reddy, J. N., *Mechanics of Laminated Composite Plates: Theory and Analysis*, CRC Press, Boca Raton, Florida (1997a).

Reddy, J. N., "On locking-free shear deformable beam finite elements", *Computer Meth. Appl. Mech. Engng.*, **149**, 113-132 (1997b).

Reddy, J. N., *Theory and Analysis of Elastic Plates*, Taylor & Francis, Philadelphia, PA (1999a).

Reddy, J. N., "On the dynamic behavior of the Timoshenko beam finite elements", *Sadhana* (Journal of the Indian Academy of Sciences), **24**, 175-198, (1999b).

Reddy, J. N. and Chao, W. C., "A comparison of closed-form and finite element solutions of thick laminated anisotropic rectangular plates", *Nuclear Engineering and Design*, **64**, 153-167 (1981).

Reddy, J. N. and Chin, C. D., "Thermomechanical Behavior of Functionally Graded Cylinders and Plates", *J. Thermal Stresses*, **26**(1), 593-626 (1998).

Reddy, J. N. and Phan, N. D., "Stability and vibration of isotropic, orthotropic and laminated plates according to a higher-order shear deformation theory", *J. Sound and Vibration*, **98**(2), 157-170 (1985).

Reddy, J. N. and Rasmussen, M. L., *Advanced Engineering Analysis*, John Wiley, New York, 1982; reprinted by Krieger, Melbourne, Florida (1990).

Reddy, J. N. and Wang, C. M., "Relationships between classical and shear deformation theories of axisymmetric circular plates", *AIAA J.*, **35**(12), 1862-1868 (1997).

Reddy, J. N. and Wang, C. M., "Deflection relationships between classical and third-order plate theories", *Acta Mechanica*, **130**(3-4), 199-208 (1998).

Reddy, J. N., Wang, C. M. and Lam, K. Y., "Unified finite elements based on the classical and shear deformation theories of beams and axisymmetric circular plates", *Communications in Numerical Methods in Engineering*, **13**, 495-510 (1997).

Reddy, J. N., Wang, C. M., and Kitipornchai, S., "Axisymmetric Bending of Functionally Graded Circular and Annular Plates", *European Journal of Mechanics, A: Solids*, **18**, 185-199 (1999).

Reismann, H., "Bending and buckling of an elastically restrained circular plate", *J. Appl. Mech.*, **19**, 167-172 (1952).

Reismann, H., *Elastic Plates: Theory and Application*. John Wiley, New York (1988).

Reissner, E., "On the theory of bending of elastic plates", *J. Mathematical Physics*, **23**, 184-191 (1944).

Reissner, E., "The effect of transverse shear deformation on the bending of elastic plates", *J. Appl. Mech.*, **12**, 69-76 (1945).

Reissner, E., "On the transverse bending of plates, including the effects of transverse shear deformation", *Int. J. Solids and Structures*, **11**, 569-573 (1975).

Reissner, E., "Reflections on the theory of elastic plates", *Applied Mechanics Reviews,* **38**(11), 1453-1464 (1985).

Roark, R. J. and Young, W. C., *Formulas for Stress and Strain.* 5th Edition, McGraw-Hill, New York (1975).

Rozvany, G. I. N and Mröz, Z., "Column design: optimization of support conditions and segmentation", *J. Structural Mechanics,* **5**, 279-290 (1977).

Rubin, C., "Stability of polar orthotropic sector plates", *J. Appl. Mech.,* **45**, 448-450 (1978).

Senthilnathan, N. R., Lim, S. P., Lee, K. H. and Chow, S. T., "Buckling of shear-deformable plates", *AIAA J.,* **25**, 1268-1271 (1987).

Senthilnathan, N. R., Lim, S. P., Lee, K. H., and Chow, S. T., "Vibration of laminated orthotropic plates using a simplified higher-order deformation theory", *Composite Structures,* **10**, 211-229 (1988).

Senthilnathan, N. R. and Lee, K. H., "Some remarks on Timoshenko beam theory", *J. Vibration and Acoustics,* **114**, 495-497 (1992).

Severn, R. T., "Inclusion of shear deflection in the stiffness matrix for a beam element", *J. Strain Analysis,* **5**, 239-241 (1970).

Soldatos, K. P., "On certain refined theories for plate bending", *J. Appl. Mech.,* **55**, 994-995 (1988).

Stephen, N. G., "The second frequency spectrum of Timoshenko beams", *J. Sound and Vibration,* **80**, 578-582 (1982).

Stephen, N. G. and Levinson, M., "A second order beam theory", *J. Sound and Vibration,* **67**, 293-305 (1979).

Szilard, R., *Theory and Analysis of Plates,* Prentice-Hall, Englewood Cliffs, NJ (1974).

Tessler, A. and Dong, S. B., "On a hierarchy of conforming Timoshenko beam elements", *Computers & Structures,* **14**(3/4), 335-344 (1981).

Timoshenko, S. P., "On the correction for shear of the differential equation for transverse vibrations of prismatic bars", *Philosophical Magazine,* **41**, 744-746 (1921).

Timoshenko, S. P., "On the transverse vibration of bars of uniform cross-section", *Philosophical Magazine*, **43**, 125-131, (1922).

Timoshenko, S. P. and Gere, J. M., *Theory of Elastic Stability*, McGraw–Hill, New York (1961).

Timoshenko, S. P. and Woinowsky-Krieger, S., *Theory of Plates and Shells*, McGraw–Hill, Singapore (1970).

Touratier, M., "An efficient standard plate theory", *Int. J. Engng. Science*, **29**, 901-916 (1991).

Ugural, A. C., *Stresses in Plates and Shells*, McGraw–Hill, New York (1981).

Vlasov, B. F., "Ob uravneniyakh teovii isgiba plastinok (On the equations of the theory of bending of plates)", *Izv. Akd. Nauk SSR, OTN*, **4**, 102-109 (1958).

Wang, C. M., "Natural frequencies formula for simply supported Mindlin plates", *J. Vibration and Acoustics*, **116**(4), 536-540 (1994).

Wang, C. M., "Timoshenko beam-bending solutions in terms of Euler–Bernoulli solutions", *J. Engng. Mech., ASCE*, **121**(6), 763-765 (1995a).

Wang, C. M., "Allowance for prebuckling deformations in buckling load relationship between Mindlin and Kirchhoff simply supported plates of general polygonal shape", *Engineering Structures*, **17**(6), 413-418 (1995b).

Wang, C. M., "Deflection of sandwich plates in terms of corresponding Kirchhoff plate solutions", *Archive of Applied Mechanics*, **65**(6), 408-414 (1995c).

Wang, C. M., "Buckling of polygonal and circular sandwich plates", *AIAA Journal*, **33**(5), 962-964 (1995d).

Wang, C. M., "Vibration frequencies of simply supported polygonal sandwich plates via Kirchhoff solutions", *J. Sound and Vibration*, **190**(2), 255-260 (1995e).

Wang, C. M., Discussion on "Postbuckling of moderately thick circular plates with edge elastic restraint", *J. Engng. Mech., ASCE*, **122**(2), 181-182 (1996).

Wang, C. M., "Relationships between Mindlin and Kirchhoff bending solutions for tapered circular and annular Plates", *Engineering Structures,* 19(3), 255-258 (1997).

Wang, C. M. and Alwis, W. A. M., "Simply supported Mindlin plate deflections using Kirchhoff plates", *J. Engng. Mech., ASCE,* 121(12), 1383-1385 (1995).

Wang, C. M., Chen, C. C., and Kitipornchai, S., "Shear deformable bending solutions for nonuniform beams and plates with elastic end restraints from classical solutions", *Archive of Applied Mechanics,* 68, 323-333 (1998).

Wang, C. M., Hong, G. M., and Tan, T. J., "Elastic buckling of Tapered circular plates", *Computers & Structures,* 55(6), 1055-1061 (1995).

Wang, C. M., Kitipornchai, S., and Xiang, Y., "Relationships between buckling loads of Kirchhoff, Mindlin and Reddy polygonal plates on Pasternak foundation", *J. Engng. Mech., ASCE,* 123(11), 1134-1137 (1997).

Wang, C. M., Kitipornchai, S., and Reddy, J. N., "Relationship between vibration frequencies of Reddy and Kirchhoff polygonal plates with simply supported edges", *J. Vibration and Acoustics,* 122(1), 77-81 (2000).

Wang, C. M. and Lee, K. H., "Deflection and stress resultants of axisymmetric Mindlin plates in terms of corresponding Kirchhoff solutions", *Int. J. Mechanical Sciences,* 38(11), 1179-1185 (1996).

Wang, C. M. and Lee, K. H., "Buckling load relationship between Reddy and Kirchhoff circular plates", *J. Franklin Institute,* 335, B(6), 989-995 (1998).

Wang, C. M., Lim, G. T., and Lee, K. H., "Relationships between Kirchhoff and Mindlin bending solutions for Lévy plates", *J. Appl. Mech.,* 66(2), 541-545 (1999).

Wang, C. M. and Reddy, J. N., "Buckling load relationship between Reddy and Kirchhoff plates of polygonal shape with simply supported edges", *Mechanics Research Communications,* 24(1), 103-108 (1997).

Wang, C. M., Tan, T. J., Hong, G. M., and Alwis, W. A. M., "Buckling of Tapered Circular Plates: Allowance for Effects of Shear and Radial Deformations", *Mechanics of Structures and Machines*, **24**(2), 135-153 (1996).

Wang, C. M., Wang, C., and Ang, K. K., "Vibration of Initially Stressed Reddy Plates on a Winkler-Pasternak Foundation", *J. Sound and Vibration*, **204**(2), 203-212 (1997).

Wang, C. M. and Xiang, Y., "Deducing Buckling Loads of Sectorial Mindlin Plates from Kirchhoff Plates", *J. Engng. Mech., ASCE*, **125**(5), 596-598 (1999).

Wang, C. M., Xiang, Y., and Kitipornchai, S., "Buckling Solutions of Rectangular Mindlin Plates Under Uniform Shear", *J. Engng Mech., ASCE*, **120**(11), 2462-2470 (1994).

Wang, C.M., Xiang, Y., Kitipornchai, S., and Liew, K.M., "Buckling solutions for Mindlin plates of various shapes," *Engineering Structures*, **16**(2), 119-127 (1994).

Wang, C. M., Yang, T. Q., and Lam, K. Y., "Viscoelastic Timoshenko beam bending solutions from viscoelastic Euler–Bernoulli solutions", *J. Engng. Mech., ASCE*, **123**(7), 746-748 (1997).

Weaver, W., Jr., Timoshenko, S. P., and Young, D. H., *Vibration Problems in Engineering*, John Wiley, New York (1990).

Wittrick, W. H., "Shear correction factors for orthotropic laminates under static load", *J. Appl. Mech.*, **40**(1), 302-304 (1973).

Wittrick, W. H., "Analytical, three-dimensional elasticity solutions to some plate problems, and some observations on Mindlin's plate theory", *Int. J. Solids and Structures*, **23**, 441-464 (1987).

Woinowsky-Krieger, S., "Berechnung der ringsum frei aufliegenden gleihseitigen dreiecksplatte", *Ing. Archiv* **4**, 254-262 (1933).

Xiang, Y., Liew, K.M., and Kitipornchai, S., "Transverse vibration of thick annular sector plates", *J. Engng. Mech., ASCE*, **119**, 1579-1599 (1993).

Xiang, Y., Wang, C. M., and Kitipornchai, S., "Exact vibration solution for initially stressed Mindlin plates on Pasternak foundations", *Int. J. Mechanical Sciences*, **36**(4), 311-316, (1994).

Yamanouchi, M., Koizumi, M., Hirai, T., and Shiota, I., (Editors), *Proceedings of the First International Symposium on Functionally Gradient Materials*, Japan, 1990.

Ye, J., "Axisymmetric buckling of homogeneous and laminated circular plates", *J. Structural Engineering, ASCE*, **121**(8), 1221-1224 (1995).

Ziegler, H., "Arguments for and against Engesser's buckling formulas", *Ingenieur-Archiv*, **52**, 105-113 (1982).

SUBJECT INDEX

Annular plates:
  boundary conditions,
    clamped, 17, 20, 24, 31
    simply supported, 20, 23, 31, 98, 114
Axisymmetric bending, 154, 156
Axisymmetric linearly varying load, 164, 251

Beam finite element: 44
  consistent, 49
  Hermite cubic, 46
  Levinson, 54
  reduced integration, 49
  unified, 48
Beam stiffness matrix, 48
Beam theory:
  Euler–Bernoulli, 5, 11, 12
  Levinson, 38
  Reddy-Bickford, 20-24, 28, 31, 36, 39, 64-74
  simplified, 42, 46
  Timoshenko, 11, 13, 17-19
  Third-order, 13, 39
  see: Reddy-Bickford
Bessel functions, 169, 183, 209, 233
Biharmonic equation, 99, 112, 152
Boundary conditions:
  annular plates, 174
  circular plates, 157-159, 168
  clamped, 17, 20, 24, 31, 80, 98, 104, 158, 186, 262
  elastically supported, 17, 20, 24,
  fixed-fixed, 59
  fixed-free, 59, 262
  free, 17, 20, 24, 31, 80, 97, 104, 157, 186, 262

  pinned-pinned, 58
  simply supported, 16, 20, 23, 31, 80, 98, 104, 113, 157, 185, 199, 238, 248, 261, 271
  hard type, 115, 227
  soft type, 115
  solid circular plate, 158, 208, 260
Buckling equations 57, 192, 196, 264, 269
Buckling load:
  beams, 55, 58, 63, 69, 72
  circular plates, 210
  polygonal plates, 200-204
  rectangular plates, 201-205
  sandwich plates, 259
  sectorial plates, 215

Characteristic polynomials, 221, 242
Circular plates:
  axisymmetric bending of, 154
  boundary conditions, 157, 168
  classical theory of, 155, 171
  governing equations, 155
  first-order theory of, 155
  third-order theory of, 155
Classical plate theory (CPT):
  circular plates, 155, 171
  displacement field, 3, 89, 171, 254
  governing equations, 94, 99, 112, 137, 155, 179, 196, 206, 225, 244, 255
  polygonal plates, 112–116
  sectorial plates, 178, 214
Consistent interpolation element, 49
Constitutive equations:
  for beams, 16, 18
  for plates, 98, 102, 107

Displacement field of:
  classical plate theory, 3, 89, 171, 254
  circular plates, 155, 171, 254
  Euler–Bernoulli beam, 12
  first-order plate theory, 90, 254
  Kirchhoff plate theory,
  see: Classical plate theory
  Mindlin plate theory,
  see: First-order plate theory
  Reddy-Bickford beam theory, 14
  Reddy plate theory,
  see: Third-order plate theory
  third-order beam theory, 14
  third-order plate theory, 91
  Timoshenko beam theory, 13

Effective shear force, 22, 30, 97, 145
Effective shear coefficient, 34
Eigenfunctions, 60
Elastic coefficients, 102
Element stiffness matrix, 47, 48, 176
Elliptical plate, 129
Energy methods, 1
Engesser-Timoshenko column, 56
Equations for buckling, 57, 192, 196, 264, 269
Equations of equilibrium:
  beams, 15, 18-20, 22-23
  circular plates, 155, 172, 206, 255
  elasticity, 6, 108
  polygonal plates, 99, 102, 106, 112, 119, 123, 196-198, 225-226, 244-248, 270
  sectorial plates, 179-181, 212-213
Equilateral triangular plate, 114
Equivalent slope, 44
Euler–Bernoulli hypothesis, 12-13
Euler–Lagrange equations, 2, 172

Finite element method, 44
Finite element model: 44, 53
  displacement, 53
First-order shear deformation theory
  bending solutions of, 112, 138-143, 156-157, 184-185, 249, 257-260
  buckling analysis of, 199-200
  displacement field of, 3, 89-91, 171, 254
  equations of equilibrium of, 102, 155, 196, 206, 225, 255, 270
  finite element models, 44,
  Navier's solution, 117, 129
  shear correction factors, 4, 19, 56, 101
  vibration analysis of, 226-228
Fixed edge: see clamped edge
Flexural rigidity: 56, 99
Force resultants, 101,
Free edge, 17, 20, 24, 31, 97
Free vibration:
  see: natural vibration
Frequency equation:
  see: characteristic equation
Functionally graded materials, 253

Generalized displacements, 46-48, 96-97, 107
Generalized forces, 48, 96-97

Hamilton's principle, 2, 223
Hermite cubic interpolation, 46, 47
  functions, 48
Higher-order theories, 4
Hooke's law, generalized, 16, 102

Independent interpolation, 51
Interdependent approximation, 52

SUBJECT INDEX

Interdependent interpolation, 50, 51
Interpolation functions:
  Hermite, 46-48
  interdependent, 52

Kinetic energy, 223
Kirchhoff assumptions, 90
Kirchhoff free-edge condition, 97
Kirchhoff hypothesis, 90
Kirchhoff plate theory,
  see: Classical plate theory

Laplace equation, 112 152
Laplace operator, 99
Levinson beam theory, 38, 54, 75
Levinson plate theory, 132
Lévy solution, 133

Marcus curvature, 110
Marcus moment,
  see: moment sum
Mindlin plate theory,
  see: First-order plate theory
Modification factor, 204
Moment resultants, 92, 119
Moment sum, 103, 112, 116, 118,
  121, 123, 131, 138, 156, 166,
  199, 213, 226, 235, 245

Navier's method, 129
Navier's solution, 118, 129

Pinned-pinned columns, 58
Plane stress state, 4,
Polar coordinates, 153
Potential energy functional, 1, 18,
  21, 93, 100, 196
Primary variables, 22, 46, 96

Principle of virtual displacements,
  2, 15, 21, 94, 101, 105, 172

Reddy plate theory,
  see: Third-order plate theory
Reddy-Bickford beam theory,
  20-24, 28, 31, 36, 39, 64-74
Reduced integration element, 49
Rotary inertia, 227-229
Rotatory inertia,
  also see: rotary inertia

Secondary variable, 22, 46, 96,
Shear correction coefficient, 4, 19,
  48, 56, 101, 231
Shear parameter, 25
Shear rigidity, 56,
Stability: see: buckling
Stiffness matrix, 44, 47
Strain energy functionals, 17, 21, 92,
  100, 195

Third-order beam theory, 6, 20-23,
  65
Third-order plate theory,
  bending solutions of, 105-107, 124,
    168
  buckling analysis of, 200, 208
  displacement field of, 4, 91, 108
  equations of equilibrium of, 106,
    119, 155, 198, 206-207, 225-226
  vibration analysis of, 236-240
Timoshenko beam theory, 11, 13,
  17-19, 56, 78
Total potential energy:
  see: potential energy functional

Unified beam element, 46
Unit-dummy-displaement method,
  47

Virtual displacements, 2, 15, 21,
    principle of, 15, 21, 94, 101, 105, 172
Virtual strain energy,
    see: Strain energy functional